本书英文版由世界卫生组织（World Health Organization）出版，书名为：

WHO TobLabNet Official Method SOP 01: Standard Operating Procedure for Intense Smoking of Cigarettes © World Health Organization 2012

WHO TobLabNet Official Method SOP 02: Standard Operating Procedure for Validation of Analytical Methods of Tobacco Product Contents and Emissions © World Health Organization 2017

WHO TobLabNet Official Method SOP 03: Standard Operating Procedure for Determination of Tobacco-Specific Nitrosamines in Mainstream Cigarette Smoke under ISO and Intense Smoking Conditions © World Health Organization 2014

WHO TobLabNet Official Method SOP 04: Standard Operating Procedure for Determination of Nicotine in Cigarette Tobacco Filler © World Health Organization 2014

WHO TobLabNet Official Method SOP 05: Standard Operating Procedure for Determination of Benzo[a]pyrene in Mainstream Cigarette Smoke under ISO and Intense Smoking Conditions © World Health Organization 2015

WHO TobLabNet Official Method SOP 06: Standard Operating Procedure for Determination of humectants in Cigarette Tobacco Filler © World Health Organization 2016

WHO TobLabNet Official Method SOP 07: Standard Operating Procedure for Determination of Ammonia in Cigarette Tobacco Filler © World Health Organization 2016

WHO TobLabNet Official Method SOP 08: Standard Operating Procedure for Determination of Aldehydes in Mainstream Cigarette Smoke under ISO and Intense Smoking Conditions © World Health Organization 2018

WHO TobLabNet Official Method SOP 09: Standard Operating Procedure for Determination of Volatile Organics in Mainstream Cigarette Smoke under ISO and Intense Smoking Conditions © World Health Organization 2018

WHO TobLabNet Official Method SOP 10: Standard Operating Procedure for Determination of Nicotine and Carbon Monoxide in Mainstream Cigarette Smoke under Intense Smoking Conditions © World Health Organization 2016

世界卫生组织（World Health Organization）授权中国科技出版传媒股份有限公司（科学出版社）翻译出版本书中文版。中文版的翻译质量和对原文的忠实性完全由科学出版社负责。当出现中文版与英文版不一致的情况时，应将英文版视作可靠和有约束力的版本。

中文版《世界卫生组织烟草实验室网络标准操作规程》
© 中国科技出版传媒股份有限公司（科学出版社） 2022

WHO TobLabNet SOP
世界卫生组织烟草实验室网络标准操作规程

胡清源　侯宏卫　等　译

科学出版社

北　京

内 容 简 介

本书为世界卫生组织（WHO）烟草实验室网络（TobLabNet）成员编写的标准操作规程（SOP）。包括卷烟深度抽吸、烟草制品成分和释放物分析方法验证、卷烟主流烟气中烟草特有亚硝胺的测定、卷烟烟丝中烟碱的测定、卷烟主流烟气中苯并[a]芘的测定、卷烟烟丝中保润剂的测定、卷烟烟丝中氨的测定、卷烟主流烟气中醛类化合物的测定、卷烟主流烟气中挥发性有机物的测定、卷烟主流烟气中烟碱和一氧化碳的测定共 10 个标准操作规程。

本书会引起吸烟与健康、烟草化学和公共卫生学等诸多领域研究人员的兴趣，可以为涉足烟草科学研究的科技工作者和烟草管制研究的决策者提供权威性参考。

图书在版编目（CIP）数据

世界卫生组织烟草实验室网络标准操作规程 / WHO 烟草实验室网络著；胡清源等译. —北京：科学出版社，2022.9
书名原文：World Health Organization Tobacco Laboratory Network Standard Operating Procedure
ISBN 978-7-03-072635-3

Ⅰ. ①世⋯　Ⅱ. ①W⋯　②胡⋯　Ⅲ. ①世界卫生组织–烟草化学–实验室管理–技术操作规程　Ⅳ. ①TS41-65

中国版本图书馆 CIP 数据核字（2022）第 113309 号

责任编辑：刘　冉 / 责任校对：杜子昂
责任印制：吴兆东 / 封面设计：北京图阅盛世

科学出版社 出版
北京东黄城根北街 16 号
邮政编码：100717
http://www.sciencep.com

北京中石油彩色印刷有限责任公司 印刷
科学出版社发行　各地新华书店经销

*

2022 年 9 月第 一 版　　开本：720 × 1000　1/16
2022 年 9 月第一次印刷　　印张：26
字数：520 000

定价：198.00 元
（如有印装质量问题，我社负责调换）

译者名单

胡清源　侯宏卫　陈　欢　刘　彤
张小涛　韩书磊　付亚宁　王红娟
冯鹏侠　苏　悦　宋凌霄

目 录

WHO TobLabNet SOP 01
卷烟深度抽吸标准操作规程 ·· 1

WHO TobLabNet SOP 02
烟草制品成分和释放物分析方法验证标准操作规程 ·· 9

WHO TobLabNet SOP 03
ISO 和深度抽吸方案下卷烟主流烟气中烟草特有亚硝胺的测定标准操作规程 ··· 45

WHO TobLabNet SOP 04
卷烟烟丝中烟碱的测定标准操作规程 ·· 64

WHO TobLabNet SOP 05
ISO 和深度抽吸方案下卷烟主流烟气中苯并[a]芘的测定标准操作规程 ············ 78

WHO TobLabNet SOP 06
卷烟烟丝中保润剂的测定标准操作规程 ·· 94

WHO TobLabNet SOP 07
卷烟烟丝中氨的测定标准操作规程 ·· 115

WHO TobLabNet SOP 08
ISO 和深度抽吸方案下卷烟主流烟气中醛类化合物的测定标准操作规程 ······· 129

WHO TobLabNet SOP 09
ISO 和深度抽吸方案下卷烟主流烟气中挥发性有机物的测定标准操作规程 ··· 151

WHO TobLabNet SOP 10
深度抽吸方案下卷烟主流烟气中烟碱和一氧化碳的测定标准操作规程 ·········· 174

CONTENTS

WHO TobLabNet Official Method SOP 01
Standard Operating Procedure for Intense Smoking of Cigarettes 189

WHO TobLabNet Official Method SOP 02
Standard Operating Procedure for Validation of Analytical Methods of Tobacco Product Contents and Emissions 196

WHO TobLabNet Official Method SOP 03
Standard Operating Procedure for Determination of Tobacco-Specific Nitrosamines in Mainstream Cigarette Smoke under ISO and Intense Smoking Conditions 240

WHO TobLabNet Official Method SOP 04
Standard Operating Procedure for Determination of Nicotine in Cigarette Tobacco Filler 261

WHO TobLabNet Official Method SOP 05
Standard Operating Procedure for Determination of Benzo[a]pyrene in Mainstream Cigarette Smoke under ISO and Intense Smoking Conditions 276

WHO TobLabNet Official Method SOP 06
Standard Operating Procedure for Determination of Humectants in Cigarette Tobacco Filler 296

WHO TobLabNet Official Method SOP 07
Standard Operating Procedure for Determination of Ammonia in Cigarette Tobacco Filler 321

WHO TobLabNet Official Method SOP 08
Standard Operating Procedure for Determination of Aldehydes in Mainstream Cigarette Smoke under ISO and Intense Smoking Conditions 338

WHO TobLabNet Official Method SOP 09
Standard Operating Procedure for Determination of Volatile Organics in Mainstream Cigarette Smoke under ISO and Intense Smoking Conditions 363

WHO TobLabNet Official Method SOP 10
Standard Operating Procedure for Determination of Nicotine and Carbon Monoxide in Mainstream Cigarette Smoke under Intense Smoking Conditions 388

WHO TobLabNet SOP 01

卷烟深度抽吸标准操作规程

方　　法：卷烟深度抽吸
分　析　物：不适用
基　　质：卷烟
更新时间：2012 年 4 月

吸烟机的抽吸方案不能代表所有人的吸烟行为：吸烟机测试对用于卷烟设计及监管目的的卷烟释放物表征非常有用，但将吸烟机的测量结果披露给吸烟者则会导致对不同品牌的暴露和风险差异的误解。吸烟机测量的烟气释放物数据可以用于产品危害评估，但并不意味着这些数据等效于人体暴露或风险的测量值。将吸烟机测量结果的差异表示为暴露或风险的差异是对世界卫生组织烟草实验室网络标准的滥用。

本方法由世界卫生组织（WHO）烟草实验室网络（TobLabNet）制订，作为卷烟深度抽吸标准操作规程（SOP）。

引言

为了在全球范围内建立具有可比性的烟草制品检测方法，需要卷烟特定成分和释放物的一致性检测方法。2008年11月在南非德班举行的世界卫生组织《烟草控制框架公约》（WHO FCTC）第三次缔约方会议回顾了 FCTC/COP1(15)和FCTC/COP2(14)号决议关于制订 WHO FCTC 第9条（烟草制品成分管制）和第10条（烟草制品披露管制）实施指南的要求，并提到第三次缔约方会议工作报告中有关其工作进展的信息，提请公约秘书处授权 WHO 无烟草行动组在5年内对检测卷烟成分和释放物的分析化学方法进行验证（FCTC/COP/3/REC/1）。

根据2006年10月在加拿大渥太华举行的第三次会议上确定的优先次序标准，WHO FCTC 第9条和第10条工作组确定优先验证以下成分的分析化学检测方法：

- 烟碱；
- 氨；
- 保润剂[1,2-丙二醇、甘油（1,2,3-丙三醇）和三甘醇（2,2′-二乙醚乙二醇）]。

需要对三种方法进行验证：一是检测烟碱的方法，二是检测氨的方法，三是检测保润剂的方法。

根据上述渥太华会议确定的优先次序标准，工作组确定了卷烟主流烟气释放物中应该验证的分析化学检测方法的下列优先级物质清单：

- 4-(*N*-甲基亚硝胺基)-1-(3-吡啶基)-1-丁酮（NNK）；
- *N*-亚硝基降烟碱（NNN）；
- 乙醛；
- 丙烯醛；
- 苯；
- 苯并[*a*]芘；
- 1,3-丁二烯；
- 一氧化碳；
- 甲醛。

这些释放物使用下述两种抽吸方案检测，对五种方法进行验证：一是检测烟草特有亚硝胺（NNK 和 NNN）的方法，二是检测苯并[*a*]芘的方法，三是检测醛类（乙醛、丙烯醛和甲醛）的方法，四是检测挥发性有机物（苯和1,3-丁二烯）的方法，五是检测一氧化碳的方法。

下表列出了验证上述方法的两种抽吸方案。

抽吸方案	抽吸容量（mL）	抽吸间隔（s）	滤嘴通风孔
ISO 方案：ISO 3308 常规分析用吸烟机 定义和标准条件	35	60	不封闭
深度抽吸方案：在 ISO 3308 基础上有所调整	55	30	所有通风孔必须如 SOP 01 **12.2** 所述 100%封闭

SOP 01 描述卷烟深度抽吸方案。

1 适用范围

本方法描述深度抽吸方案吸烟机抽吸卷烟的整个过程。

请注意，要成功操作吸烟机或其他分析设备，须对操作人员进行培训。对操作吸烟机或其他分析设备没有经验的人，在检测烟草制品成分和释放物前，需要接受专门的培训。

2 参考标准

2.1 ISO 3308：常规分析用吸烟机 定义和标准条件。

2.2 ISO 4387：卷烟 常规分析用吸烟机测定总粒相物和去烟碱干粒相物。

2.3 ISO 3402：烟草和烟草制品 用于调节和测试的大气环境。

3 术语和定义

3.1 TPM：总粒相物。

3.2 ISO 方案：抽吸过程参数为抽吸容量 35 mL，抽吸间隔 60 s，每次抽吸持续时间 2 s，不封闭滤嘴通风孔。

3.3 深度抽吸方案：抽吸过程参数为抽吸容量 55 mL，抽吸间隔 30 s，每次抽吸持续时间 2 s，100%封闭滤嘴通风孔。

3.4 烟草制品：完全或部分以烟叶为原料制造，用于抽吸、吸吮、咀嚼或吸入的产品（WHO FCTC 第 1(f)条）。

3.5 实验室样品：用于进行实验室检测的样品，由一次性或在规定时间内送到实验室的单一类型产品组成。

3.6 试样：从实验室样品中随机抽取的待测样品。所取样品数量应能代表实验室样品。

3.7 试料：从试样中随机抽取的用于单次测试的样品。所取样品数量应能代表试样。

4 方法概述

4.1 所有样品按照ISO标准规程进行处理和标记。

4.2 100%封闭通风孔。

4.3 除抽吸容量和抽吸频率外,按照ISO标准规程抽吸卷烟。

5 安全和环境预防措施

5.1 遵守所有化学实验活动的常规安全与环境预防措施。

5.2 使用本检测方法对特定产品进行检测和评估时所用材料或设备可能对环境有害。本检测方法无力解决所有与使用有关的安全问题。所有使用此方法的人有责任咨询相关机构并遵循现有监管要求,制定健康与安全规程以及环境预防措施。

5.3 应特别注意避免吸入或皮肤接触危险化学品。在制备或处理未稀释材料、标准溶液、萃取液和收集样品时,应在通风橱中进行,并穿戴合适的实验服、手套和护目镜。

6 仪器和设备

常规实验仪器,特别是:

6.1 按照ISO 3402规定调节卷烟所需设备。

6.2 按照ISO 4387规定标记烟蒂长度所需设备。

6.3 按照WHO TobLabNet SOP 01[12.2]规定采用深度抽吸方案时封闭通风孔所需设备。

6.4 按照ISO 3308规定抽吸烟草制品所需设备。

6.5 透明胶带,宽20 mm(0.75英寸),如Scotch®胶带(3M, Maplewood, Minnesota, USA)。

6.6 100%封闭通风孔所需的卷烟夹持器。

7 试剂

除非另有说明,所有试剂应至少为分析纯。试剂尽可能以CAS号来标识。

8 玻璃器皿准备

清洁、干燥的玻璃器皿，避免残留物污染。

9 溶液配制

不适用。

10 标准溶液配制

不适用。

11 抽样

按 SOP 特定方法的要求抽样。

12 卷烟准备

12.1 按照 ISO 4387 要求标记卷烟的烟蒂长度。
12.2 按照下述要求封闭所有通风孔：
 12.2.1 对于深度抽吸方案，使用 20 mm（0.75 英寸）宽的透明胶带[6.5]缠住卷烟的全部周长，以完全封闭通风孔。
 12.2.2 量出 50~55 mm 的胶带长度。
 12.2.3 将胶带的切口平行于卷烟的长轴粘贴，胶带的边缘在卷烟嘴的边缘 1 mm 以内（图 1）。
 12.2.4 小心将胶带缠绕在滤嘴上，以确保与卷烟纸紧密黏合，没有褶皱或气泡。如果出现褶皱或气泡，则应舍弃该样品并从分析中剔除。
 12.2.5 胶带应环绕卷烟滤嘴两周，且重叠部分应在 5 mm 以内（图 2）。
 12.2.6 胶带不能超过滤嘴端。
可使用卷烟夹持器[6.6]代替胶带用于封闭滤嘴通风孔。
12.3 按照 ISO 3402 要求对所有需要抽吸的卷烟进行调节。

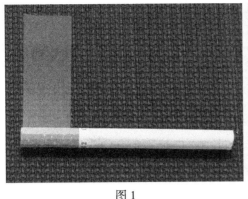
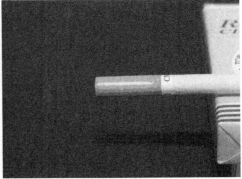

图 1   图 2

13 吸烟机设置

13.1 环境条件。

按 ISO 3308 要求设定抽吸的环境条件。

13.2 吸烟机要求。

除以下几点外，按 ISO 3308 要求设定：

    13.2.1 对于深度抽吸方案，设定吸烟机的抽吸容量为(55 ± 0.1) mL。

    13.2.2 对于深度抽吸方案，设定吸烟机的抽吸频率为每 30 s 一次。

    13.2.3 当卷烟燃烧至按[12.1]标记的位置时认为一次抽吸过程结束。

14 样品生成

在指定吸烟机上抽吸足够量的样品，样品含量应不足以使滤片发生穿透。

14.1 按照 ISO 4387 抽吸试样，并按 SOP 描述的方案收集被分析物。

14.2 至少包含一个用于质量控制的参比试样。

14.3 当样品类型为第一次检测时，评估滤片穿透的可能性，调整卷烟的数量以避免穿透。当确定焦油、烟碱和一氧化碳时，对于 92 mm 滤片和 44 mm 滤片，穿透分别发生在总粒相物水平超过 600 mg 和 150 mg 时。一旦发生穿透，必须减少每片滤片上的卷烟数量。对于不同被分析物，滤片或其他收集设备的穿透可能性不同。

15 样品制备

不适用。

16 样品分析

不适用。

17 数据分析和计算

不适用。

18 特别注意事项

无。

19 数据报告

按 SOP 要求进行。

20 质量控制

按 SOP 要求进行。

21 方法性能

不适用。

22 重复性和再现性

不适用。

23 参考文献

23.1　ISO 10185　烟草和烟草制品——词汇。

WHO TobLabNet SOP 02

烟草制品成分和释放物分析方法验证标准操作规程

方　　法：烟草制品成分和释放物分析方法验证
分　析　物：不适用
基　　质：卷烟主流烟气和卷烟烟丝成分
更新时间：2016 年 11 月

吸烟机的抽吸方案不能代表所有人的吸烟行为：吸烟机测试对用于卷烟设计及监管目的的卷烟释放物表征非常有用，但将吸烟机的测量结果披露给吸烟者则会导致对不同品牌的暴露和风险差异的误解。吸烟机测量的烟气释放物数据可以用于产品危害评估，但并不意味着这些数据等效于人体暴露或风险的测量值。将吸烟机测量结果的差异表示为暴露或风险的差异是对世界卫生组织烟草实验室网络标准的滥用。

本方法由世界卫生组织（WHO）烟草实验室网络（TobLabNet）制订，作为卷烟主流烟气和卷烟烟丝成分分析方法验证标准操作规程（SOP）。

引言

为了在全球范围内建立具有可比性的烟草制品检测方法，需要卷烟特定成分和释放物的一致性检测方法。2008年11月在南非德班举行的世界卫生组织《烟草控制框架公约》（WHO FCTC）第三次缔约方会议回顾了 FCTC/COP1(15)和FCTC/COP2(14)号决议关于制订 WHO FCTC 第 9 条（烟草制品成分管制）和第10条（烟草制品披露管制）实施指南的要求，并提到第三次缔约方会议工作报告中有关其工作进展的信息，提请公约秘书处授权 WHO 无烟草行动组在 5 年内对检测卷烟成分和释放物的分析化学方法进行验证（FCTC/COP/3/REC/1）。

根据2006年10月在加拿大渥太华举行的第三次会议上确定的优先次序标准，WHO FCTC 第9条和第10条工作组确定优先验证以下成分的分析化学检测方法：

- 烟碱；
- 氨；
- 保润剂[1,2-丙二醇、甘油（1,2,3-丙三醇）和三甘醇(2,2′-二乙醚乙二醇)]。

需要对三种方法进行验证：一是检测烟碱的方法，二是检测氨的方法，三是检测保润剂的方法。

根据上述渥太华会议确定的优先次序标准，工作组确定了卷烟主流烟气释放物中应该验证的分析化学检测方法的下列优先级物质清单：

- 4-(N-甲基亚硝胺基)-1-(3-吡啶基)-1-丁酮（NNK）；
- N-亚硝基降烟碱（NNN）；
- 乙醛；
- 丙烯醛；
- 苯；
- 苯并[a]芘；
- 1,3-丁二烯；
- 一氧化碳；
- 甲醛。

这些释放物使用下述两种抽吸方案检测，对五种方法进行验证：一是检测烟草特有亚硝胺（NNK 和 NNN）的方法，二是检测苯并[a]芘的方法，三是检测醛类（乙醛、丙烯醛和甲醛）的方法，四是检测挥发性有机物（苯和1,3-丁二烯）的方法，五是检测一氧化碳的方法。

下表列出了验证上述方法的两种抽吸方案。

抽吸方案	抽吸容量（mL）	抽吸间隔（s）	滤嘴通风孔
ISO 方案：ISO 3308 常规分析用吸烟机 定义和标准条件	35	60	不封闭
深度抽吸方案：在 ISO 3308 基础上有所调整	55	30	所有通风孔必须如 SOP 01 **12.2** 所述 100%封闭

SOP 02 描述烟草制品成分和释放物的分析方法验证。

1 适用范围

本方法描述卷烟主流烟气成分和卷烟烟丝成分分析方法验证的标准操作规程。

2 参考标准

2.1 ISO 5725-2：检测方法和结果的准确度（正确度和精密度） 第 2 部分 确定标准检测方法重复性和再现性的基本方法。

3 术语和定义

3.1 烟草制品：完全或部分以烟叶为原料制造，用于抽吸、吸吮、咀嚼或吸入的产品（WHO FCTC 第 1(f)条）。

3.2 卷烟烟丝：卷烟中含烟草的部分，包括再造烟叶、烟梗、膨胀烟丝和添加剂。

3.3 ISO 方案：抽吸过程参数为抽吸容量 35 mL，抽吸间隔 60 s，每次抽吸持续时间 2 s，不封闭滤嘴通风孔。

3.4 实验室样品：用于进行实验室检测的样品，由一次性或在规定时间内送到实验室的单一类型产品组成。

3.5 试样：从实验室样品中随机抽取的待测样品。所取样品数量应能代表实验室样品。

3.6 试料：从试样中随机抽取的用于单次测试的样品。所取样品数量应能代表试样。

3.7 重复性：对每个试样的所有测试均应在重复性条件下进行，即在短时间内由同一操作人员使用同一实验仪器完成。如果需要的话，对不同试样的测试可在不同的检测日期进行。

3.8 再现性：所有测试均应在再现性条件下进行，即可在检测日期、操作人员、实验室、实验仪器、试剂和环境条件不同的情况下，使用相同的方法对相同的材料进行测试。

4 方法概述

4.1 WHO TobLabNet 将为每种待验证的分析方法指定一个主导实验室。

4.2 为每种待验证的分析方法编写一份标准操作规程和验证方案，并发送给所有成员实验室。

4.3 为了验证分析方法，选择的产品尽可能涵盖全球范围内可找到的设计特征，同时使用特定产品以确保包装一致。一般来说，待分析的样品包括三种参比卷烟和两种市售卷烟品牌。所用市售品牌应随具体检测而不同。

4.4 用于验证目的的分析方法检测包括对单一卷烟品种的初始评估，以及对四种附加品种的全面评估。确认初始评估结果之后，再分析其他样品。主导实验室可自行决定是否包括初始评估。

4.5 数据分析和质量控制根据已经验证的标准操作规程或验证方案来进行，同时考虑每个实验室的规程。

4.6 将单个样品结果报告给 WHO 无烟草行动组。为每个实验室分配代码并且删除能够识别出实验室的信息，然后将结果发给至少两个团队。这些团队根据 ISO 规程评估数据中的异常值并独立计算重复性和再现性。完成数据评估后，将这些团队的结果进行比对。

4.7 编写验证报告，将结果包含在标准操作规程中。

5 安全与环境预防措施

5.1 遵守所有化学实验活动的常规安全与环境预防措施。

5.2 使用本检测方法对特定产品进行检测和评估时所用材料或设备可能对环境有害。本检测方法无力解决所有与使用有关的安全问题。所有使用此方法的人有责任咨询相关机构并遵循现有监管要求，制定健康与安全规程以及环境预防措施。

5.3 应特别注意避免吸入或皮肤接触危险化学品。在制备或处理未稀释材料、标准溶液、萃取液和收集样品时，应在通风橱中进行，并穿戴合适的实验服、手套和护目镜。

6 主导实验室

对每种要验证的分析方法，WHO TobLabNet 将会从其成员中指定一个实验室作为主导实验室来进行具体的验证研究。

主导实验室负有以下职责：
- 主导开发和选择合适的分析过程，编制标准操作规程和验证方案；
- 主导验证研究并跟踪研究进度；
- 如有需要，为成员实验室提供帮助；
- 编写验证研究报告。

7 标准操作规程

标准操作规程应记录所有操作的详细说明，包括试剂、耗材和所需设备，还应包含特殊预防措施和具体操作方法，并具有重复性和再现性。

标准操作规程的标准化模板见附录1。

要成功操作吸烟机或其他分析设备，须对操作人员进行培训。对操作吸烟机或其他分析设备没有经验的人员，在检测烟草制品释放物和成分前均应接受 WHO TobLabNet 的培训。

8 验证方案

验证方案应被送到验证该研究方法的所有参与者手中。验证方案应包含以下详细信息：
- 验证时间表；
- 待分析样品的数量和类型；
- 每个样品的重复次数；
- 有关如何报告结果的说明，最好使用 Excel 模板；
- 如有抽吸方案，则给出具体流程。

验证方案示例见附录2（卷烟烟丝成分）和附录3（卷烟主流烟气成分）。

9 样品

9.1 描述。

9.1.1 在寄送前，从可用制品中随机抽取样品卷烟，混合并重新包装。

9.1.2 仅用于方法验证的目的，需要检查产品是否存在物理异常。所有可看出物理异常的产品不能送到参与者手中进行后续的分析。

9.2 装运和收据。

9.2.1 烟草实验室网络将向每个参与的实验室运送每种卷烟品种的卷烟。将发送足够的样本以满足检测要求并配额外的备用样品。

9.2.2 除了测试所需的样品外，还需要在测试期间遇到问题时提供额外的样品。

9.2.3 烟草实验室网络将告知所有参与的实验室样品已发送。

9.2.4 在收据上，每个实验室将通过电子邮件告知烟草实验室网络样品已收到，并说明材料是否损坏。

9.2.5 如果在运输过程中遇到问题，烟草实验室网络将在收到告知后尽快发送替换样品。

9.2.6 收到样品后，每个实验室将完成分析并在研究方案规定时间内报告数据。

9.2.7 小心处理卷烟，避免造成损坏。

9.2.8 舍弃明显弯曲、撕裂、压碎或缺少大量烟丝的卷烟。

9.3 储存。

产品可以在室温下储存 3 个月。超过期限后，样品应在 $-20°C$ 或更低的温度下，且在原始包装或密闭容器内储存。

10 数据报告

10.1 按照 WHO 预防非传染性疾病研究方案或 SOP 规定进行结果报告。为了结果的一致性和有效的统计评估，推荐使用 Microsoft Excel 模板进行报告。如果没有 Microsoft Excel 或兼容的电子表格软件，实验室可以通过 ASCⅡ 格式进行数据报告。如果这个也不行，则可以通过打印件进行数据报告。如果可行，Excel 模板将会通过电子邮件发送到每个实验室。

10.2 为了保证机密性，每个实验室将会有相应的代码，世界卫生组织无烟草行动组（TFI）将会删除所有可能识别参与实验室的信息。进行结果报告时，指定的实验室代码会随着数据报告一起被提供。

10.3 低于报告限（LOR）的结果应表示为<LOR。当结果低于 LOR 时，不要将数据留空。

10.4 每个数据记录应包含研究方案所要求的信息，至少包括：

10.4.1 实验室代码。

10.4.2 样品 ID。

10.4.3 分析物 1 的结果。

10.4.4 分析物 2 的结果。

10.4.5 分析物 n 的结果。

10.4.6 具体分析方法。

10.4.7 意见。

不要报告质量控制或质量保证参数控制外的数据。

11 数据分析

由 WHO TFI 将编码的数据发送给至少两个团体进行统计评估。

各团体将根据 ISO 程序独立地对数据进行评估。

11.1 数据分析将遵循 ISO 5725-2 [**2.1**]规定。

11.2 按照 ISO 5725-2 规定识别异常值。在重复性和再现性的统计确定中不包括鉴定为异常值的结果。

11.3 如果数据存在任何不一致,或者数据被确定为异常值,则通过 WHO TFI 要求报告该数据的实验室检查是否存在异常并保持机密性。

11.4 根据 ISO 5725-1 [**13.2**]和 ISO 5725-2 [**13.3**]提供精确数据的要求,独立计算重复性和再现性。完成数据的评估后,比较各个团体的结果。

11.5 统计结果将会发送到 WHO 和主导实验室。

12 验证报告

验证研究的拓展报告包括:
- 研究大纲;
- 参与者和设备概述;
- 使用的数据分析方法描述;
- 数据分析总结:异常数据,丢失数据,平均值和标准差,数据一致性(曼德尔 h 检验和 k 检验),异常值测试(科克伦法则、格拉布斯法则),探索性数据分析,精度分析(重复性、再现性方差和精度限制)。

报告将会提交给 WHO TFI 出版。

13 参考文献

13.1 ISO 标准 TC 制品（ISO/TC 126）。

13.2 ISO 5725-1：检测方法和结果的准确度（正确度和精密度） 第 1 部分 一般原则和定义。

13.3 ISO 5725-2：检测方法和结果的准确度（正确度和精密度） 第 2 部分 确定标准检测方法重复性和再现性的基本方法。

附录1 示例/标准操作规程模版

ISO和深度抽吸方案下卷烟主流烟气/卷烟烟丝成分的测定标准操作规程

方　　法：ISO和深度抽吸方案下卷烟主流烟气/卷烟烟丝中某成分的测定 分析物：成分（CAS＃） 　　　　成分（CAS＃） 基　　质：卷烟主流烟气中粒相物/卷烟烟丝 上次更新：××××年×月

吸烟机的抽吸方案不能代表所有人的吸烟行为：吸烟机测试对用于卷烟设计及监管目的的卷烟释放物表征非常有用，但将吸烟机的测量结果披露给吸烟者则会导致对不同品牌的暴露和风险差异的误解。吸烟机测量的烟气释放物数据可以用于产品危害评估，但并不意味着这些数据等效于人体暴露或风险的测量值。将吸烟机测量结果的差异表示为暴露或风险的差异是对世界卫生组织烟草实验室网络标准的滥用。

本文件由世界卫生组织（WHO）烟草实验室网络（TobLabNet）制订，作为国际标准化组织（ISO）和深度抽吸方案下卷烟主流烟气/卷烟烟丝成分的测定标准操作规程（SOP）。

引言

为了在全球范围内建立具有可比性的烟草制品检测方法，需要卷烟特定成分和释放物的一致性检测方法。2008年11月在南非德班举行的世界卫生组织《烟草控制框架公约》（WHO FCTC）第三次缔约方会议回顾了 FCTC/COP1(15)和FCTC/COP2(14)号决议关于制订 WHO FCTC 第9条（烟草制品成分管制）和第10条（烟草制品披露管制）实施指南的要求，并提到第三次缔约方会议工作报告中有关其工作进展的信息，提请公约秘书处授权 WHO 无烟草行动组在5年内对检测卷烟成分和释放物的分析化学方法进行验证（FCTC/COP/3/REC/1）。

根据2006年10月在加拿大渥太华举行的第三次会议上确定的优先次序标准，第9条和第10条工作组确定了测试和测量方法（分析化学）验证优先级清单如下：
- 烟碱；
- 氨；
- 保润剂[1,2-丙二醇、甘油(1,2,3-丙三醇)和三甘醇(2,2′-二乙醚乙二醇)]。

测量这些成分需要验证三个方法：一个适用于烟碱，一个适用于氨，一个适用于保润剂。

根据上述渥太华会议确定的优先次序标准，工作组确定了在卷烟主流烟气释放中应该验证测试和测量方法（分析化学）的下列优先级物质清单：
- 4-(N-甲基亚硝胺基)-1,3-吡啶基-1-丁酮（NNK）；
- N-亚硝基去甲烟碱（NNN）；
- 乙醛；
- 丙烯醛；
- 苯；
- 苯并[a]芘；
- 1,3-丁二烯；
- 一氧化碳；
- 甲醛。

这些释放物使用下述两种抽吸方案测量，按五个方法进行验证：一个适用于烟草特有的亚硝胺（NNK 和 NNN），一个适用于苯并[a]芘，一个适用于醛类（乙醛、丙烯醛和甲醛），一个适用于挥发性有机物（苯和1,3-丁二烯），一个适用于一氧化碳。

下表列出了用于验证上述测试方法的两种抽吸方案。

抽吸方案	抽吸容量（mL）	抽吸间隔（s）	滤嘴通风孔
ISO 方案：ISO 3308 常规分析用吸烟机 定义和标准条件	35	60	不封闭
深度抽吸方案：与 ISO 3308 基础上有所调整	55	30	所有通风孔必须按 SOP 01 100%封闭

本 SOP 描述 ISO 和深度抽吸方案下卷烟主流烟气/卷烟烟丝中某成分的测定标准操作规程。

1 适用范围

本方法适用于测定卷烟主流烟气（MS）/卷烟烟丝中[成分名称]的定量测定：使用[技术名称]分析[成分名称]。如果需要，当更多的成分使用该方法进行测定时，增加：缔约方大会建议对……和……进行测试……本方法所含……和……的信息是针对那些选择这些测试的人的。

2 参考标准

2.1 ISO 3308：常规分析用吸烟机 定义和标准条件。

2.2 ISO 4387：卷烟 常规分析用吸烟机测定总粒相物和去烟碱干粒相物。

2.3 ISO 3402：烟草和烟草制品 用于调节和测试的大气环境。

2.4 ISO 5725-2：检测方法和结果的准确度（正确度和精密度） 第 2 部分 确定标准检测方法重复性和再现性的基本方法。

2.5 WHO TobLabNet SOP 01 卷烟深度抽吸标准操作规程（如需要）。

2.6 WHO TobLabNet SOP 02 烟草制品成分和释放物分析方法验证标准操作规程。

3 术语和定义

3.1

3.2

3.3

3.4

3.5 烟草制品：完全或部分以烟叶为原料制造，用于抽吸、吸吮、咀嚼或吸入的产品（WHO FCTC 第 1(f)条）。

3.6 深度抽吸方案：抽吸过程参数为抽吸容量 55 mL，抽吸间隔 30 s，每次抽吸持续时间 2 s，100%封闭滤嘴通风孔（如需要）。

3.7 ISO 方案：抽吸过程参数为抽吸容量 35 mL，抽吸间隔 60 s，每次抽吸持续时间 2 s，不封闭滤嘴通风孔（如需要）。

3.8 实验室样品：用于进行实验室检测的样品，由一次性或在规定时间内送到实验室的单一类型产品组成。

3.9 试样：从实验室样品中随机抽取的待测样品。所取样品数量应能代表实验室样品。

3.10 试料：从试样中随机抽取的用于单次测试的样品。所取样品数量应能代表试样。

4 方法概述

4.1 所选分析方法概述。

4.2

4.3

4.4

5 安全和环境预防措施

5.1 遵守所有化学实验活动的常规安全与环境预防措施。

5.2 使用本检测方法对特定产品进行检测和评估时所用材料或设备可能对环境有害。本检测方法无力解决所有与使用有关的安全问题。所有使用此方法的人有责任咨询相关机构并遵循现有监管要求，制定健康与安全规程以及环境预防措施。

5.3 应特别注意避免吸入或皮肤接触危险化学品。在制备或处理未稀释材料、标准溶液、萃取液和收集样品时，应在通风橱中进行，并穿戴合适的实验服、手套和护目镜。

5.4 如果适用，包括特定的安全和/或环境预防措施。

6 仪器和设备

6.1 按照 ISO 3402 [**2.3**]规定调节卷烟烟丝/卷烟所需设备。

6.2 按照 ISO 4387 [**2.2**]规定标记烟蒂长度所需设备（仅用于需吸烟的方法）。

6.3 按照 WHO TobLabNet SOP 01 [**2.5**]规定采用深度抽吸方案时封闭通风孔所需设备（仅用于需吸烟的方法）。

6.4 按照 ISO 3308 [**2.1**]规定抽吸烟草制品所需设备。

6.5 测量精度达小数点后×位的分析天平。

6.6

6.7

6.8 示例：配有火焰离子化检测器（FID）的毛细管气相色谱仪（GC）。

6.9 示例：能够明显分离溶剂、内标物、烟碱和其他烟草成分色谱峰的毛细管气相色谱柱，如 Varian WCOT 熔融石英色谱柱（25 m × 0.25 mm ID，涂层 CPWAX 51）。

7 试剂

除非另有说明，所有试剂应至少为分析纯。试剂尽可能以 CAS 号来识别。

7.1

7.2

7.3

7.4

7.5

8 玻璃器皿准备

清洁、干燥的玻璃器皿，避免残留物污染。

9 溶液制备

9.1 [溶液名称]

    9.1.1

    9.1.2

    9.1.3

    9.1.4

9.2 [溶液名称]

    9.2.1

    9.2.2

    9.2.3

    9.2.4

9.3 [溶液名称]

    9.3.1

    9.3.2

    9.3.3

9.3.4

10 标准溶液配制

下述标准溶液配制方法仅供参考,如有需要可进行调整。

10.1 ××内标液(如适用)。

 10.1.1 ××内标物储备液(×.× mg/mL)。

  10.1.1.1

  10.1.1.2

  10.1.1.3

 10.1.2 ××混合内标物溶液(×.× mg/mL)。

  10.1.2.1

  10.1.2.2

  10.1.2.3

10.2 ××标准溶液(× g/L)。

 10.2.1 ××标准储备液(×.× mg/mL)。

  10.2.1.1

  10.2.1.2

 10.2.2 ××混合标准中间溶液(×.× mg/mL)。

  10.2.2.1

  10.2.2.2

 10.2.3 ××最终混合标准溶液。

  10.2.3.1 根据表×制备最终混合标准溶液。

  10.2.3.2

内标物的最终浓度由下式确定:

$$\text{内标物最终浓度(ng/mL)} = x \cdot \times \times \times$$

式中:x——……中确定的标准原样的原始质量(g)。

标准溶液的最终浓度由下式确定:

$$\text{标准溶液最终浓度(ng/mL)} = x \cdot y / \times \times \times$$

式中:x——……中确定的标准原样的原始质量(g);

  y——……所述混合标准溶液的体积(μL)。

标准溶液中××的最终浓度示于表×。

表× ××标准曲线

标准	××标准溶液体积 (×g/L)(mL)	内标物溶液体积(μL)	总体积(mL)	××最终混合标准溶液的近似浓度(mg/L)	在卷烟主流烟气（mg/支）或烟丝(mg/g)中的浓度
1					
2					
3					
4					
5					

标准溶液范围可根据使用的设备和待测样品进行调整，注意方法灵敏度可能对结果产生的影响。

所有溶剂和溶液在使用前调节至室温。

11 抽样

11.1 根据 ISO 8243 [**23.4**]或替代方法，根据各实验室惯例或者要求的特殊规定或样品可用性，选取具有代表性的实验室卷烟样品。

11.2 试样的组成。

  11.2.1 若可能，将实验室样品分成单独的单元（如包、条）。

  11.2.2 从至少 $\sqrt{n}$ 个[**23.5**]单元中取出每种试样的等量产品。

  11.2.3 如果没有可用的单元，则将整个实验室样品合并为一个单元。

  11.2.4 从组合单元中随机抽取所需制品的数量。

12 卷烟准备

12.1 按照 ISO 3402 [**2.3**] 调节所有待测卷烟。

12.2 按照 ISO 4387 [**2.2**] 和 WHO TobLabNet SOP 01 [**2.5**]标记烟蒂长度。

12.3 按照 WHO TobLabNet SOP 01 [**2.5**]的深度抽吸方案或 ISO 方案准备待测样品。

或者：

12.1 从一包（例如含 20 支）卷烟和质量控制样品（若适用）中取出烟丝，或使用至少 ×× g 加工过的卷烟烟丝。

12.2 将足够的卷烟烟丝混合至每个试料至少含 ×× g 烟丝。

12.3 混合并研磨卷烟烟丝直至其足够精细，能通过 4 mm 筛网。

12.4 按照 ISO 3402 [**2.3**]对烟草制品的要求调节磨碎的卷烟烟丝（对于非挥发性成分）。除非对分析方法不利，否则建议进行调节。

13 吸烟机设置

（如果不适用，请保留标题并注明"不适用"）

13.1 环境条件。

按 ISO 3308 [**2.1**]要求设定抽吸的环境条件。

13.2 吸烟机要求。

除深度抽吸方案按 WHO TobLabNet SOP 01 [**2.5**]进行准备外，其余按照 ISO 3308 [**2.1**]的要求设置吸烟机。

14 生成样品

（如果不适用，请保留标题并注明"不适用"）

在指定吸烟机上抽吸足够量的样品，样品含量应不足以使滤片发生穿透，并且某成分（S）的浓度落在分析的标准曲线范围内。

14.1 抽吸卷烟试样并按照 ISO 4387 [**2.2**]或 WHO TobLabNet SOP 01 [**2.5**]收集 TPM。

14.2 至少包含一个用于质量控制的参比试样。

14.3 当样品类型为第一次检测时，评估滤片穿透的可能性，调整卷烟的数量以避免穿透。当确定焦油、烟碱和一氧化碳时，对于 92 mm 滤片和 44 mm 滤片，穿透分别发生在总粒相物水平超过 600 mg 和 150 mg 时。一旦发生穿透，必须减少每个滤片上的卷烟数量。

14.4 表×显示了 ISO 和深度抽吸方案中直线型和转盘型吸烟机每次抽吸的数量。

表× 直线型和转盘型吸烟机进行一次测量的卷烟数量

	ISO 方案		深度抽吸方案	
	直线型	转盘型	直线型	转盘型
每个滤片上的卷烟数量	5	20	3	10
每个结果包含的滤片数	4	1	7	2

14.5 记录每个滤片的卷烟数量和总抽吸量。

14.6 测量完成所需样品量后，进行五次空吸，然后从吸烟机上取下捕集器。

15 样品制备

对于含抽吸的方法：

15.1 滤片萃取。

从捕集器上取下滤片，将滤片对折，再对折，使滤片含有 TPM 的面向内。用不含 TPM 的另一面擦拭捕集器内表面，使残留在捕集器上的粒相物转移到滤片上。将每个滤片……

15.2

15.3

15.4

对于不含抽吸的方法：

15.1 取×.× g 充分混合、研磨和调节的试样，精确称量×.××× g 放入萃取容器中。

15.2

15.3

15.4

16 样品分析

使用[技术名称]测定卷烟主流烟气/卷烟烟丝中的某成分含量。分析物可以在所用色谱柱上与其他潜在干扰物质分离开，比较未知物的峰面积比与已知浓度标准溶液的峰面积比，得到单个分析物的浓度。

16.1 仪器操作条件示例。

本节说明 HPLC 或 GC 设备和操作条件示例。应提供足够的信息，以便不同的实验室能够重复结果。

HPLC 柱：品牌名称和柱子类型（××mm××× mm，×.× μm 粒径）或等效柱；

进样量：× μL；

柱温箱温度：(×××±×)℃（根据所使用的柱子更改）；

脱气装置：开/关；

二元泵流量：×× mL/min；

流动相 A：

流动相 B：

梯度：如表×所示；

总分析时间：×× min。

表× 某成分分离的 HPLC 梯度

时间（min）	流量（μL/min）	流动相 A（%）	流动相 B（%）

根据仪器和色谱柱条件以及色谱峰的分离度调整操作参数。

或者：

GC柱：品牌名称和柱子类型（××m××.××mm ID）；

涂层：涂层类型；

色谱柱温度：×××℃ （恒温或梯度升温程序）；

进样口温度：×××℃；

检测器温度：×××℃；

载气：载气类型，流量×.× mL/min；

进样体积：×.× μL；

进样方式：分流× ：××

根据仪器和色谱柱条件以及色谱峰的分离度调整操作参数。

16.2 预期保留时间。

 16.2.1 对于此处描述的条件，预期的洗脱顺序为[化合物名称]、[化合物名称]。

 16.2.2 对于 HPLC，流量、流动相浓度和色谱柱的使用年限等差异可能会改变保留时间。对于 GC，温度、气体流量和色谱柱的使用年限等差异可能会改变保留时间。

 16.2.3 在分析开始前必须验证洗脱顺序和保留时间。

 16.2.4 在上述条件下，预期的总分析时间约为×× min。可延长分析时间以优化性能。

 16.2.5 如果需要，可在此包括对保留时间有特定影响的特定仪器设置。

16.3 [化合物名称]的测定。

根据每个实验室的做法来确定顺序。本节给出一个确定[化合物名称]顺序的示例。

 16.3.1 在相同条件下注射两份(如需要)标准溶液和样品萃取物试样。

 16.3.2 在使用前通过注入两个× μL 样品溶液等份试样作为预试样来调节系统（如果需要的话）。

 16.3.3 在与样品相同的条件下注入检查标准品（带有内标物的空白溶液）来验证 HPLC-MS/MS 或 GC 系统的性能。

 16.3.4 注入空白溶液（不含内标物的萃取液）来检查系统或试剂中是否存在污染。

16.3.5 将每种标准溶液的等份试样注入 HPLC-MS/MS 或 GC。

16.3.6 记录某成分和内标物的峰面积。

16.3.7 计算某成分标准溶液（包括标准空白溶液）的色谱峰与内标峰的相对响应比（RF= $A_{content}/A_{IS}$）。

16.3.8 以某成分浓度（X 轴）与峰面积比（Y 轴）绘制标准曲线。

16.3.9 截距在统计上不应该与零有显著差异。

16.3.10 标准曲线应在整个标准范围内呈线性。

16.3.11 使用线性回归的斜率 b 和截距 a 来计算线性回归方程（$Y=a+bx$）。如果线性回归方程的 R^2 小于 0.99，则应重新校准。如果单个校准点与预期值（通过线性回归估计）相差超过 10%，则应该舍弃该点。

16.3.12 注入质量控制样品和试料，用适当的软件确定峰面积。

16.3.13 所有试料的信号（峰值比）必须落在标准曲线的工作范围内；否则，应调整标准溶液或试料的浓度。

代表性色谱图见附录×。

17 数据分析和计算

17.1 针对每个校准标准液计算[化合物名称]的峰面积的相对响应比：

$$RF = \frac{A_a}{A_{IS}}$$

式中：RF——相对响应比；

A_a——目标分析物的峰面积；

A_{IS}——相应内标物的峰面积。

为每个[化合物名称]绘制相对响应因子 A_a/A_{IS}（Y 轴）与浓度（X 轴）的关系图，使用线性回归的斜率（k）和截距（m）计算线性回归方程（$RF = kX + m$）。

17.2 标准曲线应在整个标准范围内呈线性。

17.3 目标分析物（ng/支）中给定组分的含量 M_{ts} 由计算的试样的相对响应比与标准曲线的斜率和截距来确定：

$$M_{ts} = \frac{Y - m}{k} \times \frac{V}{N}$$

式中：M_{ts}——每支卷烟的计算含量（ng/支）；

V——萃取液体积（40 mL）；

N——每个滤片上卷烟抽吸的数量。

如果适用，可以使用其他计算过程。

18 特别注意事项

18.1 安装新色谱柱后,按照制造商的规定,在指定的仪器条件下注入烟草样品萃取物进行调节。应重复进样,直至每个成分和内标物的峰面积(或峰高)可重复。

18.2

18.3

19 数据报告

19.1 报告每个评估样品的单独测量值。
19.2 应按照方法规范的规定报告结果。
19.3 更多信息请参阅 WHO TobLabNet SOP 02 [2.6]。

20 质量控制

20.1 控制参数。
如果控制测量值超出预期值的公差极限,则必须进行适当的调查并采取措施。如有必要,需要做额外的实验室质量保证规程,以符合各个实验室的做法。
20.2 实验室试剂空白样。
如 **16.3.4** 所述,要在样品制备和分析过程中检测潜在的污染,需包含实验室试剂空白样。空白样由分析测试样品中使用的所有试剂和材料组成,并像测试样品一样进行分析。应根据各个实验室的做法对空白样进行评估。
20.3 质量控制样品。
为了验证整个分析过程的一致性,根据各个实验室的做法分析参比卷烟(或适当的质量控制样品)。

21 方法性能

21.1 报告限。
报告限设置为所用标准曲线的最低浓度,重新计算为 mg/g 或 mg/支卷烟(×.× mg/g 对应 ×× mg/L 的最低标准浓度)。如有必要,应将在标准曲线范围外的数据报告为低于定量限(BLOQ)或高于定量限(ALOQ)。
21.2 实验室基质加标回收率。
加入到基质上的分析物的回收率用作准确度的替代量度。通过将已知含量的标准品[执行回收率确定的规程]来确定回收率。回收率由下式计算，如表×所示。

回收率（%）=100×（分析结果–未加标结果）/加标量

表× 加到基质的组分的平均值和回收率

加标量 （ng）	组分		组分		组分	
	平均值（ng）	回收率（%）	平均值（ng）	回收率（%）	平均值（ng）	回收率（%）
×.××	×.××	×.××	×.××	×.××	×.××	×.××
×.××	×.××	×.××	×.××	×.××	×.××	×.××
×.××	×.××	×.××	×.××	×.××	×.××	×.××

21.3

21.4 分析特异性。

描述所使用的方法/技术的分析特异性，例如目标分析物的保留时间用于验证分析特异性。使用 QC 卷烟主流烟气/卷烟烟丝的成分响应与内标成分响应既定比范围来验证来自未知样品结果的特异性。

21.5 线性。

建立的成分标准曲线在××~×× ng/mL 的标准浓度范围内是线性的。

21.6 潜在干扰。

××的存在可能会造成干扰，因为该成分的保留时间可能与成分/内标物的保留时间相似。含有××的样品最有可能发生这种干扰。

或者，没有已知成分因具有与成分或内标物类似的保留时间而产生的干扰。

22 重复性和再现性

根据 2012 年世界卫生组织烟草实验室网络进行的国际合作研究，烟草制品成分和释放物分析方法验证标准操作规程（WHO TobLabNet SOP 02）[2.6]对此方法给出了以下精密度限值：

常规正确操作该方法时，同一操作者使用同一设备在最短的操作时间内两个单一结果之间的差异超过重复性 r 的情况，平均 20 个样品不超过 1 次。

常规正确操作该方法时，两个实验室报告的匹配卷烟样品的单一结果超过再现性 R 的情况，平均 20 次不超过 1 次。

根据 ISO 5725-1 [23.2]和 ISO 5725-2 [2.4]对测试结果进行统计分析，得到表×所示的精密度数据。

为了计算 r 和 R，在 ISO 条件下，测试结果被定义为来自直线型吸烟机的四个滤片（每个滤片上 5 支烟）和转盘型吸烟机的一个滤片（每个滤片上 20 支烟）平均释放量的 7 次重复的平均值。在深度抽吸条件下，测试结果被定义为来自直线型吸烟机的七个滤片（每个滤片上 3 支烟）和转盘型吸烟机的两个滤片（每个滤片上 10 支烟）平均释放量的 7 次重复的平均值。

对于需要吸烟的方法：

表× 从参比卷烟主流烟气中测定成分的精密度限值（mg/cig）

参比卷烟	ISO 抽吸方案				深度抽吸方案			
	N	m_{cig}	r_{limit}	R_{limit}	N	m_{cig}	r_{limit}	R_{limit}
1R5F	×	××.××	××.××	××.××	×	××.××	××.××	××.××
3R4F	×	××.××	××.××	××.××	×	××.××	××.××	××.××
CM6	×	××.××	××.××	××.××	×	××.××	××.××	××.××

注：商品卷烟中的成分含量在参比卷烟的范围内，因此未在此表中报告。

对于无吸烟步骤的方法：

表× 从参比卷烟中测定卷烟烟丝中成分含量的精密度限值（mg/g）

参比卷烟	烟丝			
	N	m_{cig}	r_{limit}	R_{limit}
1R5F	×	××.××	××.××	××.××
3R4F	×	××.××	××.××	××.××
CM6	×	××.××	××.××	××.××

注：商品卷烟中的成分含量在参比卷烟的范围内，因此未在此表中报告。

23 参考文献

23.1 WHO TobLabNet 官方方法：卷烟主流烟气/卷烟烟丝成分的测定。

23.2 ISO 5725-1：检测方法和结果的准确度（正确度和精密度） 第 1 部分一般原则和定义。

23.3 参考该方法所引用的基本方法（如适用）。

23.4 ISO 8243：卷烟 抽样。

23.5 联合国毒品和犯罪问题办公室（UNODC）《代表性药物取样指南》（http://www.unodc.org/documents/scientific/Drug_Sampling.pdf）。

23.6 ISO 标准 TC 制品（ISO/TC 126）。

23.7

附录2　示例/成分分析方法验证规程模版

烟丝成分的测定分析方法验证规程

方　　法：卷烟烟丝中成分分析方法的验证
分　析　物：目标分析物
基　　质：烟草
上次更新：××××年×月

本文件由世界卫生组织（WHO）烟草实验室网络（TobLabNet）制订，作为测定卷烟烟丝成分分析方法验证的标准操作规程（SOP）。

引言

为了在全球范围内建立具有可比性的烟草制品检测方法，需要卷烟特定成分和释放物的一致性检测方法。2008年11月在南非德班举行的世界卫生组织《烟草控制框架公约》（WHO FCTC）第三次缔约方会议回顾了FCTC/COP1(15)和FCTC/COP2(14)号决议关于制订WHO FCTC第9条（烟草制品成分管制）和第10条（烟草制品披露管制）实施指南的要求，并提到第三次缔约方会议工作报告中有关其工作进展的信息，提请公约秘书处授权WHO无烟草行动组在5年内对检测卷烟成分和释放物的分析化学方法进行验证（FCTC/COP/3/REC/1）。

根据2006年10月在加拿大渥太华举行的第三次会议上确定的优先次序标准，WHO FCTC第9条和第10条工作组确定优先验证以下成分的分析化学检测方法：
- 烟碱；
- 氨；
- 保润剂[1,2-丙二醇、甘油（1,2,3-丙三醇）和三甘醇（2,2′-二乙醚乙二醇）]。

测量这些成分需要验证三个方法：一个适用于烟碱，一个适用于氨，一个适用于保润剂。

根据上述渥太华会议确定的优先次序标准，工作组确定了在卷烟主流烟气释放中应该验证测试和测量方法（分析化学）的下列优先级物质清单：
- 4-(N-甲基亚硝胺基)-1,3-吡啶基-1-丁酮（NNK）；
- N-亚硝基去甲烟碱（NNN）；
- 乙醛；
- 丙烯醛；
- 苯；
- 苯并[a]芘；
- 1,3-丁二烯；
- 一氧化碳；
- 甲醛。

这些释放物使用下述两种抽吸方案测量，按五个方法进行验证：一个适用于烟草特有的亚硝胺（NNK和NNN），一个适用于苯并[a]芘，一个适用于醛类（乙醛、丙烯醛和甲醛），一个适用于挥发性有机物（苯和1,3-丁二烯），一个适用于一氧化碳。

1 适用范围

引入非烟草行业附属实验室测定烟草和卷烟烟丝成分的标准方法,并确定参与实验室对成分测定的一致程度。

2 日程安排

标准方法介绍:基于方法来源描述标准方法。

2.1 标准方法将与本规程一起发送给所有烟草实验室网络成员,并邀请参加本研究,研究将在 yyyy 年 mm 月完成。

2.2 测试卷烟。

采购测试卷烟或测试物品,并由烟草实验室网络成员发送给所有参与者。本研究所需的卷烟总数将在参与者的电话会议结束后确定。yyyy 年 mm 月样本将与 SOP、研究方案和 Microsoft Excel 数据表一起分发给所有参与者以获得测试结果。请不要修改任何数据表以方便数据分析。

2.3 测量范围和时间。

成分测量应在 yyyy 年 mm 月和 yyyy 年 mm 月之间执行。参与者应在一天内对每个样品进行 7 次测定中的一次(每天可测量一个以上的样品)。

2.4 报告结果。

研究的测试结果应不迟于 yyyy 年 mm 月并采用指定数据表向 WHO TobLabNet 报告。

世界卫生组织将为每个参与的实验室分配代码以保密。

2.5 数据分析和统计评估。

烟草实验室网络将依据 WHO TobLabNet SOP 02 和 ISO 5725 检测方法和结果的准确度(正确度和精密度) 第 2 部分 确定标准检测方法重复性和再现性的基本方法对报告的数据进行统计评估。

在 yyyy 年 mm 月之前,烟草实验室网络将向所有参与者发送结果和统计评估的摘要报告。报告发送给参与者后,将在第一次烟草实验室网络会议上进行讨论。

3 测试项目

3.1 描述。

对于验证研究,样品数量已设定为 5 个,分为 3 个参比卷烟和 2 个商品卷烟(表 1)。出于效率原因,用于所有成分确定的验证研究的样本将被发送给所有参与者。

表1 烟草特有亚硝胺研究示例测试项目

样品编号	制品类型	样品名
A	参比卷烟	1R5F
B	参比卷烟	2R4F
C	参比卷烟	CM6
D	卷烟	市售品牌名
E	卷烟	市售品牌名

3.2 运输和收据。

(a) 烟草实验室网络将每种样品卷烟或烟丝样品邮寄至每个参与实验室。

(b) 烟草实验室网络会告知所有参与实验室样品已经发送。

(c) 收到后，每个参与的实验室将通过电子邮件通知烟草实验室网络收到的样品以及卷烟的任何问题或损坏情况。

(d) 如果在运输过程中遇到问题，烟草实验室网络将在被通知后尽快发送替换样品。

(e) 收到样品后，每个实验室将完成分析并在指定的时间表内报告数据。

(f) 小心处理样品以避免损坏。舍弃明显受损或缺少大量烟丝的卷烟。

3.3 储存。

样品可以在室温下储存长达3个月。为了储存更长的时间，样品应储存在原包装或密封容器中，置于-20℃或更低温度的冰箱中。

3.4 样品制备。

样品制备应根据SOP（根据需要进行研磨和筛分）进行。

3.5 要分析的测试项目数。

对于每个单独的样品，将执行7次重复测量。每个样品的7次测定中的一次都应在一天内执行（每天可测量多个样品）。

4 报告测试结果

应使用电子邮件提供的Microsoft Excel模板报告测试结果。请不要修改任何数据表以方便数据分析。

如果没有Microsoft Excel或兼容的电子表格软件，实验室可以通过ASC II格式进行数据报告。如果这个也不行，则可以通过纸质报告进行数据报告。

测试结果应根据表2中的详细说明给出。

表 2　报告的数据和格式

参数	单位	报告的小数点位数
使用烟草数量	g	#.###1 g
成分	mg/g~mg/g	#.###1 mg/g

参与者还应报告研究中使用的主要设备（类型、型号、制造商）的概况。更多详细信息，请提供可参阅的数据表。

如果出于任何原因偏离 SOP 或研究方案，请在 Microsoft Excel 数据表或单独的参与实验室报告中注明偏差及其原因。

5　预计时间表

成分方法验证计划	时间表
发送 SOP 和研究方案以及征集参与者	yyyy 年 mm 月 dd 日
意见反馈和参与调查	yyyy 年 mm 月 dd 日
将样品寄送给参与者	yyyy 年 mm 月 dd 日
进行分析并报告结果	yyyy 年 mm 月 dd 日
结果的统计评估	yyyy 年 mm 月 dd 日
向参与者报告评估摘要	yyyy 年 mm 月 dd 日
通过 TFI 向 COP 第 9 条和第 10 条工作组报告摘要结果	在烟草实验室网络会议上讨论之后

附录3 示例/释放物分析方法验证规程模版

ISO和深度抽吸方案下测定卷烟主流烟气成分含量的分析方法验证规程

方　　法：ISO和深度抽吸方案下卷烟主流烟气成分含量测定分析方法操作规程
分 析 物：不适用
基　　质：卷烟主流烟气
上次更新：××××年××月

本文件由世界卫生组织（WHO）烟草实验室网络（TobLabNet）制订，作为卷烟主流烟气/卷烟烟丝成分的测定标准操作规程（SOP）。

引言

为了在全球范围内建立具有可比性的烟草制品检测方法，需要卷烟特定成分和释放物的一致性检测方法。2008年11月在南非德班举行的世界卫生组织《烟草控制框架公约》（WHO FCTC）第三次缔约方会议回顾了FCTC/COP1(15)和FCTC/COP2(14)号决议关于制订WHO FCTC第9条（烟草制品成分管制）和第10条（烟草制品披露管制）实施指南的要求，并提到第三次缔约方会议工作报告中有关其工作进展的信息，提请公约秘书处授权WHO无烟草行动组在5年内对检测卷烟成分和释放物的分析化学方法进行验证（FCTC/COP/3/REC/1）。

根据2006年10月在加拿大渥太华举行的第三次会议上确定的优先次序标准，WHO FCTC第9条和第10条工作组确定优先验证以下成分的分析化学检测方法：
- 烟碱；
- 氨；
- 保润剂[1,2-丙二醇、甘油（1,2,3-丙三醇）和三甘醇（2,2′-二乙醚乙二醇）]。

测量这些成分需要验证三个方法：一个适用于烟碱，一个适用于氨，一个适用于保润剂。

根据上述渥太华会议确定的优先次序标准，工作组确定了在卷烟主流烟气释放中应该验证测试和测量方法（分析化学）的下列优先级物质清单：
- 4-(*N*-甲基亚硝胺基)-1,3-吡啶基-1-丁酮（NNK）；
- *N*-亚硝基去甲烟碱（NNN）；
- 乙醛；
- 丙烯醛；
- 苯；
- 苯并[*a*]芘；
- 1,3-丁二烯；
- 一氧化碳；
- 甲醛。

这些释放物使用下述两种抽吸方案测量，按五个方法进行验证：一个适用于烟草特有的亚硝胺（NNK和NNN），一个适用于苯并[*a*]芘，一个适用于醛类（乙醛、丙烯醛和甲醛），一个适用于挥发性有机物（苯和1,3-丁二烯），一个适用于一氧化碳。

下表列出了用于验证上述测试方法的两种抽吸方案。

抽吸方案	抽吸容量（mL）	抽吸间隔（s）	滤嘴通风孔
ISO 方案：ISO 3308 常规分析用吸烟机定义和标准条件	35	60	不封闭
深度抽吸方案：与 ISO 3308 基础上有所调整	55	30	所有通风孔必须按 SOP 01 100%封闭

1 适用范围

引入非烟草行业附属实验室测定烟草和卷烟烟丝成分的标准方法，并确定参与实验室对成分测定的一致程度。

2 日程安排

2.1 标准方法介绍。

基于方法来源描述标准方法。

标准方法将与本规程一起发送给所有烟草实验室网络成员，并邀请参加本研究，研究将在 yyyy 年 mm 月完成。

2.2 测试卷烟。

采购测试卷烟或测试物品，并由烟草实验室网络成员发送给所有参与者。本研究所需的卷烟总数将在参与者的电话会议结束后确定。yyyy 年 mm 月样本将与 SOP、研究协议和 Microsoft Excel 数据表一起分发给所有参与者以获得测试结果。请不要修改任何数据表以方便数据分析。

2.3 测量范围和时间。

由研究参与者确定的成分组应为主流烟草烟气中的成分。使用 ISO 和深度抽吸方案的吸烟机用于产生主流烟气。有关吸烟机类型的更多细节，包括吸烟计划，见表 3 至表 6。

成分测量应在 yyyy 年 mm 月和 yyyy 年 mm 月之间执行。参与者应在一天内对每个样品进行 7 次测定中的一次（每天可测量一个以上的样品）。

2.4 报告结果。

研究的测试结果应在不迟于 yyyy 年 mm 月并采用指定数据表向 WHO TobLabNet 报告。

世界卫生组织将为每个参与的实验室分配代码以保密。

2.5 数据分析和统计评估。

烟草实验室网络将依据 WHO TobLabNet SOP 02 和 ISO 5725 检测方法和结果的准确度（正确度和精密度） 第 2 部分 确定标准检测方法重复性和再现性的基本方法对报告的数据进行统计评估。

在 yyyy 年 mm 月之前，烟草实验室网络将向所有参与者发送结果和统计评估的摘要报告。报告发送给参与者后，将在第一次烟草实验室网络会议上进行讨论。

3 测试项目

3.1 描述。

对于验证研究，样品数量已设定为 5 个，分为 3 个参比卷烟和 2 个商品卷烟（表1）。出于效率原因，将用于所有成分确定的验证研究的样本将被发送给所有参与者。

表 1 烟草特有亚硝胺研究示例测试项目

样品编号	制品类型	样品名
A	参比卷烟	1R5F
B	参比卷烟	2R4F
C	参比卷烟	CM6
D	卷烟	市售品牌名
E	卷烟	市售品牌名

3.2 运输和收据。

（a）烟草实验室网络将每种样品卷烟或烟丝样品邮寄至每个参与实验室。

（b）烟草实验室网络会告知所有参与实验室样品已经发送。

（c）收到后，每个参与的实验室将通过电子邮件通知烟草实验室网络收到的样品以及卷烟的任何问题或损坏情况。

（d）如果在运输过程中遇到问题，烟草实验室网络将在被通知后尽快发送替换样品。

（e）收到样品后，每个实验室将完成分析并在指定的时间表内报告数据。

（f）小心处理样品以避免损坏。舍弃明显受损或缺少大量烟丝的卷烟。

3.3 储存。

样品可以在室温下储存长达 3 个月。为了储存更长的时间，样品应储存在原包装或密封容器中，置于–20℃或更低温度的冰箱中。

3.4 样品制备。

样品制备应根据 SOP（选择和储存）进行。标记样本以进一步识别。如果将烟草制品从原始包装中取出，则将其存放在密封容器中，该容器的大小足以容纳样品并将其保持在–20℃或更低温度，直至需要时为止。

3.5 要分析的测试项目数。

对于每个单独的样品，将执行 7 次重复测量。每个样品的 7 次测定中的一次都应在一天内执行（每天可测量多个样品）。

4 报告测试结果

应使用电子邮件提供的 Microsoft Excel 模板报告测试结果。请不要修改任何数据表以方便数据分析。

如果没有 Microsoft Excel 或兼容的电子表格软件，实验室可以通过 ASC II 格式进行数据报告。如果这个也不行，则可以通过纸质报告进行数据报告。

测试结果应根据表 2 中的详细说明给出。

表 2　报告的数据和格式

参数	单位	报告的小数点位数
使用烟草数量	g	#.###1 g
成分	ng/支	#.##1 ng/支
成分	ng/支	#.##1 ng/支
成分	ng/支	#.##1 ng/支

参与者还应报告研究中使用的主要设备（类型、型号、制造商）的概况。更多详细信息，请提供可参阅的数据表。

如果出于任何原因偏离 SOP 或研究方案，请在 Microsoft Excel 数据表或个别参与实验室报告中注明偏差及其原因。

5 预计时间表

成分方法验证计划	时间表
发送 SOP 和研究方案以及征集参与者	yyyy 年 mm 月 dd 日
意见反馈和参与调查	yyyy 年 mm 月 dd 日
将样品寄送给参与者	yyyy 年 mm 月 dd 日
进行分析并报告结果	yyyy 年 mm 月 dd 日
结果的统计评估	yyyy 年 mm 月 dd 日
向参与者报告评估摘要	yyyy 年 mm 月 dd 日
通过 TFI 向 COP 第 9 条和第 10 条工作组报告摘要结果	在烟草实验室网络会议上讨论之后

6 抽吸计划

所描述的抽吸计划基于5种不同的样品（卷烟品牌/类型）和每个样品7次重复。如果使用不同数量的样品或重复样品，则应相应调整计划。

6.1 ISO方案。

表3 转盘型吸烟机的抽吸计划示例（每个滤片上20支卷烟）

天	实验编号	样品	天	实验编号	样品
1	1	A	5	21	C
1	2	B	5	22	D
1	3	C	5	23	E
1	4	D	5	24	A
1	5	E	5	25	B
2	6	D	6	26	D
2	7	E	6	27	C
2	8	A	6	28	B
2	9	B	6	29	A
2	10	C	6	30	E
3	11	B	7	31	C
3	12	C	7	32	B
3	13	D	7	33	A
3	14	E	7	34	E
3	15	A	7	35	D
4	16	E			
4	17	A			
4	18	B			
4	19	C			
4	20	D			

注：每个滤片应使用××mL萃取溶液单独萃取，然后进一步分析。一个结果是每个滤片/提取液（20支卷烟）的结果。

表4 20通道直线型吸烟机的抽吸计划示例（每个滤片5支卷烟）

天	通道编号																			
	1	2	3	4	5	6	7	8	9	10	11	12	13	14	15	16	17	18	19	20
1	A	A	A	A	B	B	B	B	C	C	C	C	D	D	D	D	E	E	E	E
2	D	D	D	D	E	E	E	E	A	A	A	A	B	B	B	B	C	C	C	C
3	B	B	B	B	C	C	C	C	D	D	D	D	E	E	E	E	A	A	A	A
4	E	E	E	E	A	A	A	A	B	B	B	B	C	C	C	C	D	D	D	D
5	C	C	C	C	D	D	D	D	E	E	E	E	A	A	A	A	B	B	B	B
6	D	D	D	D	C	C	C	C	B	B	B	B	A	A	A	A	E	E	E	E
7	C	C	C	C	B	B	B	B	A	A	A	A	E	E	E	E	D	D	D	D

注：应将四个滤片组合并用××mL萃取溶液萃取，并进一步分析。结果是四个滤片组合（20支卷烟）萃取液的结果。

6.2 深度抽吸方案。

表 5 转盘型吸烟机的抽吸计划示例（每个滤片上 10 支卷烟）

天	实验编号	样品	天	实验编号	样品
8	36	A	12	76	C
	37	B		77	D
	38	C		78	E
	39	D		79	A
	40	E		80	B
	41	A		81	C
	42	B		82	D
	43	C		83	E
	44	D		84	A
	45	E		85	B
8	46	D	13	86	D
	47	E		87	C
	48	A		88	B
	49	B		89	A
	50	C		90	E
	51	D		91	D
	52	E		92	C
	53	A		93	B
	54	B		94	A
	55	C		95	E
10	56	B	14	96	C
	57	C		97	B
	58	D		98	A
	59	E		99	E
	60	A		100	D
	61	B		101	C
	62	C		102	B
	63	D		103	A
	64	E		104	E
	65	A		105	D
11	66	E			
	67	A			
	68	B			
	69	C			
	70	D			
	71	E			
	72	A			
	73	B			
	74	C			
	75	D			

注：每个滤片应使用×× mL 萃取溶液单独萃取，然后进一步分析。结果是两个单独的滤片/萃取液（20 支卷烟）的平均值。

表6 20通道直线型吸烟机的抽吸计划示例（每个滤片3支卷烟）

天	实验号	通道编号																			
		1	2	3	4	5	6	7	8	9	10	11	12	13	14	15	16	17	18	19	20
8	8	A	A	A		B	B	B		C	C	C		D	D	D			E	E	E
	9			B	B	B	C	C	C	D	D	D	E	E	E		A	A	A		
9	10	D	D	D		E	E	E		A	A	A		B	B	B			C	C	C
	11		E	E	E	A	A	A	B	B	B	C	C	C	D	D	D				
10	12	B	B	B		C	C	C		D	D	D		E	E	E			A	A	A
	13			C	C	C	D	D	D	E	E	E	A	A	A	B	B	B			
11	14	E	E	E		A	A	A		B	B	B		C	C	C			D	D	D
	15		A	A	A	B	B	B	C	C	C	D	D	D	E	E	E				
12	16	C	C	C		D	D	D		E	E	E		A	A	A			B	B	B
	17		D	D	D	E	E	E	A	A	A	B	B	B	C	C	C				
13	18	D	D	D		C	C	C		B	B	B		A	A	A			E	E	E
	19			C	C	C	B	B	B	A	A	A	E	E	E	D	D	D			
14	20	C	C	C		B	B	B		A	A	A		E	E	E			D	D	D
	21		B	B	B	A	A	A	E	E	E	D	D	D	C	C	C				

注：将三个滤片组合并用××mL 萃取溶液萃取，然后进一步分析。结果是两套组合滤片/萃取液（18支卷烟＝2×9支卷烟）的平均值。

WHO TobLabNet SOP 03

ISO 和深度抽吸方案下卷烟主流烟气中烟草特有亚硝胺的测定标准操作规程

方　　法：	ISO 和深度抽吸方案下卷烟主流烟气中烟草特有亚硝胺的测定
分 析 物：	N-亚硝基降烟碱（CAS 号：16543-55-8）
	4-(N-甲基亚硝胺基)-1-(3-吡啶基)-1-丁酮（CAS 号：64091-91-4）
	N-亚硝基新烟草碱（CAS 号：71267-22-6）
	N-亚硝基假木贼碱（CAS 号：37620-20-5）
基　　质：	卷烟主流烟气粒相物
更新时间：	2014 年 6 月

> 吸烟机的抽吸方案不能代表所有人的吸烟行为：吸烟机测试对用于卷烟设计及监管目的的卷烟释放物表征非常有用，但将吸烟机的测量结果披露给吸烟者则会导致对不同品牌的暴露和风险差异的误解。吸烟机测量的烟气释放物数据可以用于产品危害评估，但并不意味着这些数据等效于人体暴露或风险的测量值。将吸烟机测量结果的差异表示为暴露或风险的差异是对世界卫生组织烟草实验室网络标准的滥用。

提到的公司和制造商的产品，并不意味着 WHO 优先推荐这些产品。除错误和疏漏外，专有产品的名称以首字母大写表示。

本方法由世界卫生组织（WHO）烟草实验室网络（TobLabNet）制订，作为国际标准化组织（ISO）和深度抽吸方案下卷烟主流烟气中烟草特有亚硝胺的测定标准操作规程（SOP）。

引言

为了在全球范围内建立具有可比性的烟草制品检测方法，需要卷烟特定成分和释放物的一致性检测方法。2008年11月在南非德班举行的世界卫生组织《烟草控制框架公约》（WHO FCTC）第三次缔约方会议回顾了 FCTC/COP1(15)和FCTC/COP2(14)号决议关于制订 WHO FCTC 第9条（烟草制品成分管制）和第10条（烟草制品披露管制）实施指南的要求，并提到第三次缔约方会议工作报告中有关其工作进展的信息，提请公约秘书处授权 WHO 无烟草行动组在5年内对检测卷烟成分和释放物的分析化学方法进行验证（FCTC/COP/3/REC/1）。

根据2006年10月在加拿大渥太华举行的第三次会议上确定的优先次序标准，WHO FCTC 第9条和第10条工作组确定优先验证以下成分的分析化学检测方法：
- 烟碱；
- 氨；
- 保润剂[1,2-丙二醇、甘油（1,2,3-丙三醇）和三甘醇（2,2′-二乙醚乙二醇）]。

需要对三种方法进行验证：一是检测烟碱的方法，二是检测氨的方法，三是检测保润剂的方法。

根据上述渥太华会议确定的优先次序标准，工作组确定了卷烟主流烟气释放物中应该验证的分析化学检测方法的下列优先级物质清单：
- 4-(*N*-甲基亚硝胺基)-1-(3-吡啶基)-1-丁酮（NNK）；
- *N*-亚硝基降烟碱（NNN）；
- 乙醛；
- 丙烯醛；
- 苯；
- 苯并[*a*]芘；
- 1,3-丁二烯；
- 一氧化碳；
- 甲醛。

这些释放物使用下述两种抽吸方案检测，对五种方法进行验证：一是检测烟草特有亚硝胺（NNK 和 NNN）的方法，二是检测苯并[*a*]芘的方法，三是检测醛类（乙醛、丙烯醛和甲醛）的方法，四是检测挥发性有机物（苯和1,3-丁二烯）的方法，五是检测一氧化碳的方法。

下表列出了验证上述方法的两种抽吸方案。

抽吸方案	抽吸容量（mL）	抽吸间隔（s）	滤嘴通风孔
ISO 方案：ISO 3308 常规分析用吸烟机 定义和标准条件	35	60	不封闭
深度抽吸方案：在 ISO 3308 基础上有所调整	55	30	所有通风孔必须如 SOP 01 **12.2** 所述 100%封闭

SOP 03 描述 ISO 和深度抽吸方案下卷烟主流烟气中烟碱特有亚硝胺的测定标准操作规程。

1 适用范围

1.1 本方法适用于高效液相色谱-质谱联用（HPLC-MS-MS）法定量测定卷烟主流烟气中以下四种烟草特有亚硝胺（TSNA）：N-亚硝基降烟碱（NNN），4-(N-甲基亚硝胺基)-1-(3-吡啶基)-1-丁酮（NNK），N-亚硝基新烟草碱（NAT），N-亚硝基假木贼碱（NAB）。

请注意，缔约方大会建议仅测量 NNN 和 NNK。对于选择测量 NAT 和 NAB 的实验室，包括这些物质的分析信息。

1.2 NNN 和 NNK 是潜在致癌物质，NAB 是一种较弱的致癌物质，而 NAT 则不致癌。NNN 和 NNK 最初并不存在于烟叶中，而是烟草在调制和储藏过程中烟碱发生亚硝化作用形成的，NAB 由亚硝化的假木贼碱形成，而 NAT 来自新烟草碱亚硝化。被人体吸收后，NNN 和 NNK 可羟基化，与血红蛋白或 DNA 形成加合物。

2 参考标准

2.1 ISO 3402：烟草和烟草制品 用于调节和测试的大气环境。

2.2 ISO 4387：卷烟 常规分析用吸烟机测定总粒相物和去烟碱干粒相物。

2.3 WHO TobLabNet SOP 01：卷烟深度抽吸标准操作规程。

2.4 ISO 3308：常规分析用吸烟机 定义和标准条件。

2.5 ISO 8243：卷烟 抽样。

2.6 联合国毒品和犯罪问题办公室（UNODC）《代表性药物取样指南》（http://www.unodc.org/documents/scientific/Drug_Sampling.pdf）。

2.7 WHO TobLabNet SOP 02：烟草制品成分和释放物分析方法验证标准操作规程。

2.8 WHO TobLabNet 官方方法：ISO 和深度抽吸方案下卷烟主流烟气中烟草特有亚硝胺的测定。

2.9 ISO 5725-1：检测方法和结果的准确度（正确度和精密度） 第 1 部分 一般原则和定义。

2.10 ISO 5725-2：检测方法和结果的准确度（正确度和精密度） 第 2 部分 确定标准检测方法重复性和再现性的基本方法。

3 术语和定义

3.1 TPM：总粒相物。

3.2 TSNA：烟草特有亚硝胺。

3.3 NNN：N-亚硝基降烟碱。

3.4 NNK：4-(N-甲基亚硝胺基)-1-(3-吡啶基)-1-丁酮。

3.5 NAT：N-亚硝基新烟草碱。

3.6 NAB：N-亚硝基假木贼碱。

3.7 烟草制品：完全或部分以烟叶为原料制造，用于抽吸、吸吮、咀嚼或吸入的产品（WHO FCTC 第 1(f)条）。

3.8 深度抽吸方案：抽吸过程参数为抽吸容量 55 mL，抽吸间隔 30 s，每次抽吸持续时间 2 s，100%封闭滤嘴通风孔。

3.9 ISO 方案：抽吸过程参数为包括抽吸容量 35 mL，抽吸间隔 60 s 和每每次抽吸持续时间 2 s，不封闭滤嘴通风孔。

3.10 实验室样品：用于进行实验室检测的样品，由一次性或在规定时间内送到实验室的单一类型产品组成。

3.11 试样：从实验室样品中随机抽取的待测样品。所取样品数量应能代表实验室样品。

3.12 试料：从试样中随机抽取的用于单次测试的样品。所取样品数量应能代表试样。

4 方法概述

4.1 来自卷烟试样的主流烟气被捕集到玻璃纤维滤片上。

4.2 可能需要调整卷烟数量来防止滤片发生穿透。若 92 mm 的滤片 TPM 超过 600 mg 或者 44 mm 的滤片 TPM 超过 150 mg，则必须减少每个滤片上卷烟的数量。

4.3 将含有两种（或四种）同位素标记的内标的溶液加在滤片上，然后用乙酸铵进行萃取。

4.4 过滤萃取液，采用配备电喷雾电离源的高效液相色谱-质谱联用仪（HPLC-MS-MS）进行分析。在 MS-MS 模式下检测分析物离子。

4.5 通过将分析物与同位素标记内标物的重建离子色谱图峰面积比进行作图，建立已知浓度 TSNA 的校准曲线，从而确定试料的 TSNA 含量。

5 安全和环境预防措施

亚硝胺类是强致癌物质，应采取措施避免人体接触。
亚硝胺类及其溶液应在通风橱或手套箱中进行处理。
实验室应建立处理含亚硝胺类溶液的规程。

5.1 遵守所有化学实验活动的常规安全与环境预防措施。

5.2 使用本检测方法对特定产品进行检测和评估时所有材料或设备可能对环境有害。本检测方法无力解决所有与使用有关的安全问题。所有使用此方法的人有责任咨询相关机构并遵循现有监管要求，制定健康与安全规程以及环境预防措施。

5.3 应特别注意避免吸入或皮肤接触危险化学品。在制备或处理未稀释材料、标准溶液、萃取液和收集样品时，应在通风橱中进行，并穿戴合适的实验服、手套和护目镜。

6 仪器和设备

常规实验仪器，特别是：

6.1 按照 ISO 3402 [**2.1**]规定调节卷烟所需设备。

6.2 按照 ISO 4387 [**2.2**]规定标记烟蒂长度所需设备。

6.3 按照 WHO TobLabNet SOP 01 [**2.3**]规定采用深度抽吸方案时封闭通风孔所需设备。

6.4 按照 ISO 3308 [**2.4**]规定抽吸烟草制品所需设备。

6.5 测量精度为 0.0001 g 的分析天平。

6.6 涡旋振荡器，如腕式振动器。

6.7 移液枪及能精确移取 100~1000 μL 体积的枪头。

6.8 可量取 2 mL 的移液管。

6.9 25 mL 和 2 L 的玻璃量筒。

6.10 聚四氟乙烯垫片。

6.11 10 mL、50 mL、100 mL、200 mL、1 L 和 2 L 的容量瓶。

6.12 60 mL、1 L 的棕色玻璃瓶或合适的烧瓶。

6.13 25 mL × 0.45 μm GD/X 尼龙注射式过滤器。

6.14 5 mL 注射器。

6.15 47 mm × 0.45 μm 尼龙膜过滤器。

6.16 玻璃移液管。

6.17 2 mL 自动进样瓶。

6.18 HPLC 联用 MS-MS。

6.19 MS-MS 三重四极杆质谱仪。

6.20 能够明显分离卷烟释放物中 TSNA 和同位素标记 TSNA 峰的 HPLC 柱，如 Agilent Zorbax Eclipse XDB-C_{18} (2.1×150 mm, 粒径 3.5 μm)。

6.21 Opti-Guard RP C_{18} 保护柱（可选）。

7 试剂

除非另有说明，所有试剂应至少为分析纯。试剂尽可能以 CAS 号来识别。

7.1 N-亚硝基降烟碱（NNN）[16543-55-8]。

7.2 氘代 N-亚硝基降烟碱（NNN-d_4）。

7.3 4-(N-甲基亚硝胺基)-1-(3-吡啶基)-1-丁酮（NNK）[64091-91-4]。

7.4 氘代 4-(N-甲基亚硝胺基)-1-(3-吡啶基)-1-丁酮（NNK-d_4）。

7.5 N-亚硝基新烟草碱（NAT）[71267-22-6]。

7.6 氘代 N-亚硝基新烟草碱（NAT-d_4）。

7.7 N-亚硝基假木贼碱（NAB）[37620-20-5]。

7.8 氘代 N-亚硝基假木贼碱（NAB-d_4）。

7.9 乙酸铵 [631-61-8]（色谱纯）。

7.10 冰醋酸[64-19-7]。

7.11 去离子水[7732-18-5]。

7.12 乙腈[75-05-8]（色谱纯）。

7.13 甲醇[67-56-1]（色谱纯）。

8 玻璃器皿准备

清洁、干燥的玻璃器皿，避免残留物污染。

9 溶液配制

下述溶液配制方法仅供参考，如有需要可进行调整。

9.1 萃取液（100 mmol/L 的乙酸铵）：

    9.1.1 称取约 7.7 g 乙酸铵。

    9.1.2 将称好的乙酸铵溶于水中，置于 1 L 的容量瓶或其他合适的烧瓶。

    9.1.3 盖紧盖子，摇匀，确保混合均匀。

9.1.4 转移到 1 L 棕色玻璃瓶中，在 4~10℃下保存。

9.2 流动相 A（体积分数 0.1%的乙酸水溶液）：

9.2.1 将约 1.4 L 的水转移到 2 L 的容量瓶或者合适的烧瓶中。

9.2.2 移取 2 mL 冰醋酸到容量瓶中。

9.2.3 混合，加水至刻度处。

9.2.4 盖紧盖子并混合均匀，在 4~10℃下保存。

9.3 流动相 B（体积分数 0.1%的乙酸甲醇溶液）：

9.3.1 将约 1.4 L 的甲醇转移到 2 L 的容量瓶或者合适的烧瓶中。

9.3.2 移取 2 mL 冰醋酸到容量瓶中。

9.3.3 混合，加甲醇至刻度处。

9.3.4 盖紧盖子并混合均匀，在 4~10℃下保存。

10 标准溶液配制

下述标准溶液配制方法仅供参考，如有需要可进行调整。

10.1 同位素内标物。

10.1.1 TSNA 内标物储备液：

10.1.1.1 用测量精度为 0.0001 g 的天平称取约 4 mg（m_{IS}）同位素标记的 TSNA 置于 10 mL 的容量瓶中，并记录质量，精确至 0.1 mg。

10.1.1.2 将每个称量好的 TSNA 内标物分别溶解在 10 mL 乙腈中，混合均匀。

10.1.1.3 贴上标签，用棕色瓶在–20℃±5℃保存。该溶液可以稳定保存两年。

10.1.2 TSNA 混合内标物溶液（5 μg/mL）：

10.1.2.1 移取 1.25 mL（V_{IS}）内标物储备液[10.1.1]置于 100 mL 的容量瓶中。

10.1.2.2 用 100 mmol/L 乙酸铵稀释至刻度，并混合均匀。

10.1.2.3 贴上标签，用棕色瓶在–20℃±5℃保存。该溶液可以稳定保存两周。

10.2 标准溶液。

各标准储备液用乙腈制备，在–20℃±5℃储存。混合标准中间溶液由各标准储备液配制。

10.2.1 TSNA 标准储备液（0.4 mg/mL）：

10.2.1.1 用测量精度为 0.0001 g 的天平分别称取约 4 mg（m_S）的 4 种 TSNA，并记录质量。

10.2.1.2 将每个TSNA标样用约8 mL的乙腈溶解在10 mL的容量瓶中。

10.2.1.3 用乙腈定容至刻度,混合均匀。

10.2.1.4 贴上标签,用棕色瓶在–20℃±5℃保存。该溶液可以稳定保存两年。

10.2.2 TSNA的1 μg/mL混合标准中间溶液:

10.2.2.1 每个标准储备液[**10.2.1**]量取25 μL(V_{SS})置于10 mL容量瓶中。

10.2.2.2 用100 mmol/L乙酸铵稀释至刻度,并混合均匀。

10.2.2.3 贴上标签,用棕色瓶在–20℃±5℃保存。该溶液可以稳定保存两周。

10.2.3 TSNA的最终混合标准溶液:

10.2.3.1 根据表1制备最终混合标准溶液。

10.2.3.2 移取不同体积的混合标准中间溶液[**10.2.2**]至10 mL容量瓶中。

10.2.3.3 加100 μL的内标物溶液[**10.1.2**],用100 mmol/L乙酸铵定容至刻度,并混合均匀。

10.2.3.4 内标物的最终浓度如下所示:

$$最终浓度(IS)(ng/mL) = \frac{m_{IS}}{10} \times \frac{V_{IS}}{100} \times \frac{0.1}{10} \times 10^6$$

式中,m_{IS} 为内标物[**10.1.1.1**]的原始质量(mg);V_{IS} 为移取的混合内标物溶液[**10.1.2.1**]的体积(mL);10^6 为将mg转换为ng的系数。

10.2.3.5 标准溶液的最终浓度如下所示:

$$标准溶液最终浓度(ng/mL) = \frac{m_S}{10} \times \frac{V_{SS}}{10} \times \frac{V_{MIX}}{10} \times 10^6$$

式中,m_S 为标准原样[**10.2.1.1**]的原始质量(mg);V_{SS} 为移取的混合标准溶液[**10.2.2.1**]的体积(mL);V_{MIX} 为表1所示的混合标准溶液的体积(mL)。

10.2.3.6 另外,可能需要更高浓度的标准溶液用来扩展校准范围,以涵盖具有高含量TSNA的烟草制品。

表1 TSNA标准曲线

标准	TSNA混合标准中间溶液体积 (1 μg/mL) (mL) (V_{MIX})	混合TSNA内标物溶液体积 (mL) (V_{IS})	总体积 (mL)	TSNA的近似浓度 (ng/mL)
1	0.005	0.100	10	0.5
2	0.010	0.100	10	1.0
3	0.020	0.100	10	2.0

续表

标准	TSNA 混合标准中间溶液体积 (1 μg/mL) (mL) (V_{MIX})	混合 TSNA 内标物溶液体积 (mL) (V_{IS})	总体积 (mL)	TSNA 的近似浓度 (ng/mL)
4	0.050	0.100	10	5.0
5	0.200	0.100	10	20.0
6	0.500	0.100	10	50.0
7	1.000	0.100	10	100.0
8	2.000	0.100	10	200.0

标准溶液范围可根据使用的设备和待测样品进行调整，注意方法灵敏度可能对结果产生的影响。

所有溶剂和溶液在使用前调节至室温。

11 抽样

11.1 根据 ISO 8243 [**2.5**]或替代方法，根据各实验室惯例或者要求的特殊规定或样品可用性，选取具有代表性的实验室卷烟样品。

11.2 试样的组成。

  11.2.1 若可能，将实验室样品分成独立的单元（如包、条）。

  11.2.2 从至少 $\sqrt{n}$ 个[**2.6**]单元中取出每种试样的等量产品。

12 卷烟准备

12.1 按照 ISO 3402 [**2.1**]调节所有待测卷烟。

12.2 按照 ISO 4387 [**2.2**]和 WHO TobLabNet SOP 01 [**2.3**]标记烟蒂长度。

12.3 按照 WHO TobLabNet SOP 01 [**2.3**]的深度抽吸方案或 ISO 方案准备待测样品。

13 吸烟机设置

13.1 环境条件。

按 ISO 3308 [**2.4**]要求设定抽吸的环境条件。

13.2 吸烟机要求。

除深度抽吸方案按 WHO TobLabNet SOP 01 [**2.3**]进行准备外，其余按照 ISO 3308 [**2.4**]的要求设置吸烟机。

14 样品生成

在规定的吸烟机上抽吸足够数量的卷烟，不使滤片发生穿透，并确保 NNN、NNK、NAT 和 NAB 的浓度落在分析的校准范围内。

14.1 抽吸卷烟试样并按照 ISO 4387 [2.2]或 WHO TobLabNet SOP 01 [2.3]收集 TPM。

14.2 至少包含一个用于质量控制的参比试样。

14.3 当样品类型为第一次检测时，评估滤片穿透的可能性，调整卷烟的数量以避免穿透。当测定焦油、烟碱和一氧化碳时，对于 92 mm 滤片和 44 mm 滤片，穿透分别发生在总粒相物水平超过 600 mg 和 150 mg 时。一旦发生穿透，必须减少每个滤片上的卷烟数量。

14.4 表 2 显示了 ISO 和深度抽吸方案中直线型和转盘型吸烟机每次抽吸的数量。

表2 直线型和转盘型吸烟机进行一次测量的卷烟数量

	ISO 方案		深度抽吸方案	
	直线型	转盘型	直线型	转盘型
每个滤片上的卷烟数量	5	20	3	10
每个结果包含的滤片数	4	1	7	2

14.5 记录每个滤片的卷烟数量和总抽吸量。

14.6 测量完成所需的样品量后，进行五次空吸，然后从吸烟机上取下滤片捕集器。

15 样品制备

15.1 滤片萃取。

    15.1.1 从捕集器上取下滤片，将滤片对折，再对折，使滤片含有 TPM 的面向内。用不含 TPM 的另一面擦拭捕集器内表面，使残留在捕集器上的粒相物转移到滤片上。将每个滤片置于 60 mL 的棕色瓶中。

    15.1.2 将含有同位素标记 TSNA 的混合内标物溶液加到每个滤片上（44 mm 的滤片加 200 μL，92 mm 的滤片加 500 μL）。

    15.1.3 移取 100 mmol/L 乙酸铵水溶液置于棕色瓶中（44 mm 滤片用 20 mL，92 mm 滤片用 50 mL）。

15.1.4 使用腕式或环形振荡器以 250 r/min 振动 30 min，从滤片上萃取 TSNA。

15.1.5 用最大为 0.45 μm 的滤膜过滤，并置于自动进样瓶中。

16 样品分析

使用 HPLC 三重四极杆质谱联用仪测定卷烟主流烟气中 TSNA。在 HPLC 柱上从其他潜在干扰物质中分离分析物。在 MS-MS 模式下，使用带有电喷雾电离的三重四极杆质谱仪可以获得进一步的选择性，比较未知物的峰面积比（分析物峰面积与同位素标记的分析物峰面积）与已知浓度标准液的峰面积比（分析物峰面积与同位素标记的分析物峰面积），得到单个分析物的浓度。

16.1 HPLC 和质谱仪操作条件。
HPLC 操作条件示例：
HPLC 柱：Agilent Zorbax Eclipse XDB-C_{18}（2.1×150 mm, 3.5 μm）；
进样量：5 μL；
柱温箱温度：（40±1）℃（根据所使用的柱子进行更改）；
脱气装置：开；
泵流量：0.2 mL/min；
流动相 A：0.1%乙酸水溶液；
流动相 B：0.1%乙酸甲醇溶液；
梯度：如表 3 所示；
总分析时间：12 min。

表 3 TSNA 分离的 HPLC 梯度

时间（min）	流量（μL/min）	流动相 A（%）	流动相 B（%）
0	200	50	50
3.0	200	10	90
4.0	200	0	100
5.0	200	0	100
5.5	200	50	50
12.0	200	50	50

质谱操作条件：
驻留时间：40 ms；
电离/模式：电喷雾电离源/正离子模式；
离子电压：1500 V；
电喷雾离子源温度：450℃；

气帘气：氮气；
碰撞诱导的解离气体：氮气；
雾化气体：氮气。
根据仪器和色谱柱条件以及色谱峰的分离度调整操作参数。
NNN、NNK、NAT 和 NAB 定量和确认离子对如表 4 所示。

表4 每个 TSNA 的定量和确认离子对

分析物	离子对 Q1/Q3 质荷比（m/z）
NNN 定量	178/148
NNN 确认	178/120
NNN-d_4	182/152
NNK 定量	208/122
NNK 确认	208/106
NNK-d_4	212/126
NAT 定量	190/160
NAT 确认	190/106
NAT-d_4	194/164
NAB 定量	192/162
NAB 确认	192/133
NAB-d_4	196/166

注：确认离子比可以作为确定单个分析物真实性的有效指南。

16.2 预计保留时间。

  16.2.1 对于此处的描述条件，预期的洗脱顺序为 NNN、NNN-d_4、NNK、NNK-d_4、NAT、NAT-d_4、NAB、NAB-d_4。

  16.2.2 流速、流动相浓度和色谱柱的老化等差异可能会改变保留时间。

16.3 TSNA 的测定。

根据每个实验室的做法来确定 TSNA 的测定序列，举例说明如下：

  16.3.1 注入空白溶液（不含内标物的萃取液）来检查系统或试剂中是否存在污染。

  16.3.2 注入标准空白溶液（含内标物的空白溶液）来验证 HPLC-MS-MS 系统的性能。

  16.3.3 注入校准标准液、质量控制溶液和样品。

  16.3.4 记录每个 TSNA 和内标物的峰面积。

  16.3.5 计算每个 TSNA 的每种组分标准液（包括标准空白溶液）的峰与内标峰的相对响应比 RF 值（RF=A_{TSNA}/A_{IS}）。

  16.3.6 以每个 TSNA 的浓度为 X 轴，以相应的峰面积比为 Y 轴绘制标准曲线。

16.3.7 截距在统计上与零不应该有显著差异。

16.3.8 标准曲线应在整个标准范围内呈线性。

16.3.9 使用线性回归的斜率 b 和截距 a 来计算线性回归方程（$Y=a+bx$），如果线性回归方程的 R^2 小于 0.99，则应重新校准。如果单个校准点与预期值（通过线性回归估计）相差超过 10%，则应该舍弃该点。

16.3.10 注入质量控制溶液和样品，用合适的软件确定峰面积。

16.3.11 所有试样的峰值比必须落在标准曲线的工作范围内；否则，应调整标准溶液或试样的浓度。

代表性色谱图见附录 1。

17 数据分析和计算

斜率和截距是由分析物和内标物的相对响应与分析物和内标物的相对浓度确定的。

17.1 针对每个校准标准液计算每个 TSNA 的峰面积的相对响应比[10.2.3]：

$$RF = \frac{A_a}{A_{IS}}$$

式中：RF——相对响应比；

A_a——目标分析物的峰面积；

A_{IS}——相应内标物的峰面积。

为每个 TSNA 绘制相对响应因子 A_a/A_{IS}（Y 轴）与浓度（X 轴）的关系图，使用线性回归的斜率 b 和截距 a 计算线性回归方程（$Y=a+bX$）。

17.2 标准曲线应在整个标准范围内呈线性。

17.3 目标分析物中 TSNA 的含量 M_{ts} 由计算的试样的相对响应比与标准曲线的斜率和截距来确定：

$$M_{ts} = \frac{Y-a}{b} \times \frac{V}{n}$$

式中：M_{ts}——每支卷烟的计算含量（ng/支）；

Y——相对响应比（A_a/A_{IS}）；

a——从标准曲线获得的线性回归方程的截距；

b——从标准曲线获得的线性回归方程的斜率；

V——萃取液的体积（对于 44 mm 滤片是 20 mL，对于 92 mm 滤片是 50 mL）；

n——每个滤片上卷烟抽吸的数量（表 2）。

如果适用，可以使用替代计算方程。

18 特别注意事项

18.1 安装新色谱柱后，按照制造商的规定，在指定的仪器条件下注入烟草样品萃取物进行调节。应重复进样，直至每个 TSNA 和内标物的峰面积（或峰高）可重复。

19 数据报告

19.1 报告每个评估样品的单独测量值。
19.2 以 ng/支或根据需要报告结果。

20 质量控制

20.1 控制参数。
如果控制测量值超出预期值的公差极限，则必须进行适当的调查并采取措施。如有必要，需要做额外的实验室质量保证规程，以符合各个实验室的做法。
20.2 实验室试剂空白样。
如 16.3.2 所述，要在样品制备和分析过程中检测潜在的污染，需包含实验室试剂空白样。空白样由分析测试样品中使用的所有试剂和材料组成，并像测试样品一样进行分析。应根据各个实验室的做法对空白样进行评估。
20.3 质量控制样品。
为了验证整个分析过程的一致性，根据各个实验室的做法分析参比卷烟（或适当的质量控制样品）。
20.4 每运行 20 次后制作新的标准曲线（20 个样品萃取物）。

21 方法性能

21.1 报告限。
报告限设置为所用标准曲线的最低浓度，重新计算为 ng/支卷烟（例如 20 ng/支卷烟对应 0.5 ng/mL 的最低标准浓度）。
21.2 实验室基质加标回收率。
加入到基质上的分析物的回收率用作准确度的替代量度。通过将已知含量的标准品（抽吸后）加到带有卷烟烟气的滤片上并通过用与样品相同的方法萃取滤片确定回收率，未加标的滤片也要进行分析。回收率由下式计算，如表 5 所示。

回收率=100×（分析结果－未加标结果/加标量）

表 5 TSNA 加标的平均值和回收率

加标量 （ng）	NNN		NNK		NAT		NAB	
	平均值 （ng）	回收率 （%）	平均值 （ng）	回收率 （%）	平均值 （ng）	回收率 （%）	平均值 （ng）	回收率 （%）
150	146.3	97.5	145.3	96.8	161.3	107.5	149.1	99.4
500	452.8	90.6	473.8	94.8	516.5	103.3	509.4	101.9
1000	962.5	96.3	973.8	97.4	1015.8	101.6	1001.9	100.1

21.3 分析特异性。

LC-MS-MS 具有出色的分析特异性。保留时间和确认离子比，定量离子与确认离子和质量控制样品的响应比的范围，用于验证未知样品的结果的特异性。

21.4 线性。

建立的 TSNA 标准曲线在 0.5~200 ng/mL 的标准浓度范围内是线性的。

21.5 潜在干扰。

没有已知的组分具有与 TSNA 或内标物相似的保留时间和相同的确认离子对比。

22 重复性和再现性

2009~2012 年间进行的一项国际合作研究[2.8]涉及 9 个实验室和 3 种参比卷烟（1R5F、3R4F 和 CM6）以及 2 种商品卷烟，对测试结果进行统计分析给出了表 6 至表 9 所示方法的精确度限值。

常规正确操作该方法时，同一操作者使用同一设备在最短的操作时间内两个单一结果之间的差异超过重复性 r 的情况，平均 20 个样品不超过 1 次。

常规正确操作该方法时，两个实验室报告的匹配卷烟样品的单一结果超过再现性 R 的情况，平均 20 次不超过 1 次。

根据 ISO 5725-1 [2.9]和 ISO 5725-2 [2.10]对测试结果进行统计分析，得到表 6 至表 9 所示的精密度数据。

为了计算 r 和 R，在 ISO 条件下，测试结果被定义为来自直线型吸烟机的四个滤片（每个滤片上 5 支烟）和转盘型吸烟机的一个滤片（每个滤片上 20 支烟）平均释放量的 7 次重复的平均值。在深度抽吸下，测试结果被定义为来自直线型吸烟机的七个滤片（每个滤片上 3 支烟）和转盘型吸烟机的两个滤片（每个滤片上 10 支烟）平均释放量的 7 次重复的平均值。更多信息请参考[2.8]。

表6 参比卷烟主流烟气中NNN（ng/支）的精密度限值

参比卷烟	ISO方案				深度抽吸方案			
	N	m_{cig}	r_{limit}	R_{limit}	N	m_{cig}	r_{limit}	R_{limit}
1R5F	9	45.55	6.15	11.22	9	256.72	23.44	53.37
3R4F	9	113.76	12.24	20.02	9	302.24	26.97	60.77
CM6	8	21.02	3.97	8.05	8	40.26	6.77	18.18

表7 参比卷烟主流烟气中NNK（ng/支）的精密度限值

参比卷烟	ISO方案				深度抽吸方案			
	N	m_{cig}	r_{limit}	R_{limit}	N	m_{cig}	r_{limit}	R_{limit}
1R5F	8	23.31	6.17	11.90	9	128.28	21.23	49.05
3R4F	8	95.81	12.65	33.78	9	259.25	34.27	84.70
CM6	9	28.47	6.63	13.46	9	55.91	14.03	29.35

表8 参比卷烟主流烟气中NAT（ng/支）的精密度限值

参比卷烟	ISO方案				深度抽吸方案			
	N	m_{cig}	r_{limit}	R_{limit}	N	m_{cig}	r_{limit}	R_{limit}
1R5F	7	102.43	11.52	48.66	7	264.43	24.41	129.21
3R4F	6	39.7	6.95	12..97	7	225.58	19.36	116.45
CM6	7	32.03	4.45	15.41	7	62.48	8.84	30.59

表9 参比卷烟主流烟气中NAB（ng/支）的精密度限值

参比卷烟	ISO方案				深度抽吸方案			
	N	m_{cig}	r_{limit}	R_{limit}	N	m_{cig}	r_{limit}	R_{limit}
1R5F	6	12.85	1.87	3.46	6	31.95	5.65	10.38
3R4F	6	6.36	1.87	2.12	6	29.48	3.49	7.62
CM6	7	4.67	1.55	6.21	7	8.98	2.88	10.93

注：商品卷烟的TSNA含量在参比卷烟的范围内，因此未在此表中报告。
  1R5F，3R4F和CM6是本研究中分析的三种参比卷烟；
  N为参与实验室数量；
  m_{cig}为各亚硝胺的平均值；
  r_{limit}为各亚硝胺的重复性限值；
  R_{limit}为各亚硝胺的再现性限值。

23 检测报告

检测报告应包含以下内容：
（a）相关方法即 WHO TobLabNet SOP 03；
（b）收到样品的时间；
（c）结果及其单位。

附录1 卷烟主流烟气中烟草特有亚硝胺测定的典型色谱图

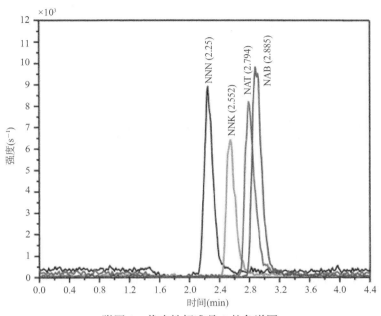

附图1 代表性标准品2的色谱图

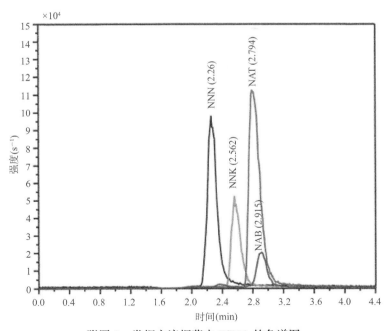

附图2 卷烟主流烟草中TSNA的色谱图

WHO TobLabNet SOP 04

卷烟烟丝中烟碱的测定标准操作规程

方　　法：卷烟烟丝中烟碱的测定
分 析 物：烟碱(3-[(2*S*)-1-甲基吡咯烷基-2-基]吡啶)（CAS 号：54-11-5）
基　　质：卷烟烟丝
更新时间：2014 年 6 月

> 吸烟机的抽吸方案不能代表所有人的吸烟行为：吸烟机测试对用于卷烟设计及监管目的的卷烟释放物表征非常有用，但将吸烟机的测量结果披露给吸烟者则会导致对不同品牌的暴露和风险差异的误解。吸烟机测量的烟气释放物数据可以用于产品危害评估，但并不意味着这些数据等效于人体暴露或风险的测量值。将吸烟机测量结果的差异表示为暴露或风险的差异是对世界卫生组织烟草实验室网络标准的滥用。

提到的公司和制造商的产品，并不意味着 WHO 优先推荐这些产品。除错误和疏漏外，专有产品的名称以首字母大写表示。

本方法由世界卫生组织（WHO）烟草实验室网络（TobLabNet）制订，作为卷烟烟丝中烟碱测定的标准操作规程（SOP）。

引言

为了在全球范围内建立具有可比性的烟草制品检测方法，需要卷烟特定成分和释放物的一致性检测方法。2008年11月在南非德班举行的世界卫生组织《烟草控制框架公约》（WHO FCTC）第三次缔约方会议回顾了FCTC/COP1(15)和FCTC/COP2(14)号决议关于制订WHO FCTC第9条（烟草制品成分管制）和第10条（烟草制品披露管制）实施指南的要求，并提到第三次缔约方会议工作报告中有关其工作进展的信息，提请公约秘书处授权WHO无烟草行动组在5年内对检测卷烟成分和释放物的分析化学方法进行验证（FCTC/COP/3/REC/1）。

根据2006年10月在加拿大渥太华举行的第三次会议上确定的优先次序标准，WHO FCTC第9条和第10条工作组确定优先验证以下成分的分析化学检测方法：

- 烟碱；
- 氨；
- 保润剂[1,2-丙二醇、甘油（1,2,3-丙三醇）和三甘醇（2,2′-二乙醚乙二醇）]。

测量这些成分需要验证三个方法：一个适用于烟碱，一个适用于氨，一个适用于保润剂。

根据上述渥太华会议确定的优先次序标准，工作组确定了卷烟主流烟气释放物中应该验证测试和测量方法（分析化学）的下列优先级物质清单：

- 4-(N-甲基亚硝胺基)-1,3-吡啶基-1-丁酮（NNK）；
- N-亚硝基去甲烟碱（NNN）；
- 乙醛；
- 丙烯醛；
- 苯；
- 苯并[a]芘；
- 1,3-丁二烯；
- 一氧化碳；
- 甲醛。

这些释放物使用下述两种抽吸方案测量，按五个方法进行验证：一个适用于烟草特有的亚硝胺（NNK和NNN），一个适用于苯并[a]芘，一个适用于醛类（乙醛、丙烯醛和甲醛），一个适用于挥发性有机物（苯和1,3-丁二烯），一个适用于一氧化碳。

下表列出了两种用于验证上述测试方法的吸烟方案。

抽吸方案	抽吸容量（mL）	抽吸间隔（s）	滤嘴通风孔
ISO 方案：ISO 3308 常规分析用吸烟机 定义和标准条件	35	60	不封闭
深度抽吸方案：在 ISO 3308 基础上有所调整	55	30	所有通风孔必须如 SOP 01 **12.2** 所述 100%封闭

SOP 04 描述卷烟烟丝中烟碱的测定标准操作规程。

1 适用范围

本方法适用于气相色谱（GC）法定量测定卷烟烟丝中的烟碱含量。

2 参考标准

2.1 ISO 3402：烟草和烟草制品 用于调节和测试的大气环境。

2.2 ISO 13276：烟草和烟草制品 钨硅酸重量法测定烟碱纯度。

2.3 ISO 8243：卷烟 抽样。

2.4 联合国毒品和犯罪问题办公室（UNODC）《代表性药物取样指南》（http://www.unodc.org/documents/scientific/Drug_Sampling.pdf）。

2.5 WHO TobLabNet SOP 02：烟草制品成分和释放物分析方法验证标准操作规程。

2.6 WHO TobLabNet 官方方法：卷烟烟丝中烟碱的测定。

2.7 ISO 5725-1：检测方法和结果的准确度（正确度和精密度） 第 1 部分 一般原则和定义。

2.8 ISO 5725-2：检测方法和结果的准确度（正确度和精密度） 第 2 部分 确定标准检测方法重复性和再现性的基本方法。

3 术语和定义

3.1 烟碱含量：卷烟烟丝中的烟碱总量，以 mg/g 卷烟烟丝表示。

3.2 卷烟烟丝：卷烟中含烟草的部分，包括再造烟叶、烟梗、膨胀烟丝和添加物。

3.3 烟草制品：完全或部分以烟叶为原料制造，用于抽吸、吸吮、咀嚼或吸入的产品（WHO FCTC 第 1(f)条）。

3.4 实验室样品：用于进行实验室检测的样品，由一次性或在规定时间内送到实验室的单一类型产品组成。

3.5 试样：从实验室样品中随机抽取的待测样品。所取样品数量应能代表实验室样品。

3.6 试料：从试样中随机抽取的用于单次测试的样品。所取样品数量应能代表试样。

4 方法概述

4.1 经过调制后，将卷烟烟丝研磨并混合。
4.2 用正己烷、氢氧化钠溶液和水的混合物从卷烟烟丝中萃取烟碱。
4.3 用氢火焰离子化检测器通过 GC 分析有机层。
4.4 通过分析物与同位素标记内标物的重建离子色谱图峰面积比进行作图，建立已知浓度烟碱的校准曲线，从而确定试料的烟碱含量。

5 安全和环境预防措施

5.1 遵守所有化学实验活动的常规安全与环境预防措施。
5.2 使用本检测方法对特定产品进行检测和评估时所用材料或设备可能对环境有害。本检测方法无力解决所有与使用有关的安全问题。所有使用此方法的人有责任咨询相关机构并遵循现有监管要求，制定健康与安全规程以及环境预防措施。
5.3 应特别注意避免吸入或皮肤接触危险化学品。在制备或处理未稀释材料、标准溶液、萃取液和收集样品时，应在通风橱中进行，并穿戴合适的实验服、手套和护目镜。

6 仪器和设备

常规实验仪器，特别是：
6.1 按照 ISO 3402 [2.1]规定调节卷烟所需设备。
6.2 萃取瓶：带有塞子的锥形瓶（250 mL），带有波纹密封盖和隔垫的隔热玻璃瓶(100 mL)，带有聚四氟乙烯塞子的 100 mL 培养管或其他合适的烧瓶。
6.3 使萃取瓶保持在适当位置的线性振荡器。
6.4 配有氢火焰离子化检测器的毛细管气相色谱仪。
6.5 能够明显分离溶剂、内标物、烟碱和其他烟草成分色谱峰的毛细管气相色谱柱，如 Varian WCOT 熔融石英色谱柱（25 m×0.25 mm ID, 涂层 CP-WAX 51）。
6.6 超声波水浴。

7 试剂

除非另有说明，所有试剂应至少为分析纯。试剂尽可能以 CAS 号来识别。

7.1 载气：高纯度（99.999%）的氦气[7440-59-7]。

7.2 辅助气体：用于氢火焰离子化检测器的高纯度（99.999%）空气和氢气[1333-74-0]。

7.3 最大含水量为 1.0 g/L 的正己烷[110-54-3]（色谱纯）。

7.4 纯度不低于 98%的烟碱[54-11-5]。纯度不低于 98%的烟碱水杨酸盐。如有必要，实验室应检验烟碱的纯度。

7.5 氢氧化钠[1310-73-2]。

7.6 内标物：正十七烷（纯度质量分数不低于98%）[629-78-7]。喹哪啶[91-63-4]、异喹啉[119-65-3]、喹啉[91-22-5]或其他合适的替代物。

8 玻璃器皿准备

8.1 清洁、干燥的玻璃器皿，避免残留物污染。

9 溶液配制

9.1 2 mol/L 的氢氧化钠溶液：

    9.1.1 称取约 80 g 氢氧化钠。

    9.1.2 将称好的氢氧化钠溶解在水中并用水稀释至 1 L。

9.2 0.5 mg/mL 的萃取液：

    9.2.1 称取质量约为 0.5 g（精确至 0.001 g）的正十七烷或其他可替代物作内标物。

    9.2.2 将称好的正十七烷或替代内标物溶于正己烷中，并用正己烷稀释至 1 L。

10 标准溶液配制

下述标准溶液配制方法仅供参考，如有需要可进行调整。

10.1 2 g/L 的烟碱标准储备液：

    10.1.1 用精度为 0.0001 g 的天平称取 200 mg 烟碱或者 370 mg 烟碱水杨酸盐置于 250 mL 的烧瓶中。

    10.1.2 用 50 mL 水溶解称量好的烟碱。

10.1.3 移取 100 mL 萃取液[9.2.2]，并加入 25 mL 2 mol/L 的氢氧化钠溶液。

10.1.4 将获得的双相混合物在振荡器中剧烈振动 60 min±2 min，混合均匀。

10.1.5 分离上层有机相，在 4~8℃下避光储存。

10.2 烟碱标准溶液：

10.2.1 移取标准储备液[10.1.5] 0.5 mL、2.5 mL、5.0 mL、7.5 mL 和 10.0 mL 分别置于 20 mL 的容量瓶中。

10.2.2 用萃取液[9.2.2]定容至刻度处。

10.2.3 在 4~8℃下避光储存标准溶液。

10.2.4 确定标准溶液中的最终烟碱浓度：

$$最终浓度（mg/L）= x \times y \times \frac{1000}{100 \times 20} \times 标准品纯度$$

式中：x——烟碱[10.1.1]的原始质量（mg）；

　　　y——移取标准储备液[10.2.1]的体积。

标准溶液中的最终烟碱浓度如表1所示。

表 1 标准溶液中烟碱的浓度

标准	烟碱标准储备液(2 g/L)的体积(mL)(y)	内标液的体积	总体积(mL)	最终标准溶液中的近似烟碱浓度（mg/L）	相当于卷烟烟丝中的近似水平（mg/g）
1	0.5	不适用，包含于萃取液	20	50	1.3
2	2.5		20	250	6.7
3	5.0		20	500	13.3
4	7.5		20	750	20.0
5	10.0		20	1000	26.7

标准溶液范围可根据使用的设备和待测样品进行调整，注意方法灵敏度可能对结果产生的影响。

所有溶剂和溶液在使用前调节至室温。

11 抽样

11.1 根据 ISO 8243 [2.5]或替代方法，根据各实验室惯例或者要求的特殊规定或样品可用性，选取具有代表性的实验室卷烟样品。

11.2 试样的组成。

11.2.1 若可能，将实验室样品分成独立的单元（如包、条）。

11.2.2 从至少 $\sqrt{n}$ 个[2.4]单元中取出每种试样的等量产品。

12 卷烟准备

12.1 从一包（例如含 20 支）卷烟和质量控制样品（若适用）中取出烟丝，或使用至少 15 g 加工过的卷烟烟丝。

12.2 将足够的卷烟烟丝混合至每个试料至少含 10 g 烟丝，制备至少三个重复的试料。

12.3 混合并研磨卷烟烟丝直至其足够精细，能通过 4 mm 筛网。

12.4 按照 ISO 3402 [**2.1**]对烟草制品的要求调节磨碎的卷烟烟丝。

13 吸烟机设置

不适用。

14 样品生成

不适用。

15 样品制备

15.1 对于每个试料，用精确度为 0.001 g 的天平称取 1.5 g 经充分混合、研磨、调节的试料，置于萃取容器中。

15.2 将试料与 20 mL 水，40 mL 萃取液（V_e）[**9.2.2**]和 10 mL 2 mol/L 氢氧化钠溶液混合。

15.3 在振荡器上振动烧瓶(60±2) min。

15.4 将样品瓶静置 20 min 后，使两相澄清分离。分离各相，尽快用 GC 分析有机相（上层）。为了便于分离各相，可将锥形瓶置于超声波水浴中。

15.5 如果要保存萃取的样品，于 4~8℃下避光保存。

16 样品分析

使用配备氢火焰离子化检测器的气相色谱仪测定卷烟烟丝中的烟碱。分析物可以在所用气相色谱柱上与其他潜在干扰物分离开，比较未知物的峰面积比与已知浓度标准溶液的峰面积比，得到单个分析物的浓度。

16.1 GC 操作条件示例。

GC 柱：Varian WCOT 熔融石英毛细管柱，25 m×0.25 mm ID；

涂层：CP-WAX 51；

色谱柱温度：170℃（恒温）；

进样口温度：270℃；

检测器温度：270℃；

载气：流速为 1.5 mL/min 的氦气；

进样体积：1.0 μL；

进样方式：分流 1∶10。

根据仪器和色谱柱条件以及色谱峰的分离度调整操作参数。

16.2 预计保留时间。

  16.2.1 对于此处描述的条件，预期的洗脱顺序是正十七烷、烟碱。

  16.2.2 温度、气体流速和色谱柱的使用年限等差异可能会改变保留时间。

  16.2.3 在上述条件下，预期的总分析时间约为 10 min。延长分析时间可以优化性能。

16.3 烟碱的测定。

根据每个实验室的做法来确定序列。本节给出一个确定卷烟烟丝中烟碱测定顺序的示例。

  16.3.1 注入空白溶液（不含内标物的萃取液）来检查系统或试剂中是否存在污染。

  16.3.2 注入标准空白溶液（带有内标物的空白溶液）来验证气相色谱系统的性能。

  16.3.3 注入标准溶液、质量控制溶液和样品。

  16.3.4 评估标准品的保留时间和响应强度（峰面积）。如果保留时间与先前的进样的保留时间相似（±0.2 min），并且响应与先前进样中的典型响应差异在 20%以内，则说明系统已准备好进行分析。如果响应超出规定，请根据实验室做法寻求纠正的措施。

  16.3.5 记录烟碱和内标物的峰面积。

  16.3.6 计算每种烟碱标准液（包括标准空白溶液）的烟碱峰与内标峰的相对响应比（$RF = A_{nicotine}/A_{IS}$）。

  16.3.7 以烟碱浓度（X 轴）与峰面积比（Y 轴）绘制标准曲线。

  16.3.8 截距在统计上与零不应该有显著差异。

  16.3.9 标准曲线应在整个标准范围内呈线性。

  16.3.10 使用线性回归的斜率 b 和截距 a 来计算线性回归方程（$Y=a+bx$），如果线性回归方程的 R^2 小于 0.99，则应重新校准。如果单个校准点与预期值（通过线性回归估计）相差超过 10%，则应该舍弃该点。

16.3.11 注入 1 μL 每种质量控制样品和试料,用适当的软件确定峰面积。

16.3.12 所有试料的信号(峰值比)必须落在标准曲线的工作范围内;否则,应调整标准溶液或试料的浓度。

代表性色谱图见附录1。

17 数据分析和计算

17.1 对于每个试料,计算烟碱响应与内标物响应的峰面积比(Y_t)。

17.2 使用线性方程的回归系数来计算每个试料等份试样的烟碱浓度:

$$m_t = \frac{Y_t - a}{b}$$

式中:m_t——试样溶液中烟碱的浓度(mg/L);
Y_t——烟碱峰面积与内标物峰面积的比;
a——从标准曲线获得的线性回归方程的截距;
b——从标准曲线获得的线性回归方程的斜率。

17.3 使用下式计算烟草样品中烟碱含量 m_n:

$$m_n = \frac{m_t \times V_e}{m_o \times 1000}$$

式中:m_t——试样溶液中烟碱的浓度(mg/L);
V_e——所用萃取液的体积(mL);
m_o——试料的质量[15.1](g)。

18 特别注意事项

18.1 安装新色谱柱后,在所述 GC 条件下通过注入烟草样品萃取物进行调节。应重复进样,直至烟碱和内标物的峰面积(或峰高)可重复。

18.2 建议在每个样品组(系列)后,通过将色谱柱温度升至 220 ℃并持续 30 min,从 GC 色谱柱中清除出高沸点组分。

18.3 当内标物的峰面积(或峰高)显著高于预期时,建议在萃取液不含内标物的情况下萃取烟草样品,这样可以确定某组分是否与内标物共同洗脱,从而导致烟碱值偏低。

19 数据报告

19.1 报告每个评估样品的单独测量值。

19.2 以 mg/g 烟草或根据需要报告结果。

20 质量控制

20.1 控制参数。

如果控制测量值超出预期值的公差极限,则必须进行适当的调查并采取措施。如有必要,需要做额外的实验室质量保证规程,以符合各个实验室的做法。

20.2 实验室试剂空白样。

如 **16.3.1** 所述,要在样品制备和分析过程中检测潜在的污染,需包含实验室试剂空白样。空白样由分析测试样品中使用的所有试剂和材料组成,并像测试样品一样进行分析。应根据各个实验室的做法对空白样进行评估。

20.3 质量控制样品。

为了验证整个分析过程的一致性,根据各个实验室的做法分析参比卷烟(或适当的质量控制样品)。

20.4 每运行 20 次后制作新的标准曲线(20 个样品萃取物)。

21 方法性能

21.1 报告限。

报告限设置为所用标准曲线的最低浓度,重新计算为 mg/g(例如 1.3 mg/g 对应 50 mg/L 的最低标准浓度)。

21.2 实验室基质加标回收率。

加入到基质上的分析物的回收率用作准确度的替代量度。通过将已知含量的标准品(萃取前)加到放置烟草的锥形瓶中并通过与样品相同的方法萃取烟碱来确定回收率,未加标的烟草也要进行分析。回收率由下式计算,如表 2 所示。

回收率(%)=100×(分析结果 – 未加标结果) / 加标量

表 2 加标到基质上烟碱的平均值和回收率

加标量(mg/g)	烟碱(mg/g)	
	平均值(mg/g)	回收率(%)
4.92	4.81	97.8
5.34	5.48	102.7

续表

加标量（mg/g）	烟碱（mg/g）	
	平均值（mg/g）	回收率（%）
8.01	7.75	96.8
9.84	9.72	98.8
13.37	12.72	95.1
19.68	19.60	99.6

21.3 分析特异性。

目标分析物的保留时间用于验证分析特异性。使用质量控制卷烟烟丝的成分响应与内标成分的响应比范围来验证未知样品的结果的特异性。

21.4 线性。

建立的烟碱标准曲线在 50~1000 mg/L（1.3~26.7 mg/g）的标准浓度范围内是线性的。

21.5 潜在干扰。

丁香酚的存在会引起干扰，因为其保留时间与烟碱的保留时间相似。含有丁香的样品最容易产生干扰。实验室可能需要通过调整分析仪器参数来排除干扰。

22 重复性和再现性

2010年进行的一项国际合作研究[2.6]涉及18个实验室和3种参比卷烟（1R5F、3R4F 和 CM6）以及2种商品卷烟，给出了表3所示方法的精密度限值。

常规正确操作该方法时，同一操作者使用同一设备在最短的操作时间内两个单一结果之间的差异超过重复性 r 的情况，平均20个样品不超过1次。

常规正确操作该方法时，两个实验室报告的匹配卷烟样品的单一结果超过再现性 R 的情况，平均20次不超过1次。

根据 ISO 5725-1 [2.7]和 ISO 5725-2 [2.8]对测试结果进行统计分析，得到表3所示的数据。

表3 测定卷烟烟丝中烟碱的精密度限值（mg/g）

参比卷烟	N	m_{cig}	r_{limit}	R_{limit}
1R5F	15	15.92	0.977	2.243
3R4F	17	17.16	1.152	2.414
CM6	17	18.77	1.378	2.635

注：商品卷烟的 TSNA 含量在参比卷烟的范围内，因此未在此表中报告。
1R5F，3R4F 和 CM6 是本研究中分析的三种参比卷烟；
N 为参与实验室数量；
m_{cig} 为卷烟烟丝中烟碱含量的平均值；
r_{limit} 为卷烟烟丝中烟碱含量的重复性限值；
R_{limit} 为卷烟烟丝中烟碱含量的再现性限值。

23 检测报告

检测报告应包含以下内容：
（a）相关方法即 WHO TobLabNet SOP 04；
（b）收到样品的时间；
（c）结果及其单位。

附录1 卷烟烟丝中烟碱的典型色谱图

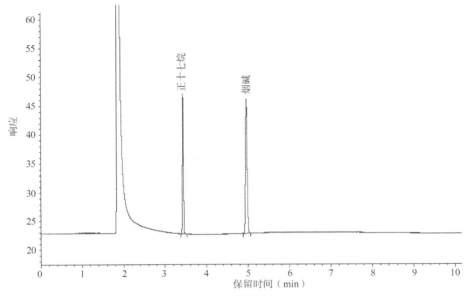

附图1 50 mg/L 烟碱标准液的色谱图

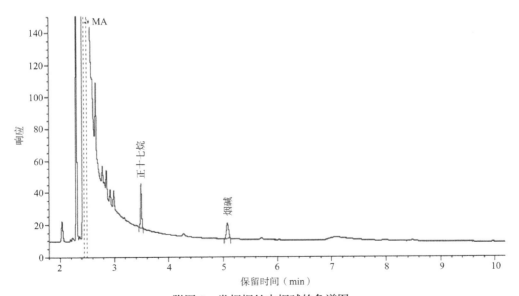

附图2 卷烟烟丝中烟碱的色谱图

WHO TobLabNet SOP 05

ISO 和深度抽吸方案下卷烟主流烟气中苯并[a]芘的测定标准操作规程

方　　法：ISO 和深度抽吸方案下卷烟主流烟气中苯并[a]芘的测定
分 析 物：苯并[a]芘（CAS 号：50-32-8）
基　　质：卷烟主流烟气粒相物
更新时间：2015 年 2 月

> 吸烟机的抽吸方案不能代表所有人的吸烟行为：吸烟机测试对用于卷烟设计及监管目的的卷烟释放物表征非常有用，但将吸烟机的测量结果披露给吸烟者则会导致对不同品牌的暴露和风险差异的误解。吸烟机测量的烟气释放物数据可以用于产品危害评估，但并不意味着这些数据等效于人体暴露或风险的测量值。将吸烟机测量结果的差异表示为暴露或风险的差异是对世界卫生组织烟草实验室网络标准的滥用。

提到的公司和制造商的产品，并不意味着 WHO 优先推荐这些产品。除错误和疏漏外，专有产品的名称以首字母大写表示。

本方法由世界卫生组织（WHO）烟草实验室网络（TobLabNet）制订，作为国际标准化组织（ISO）和深度抽吸方案下卷烟主流烟气中的苯并[a]芘（B[a]P）的测定标准操作规程（SOP）。

引言

为了在全球范围内建立具有可比性的烟草制品检测方法，需要卷烟特定成分和释放物的一致性检测方法。2008 年 11 月在南非德班举行的世界卫生组织《烟草控制框架公约》（WHO FCTC）第三次缔约方会议回顾了 FCTC/COP1(15)和 FCTC/COP2(14)号决议关于制订 WHO FCTC 第 9 条（烟草制品成分管制）和第 10 条（烟草制品披露管制）实施指南的要求，并提到第三次缔约方会议工作报告中有关其工作进展的信息，提请公约秘书处授权 WHO 无烟草行动组在 5 年内对检测卷烟成分和释放物的分析化学方法进行验证（FCTC/COP/3/REC/1）。

根据 2006 年 10 月在加拿大渥太华举行的第三次会议上确定的优先次序标准，WHO FCTC 第 9 条和第 10 条工作组确定优先验证以下成分的分析化学检测方法：

- 烟碱；
- 氨；
- 保润剂[1,2-丙二醇、甘油（1,2,3-丙三醇）和三甘醇（2,2′-二乙醚乙二醇）]。

测量这些成分需要验证三个方法：一个适用于烟碱，一个适用于氨，一个适用于保润剂。

根据上述渥太华会议确定的优先次序标准，工作组确定了在卷烟主流烟气释放中应该验证测试和测量方法（分析化学）的下列优先级物质清单：

- 4-(N-甲基亚硝胺基)-1,3-吡啶基-1-丁酮（NNK）；
- N-亚硝基去甲烟碱（NNN）；
- 乙醛；
- 丙烯醛；
- 苯；
- 苯并[a]芘；
- 1,3-丁二烯；
- 一氧化碳；
- 甲醛。

这些释放物使用下述两种抽吸方案测量，按五个方法进行验证：一个适用于烟草特有的亚硝胺（NNK 和 NNN），一个适用于苯并[a]芘，一个适用于醛类（乙醛、丙烯醛和甲醛），一个适用于挥发性有机物（苯和 1,3-丁二烯），一个适用于一氧化碳。

下表列出了用于验证上述测试方法的两种抽吸方案。

抽吸方案	抽吸容量（mL）	抽吸间隔（s）	滤嘴通风孔
ISO 方案：ISO 3308 常规分析用吸烟机 定义和标准条件	35	60	不封闭
深度抽吸方案：在 ISO 3308 基础上有所调整	55	30	所有通风孔必须如 SOP 01 **12.2** 所述 100%封闭

SOP 05 描述 ISO 和深度抽吸方案下卷烟主流烟气中苯并[a]芘的测定标准操作规程。

1 适用范围

本方法适用于气相色谱-质谱（GC-MS）法定量测定卷烟主流烟气中的苯并[a]芘。

苯并[a]芘是多环芳烃，2012 年被国际癌症研究机构（IARC）列为第 1 类人体致癌物，由有机物质不完全燃烧形成，主要通过吸入、口服和皮肤吸收进入人体。

2 参考标准

2.1 ISO 3402：烟草和烟草制品 用于调节和测试的大气环境。
2.2 ISO 4387：卷烟 常规分析用吸烟机测定总粒相物和去烟碱干粒相物。
2.3 TobLabNet SOP 01：卷烟深度抽吸标准操作规程。
2.4 ISO 3308：常规分析用吸烟机 定义和标准条件。
2.5 联合国毒品和犯罪问题办公室（UNODC）《代表性药物取样指南》（http://www.unodc.org/documents/scientific/Drug_Sampling.pdf）。
2.6 ISO 8243：卷烟 抽样。
2.7 ISO 5725-1：检测方法和结果的准确度（正确度和精密度） 第 1 部分 一般原则和定义。
2.8 ISO 5725-2：检测方法和结果的准确度（正确度和精密度） 第 2 部分 确定标准检测方法重复性和再现性的基本方法。
2.9 WHO TobLabNet 官方方法：ISO 和深度抽吸方案下卷烟主流烟气中苯并[a]芘的测定。

3 术语和定义

3.1 TPM：总粒相物。
3.2 B[a]P：苯并[a]芘。

3.3 B[*a*]P-d12：氘代苯并[*a*]芘。

3.4 烟草制品：完全或部分以烟叶为原料制造，用于抽吸、吸吮、咀嚼或吸入的产品（WHO FCTC 第 1(f)条）。

3.5 深度抽吸方案：抽吸过程参数为抽吸容量 55 mL，抽吸间隔 30 s，每次抽吸持续时间 2 s，100%封闭滤嘴通风孔。

3.6 ISO 方案：抽吸过程参数为抽吸容量 35 mL，抽吸间隔 60 s，每次抽吸持续时间 2 s，不封闭滤嘴通风孔。

3.7 实验室样品：用于进行实验室检测的样品，由一次性或在规定时间内送到实验室的单一类型产品组成。

3.8 试样：从实验室样品中随机抽取的待测样品。所取样品数量应能代表实验室样品。

3.9 试料：从试样中随机抽取的用于单次测试的样品。所取样品数量应能代表试样。

4 方法概述

4.1 来自卷烟试样的主流烟气被捕集到玻璃纤维滤片上。

4.2 可能需要调整卷烟数量来防止发生滤片发生穿透。若 92 mm 的滤片 TPM 超过 600 mg 或者 44 mm 的滤片 TPM 超过 150 mg，则必须减少每个滤片上卷烟的数量。

4.3 将含有同位素标记的内标物的溶液加在滤片上，然后用环己烷进行萃取。

4.4 环己烷萃取物用二氧化硅固相萃取柱进行洗脱，采用配备电子电离源的气相色谱-质谱（GC-MS）分析收集的洗脱液。

4.5 将 B[*a*]P 与 B[*a*]P-d12 的离子色谱峰面积比对标准品中已知的 B[*a*]P 浓度进行作图，建立标准曲线，从而确定样品的 B[*a*]P 含量。

5 安全和环境预防措施

注意：苯并[*a*]芘是已知的人体致癌物，应采取措施避免人体接触。B[*a*]P 及其溶液应在通风橱或手套箱中进行处理。
实验室应建立处理含 B[*a*]P 溶液的规程。

5.1 遵守所有化学实验活动的常规安全与环境预防措施。

5.2 使用本检测方法对特定产品进行检测和评估时所用材料或设备可能对环境有害。本检测方法无力解决所有与使用有关的安全问题。所有使用此方法的人有责任咨询相关机构并遵循现有监管要求，制定健康与安全规程以及环境预防措施。

5.3 应特别注意避免吸入或皮肤接触危险化学品。在制备或处理未稀释材料、标准溶液、萃取液和收集样品时，应在通风橱中进行，并穿戴合适的实验服、手套和护目镜。

6 仪器和设备

常规实验仪器，特别是：

6.1 按照 ISO 3402 [**2.1**]规定调节卷烟所需设备。

6.2 按照 ISO 4387 [**2.2**]规定标记烟蒂长度所需设备。

6.3 按照 WHO TobLabNet SOP 01 [**2.3**]规定采用深度抽吸方案时封闭通风孔所需设备。

6.4 按照 ISO 3308 [**2.4**]规定抽吸烟草制品所需设备。

6.5 测量精度为 0.0001 g 的分析天平。

6.6 涡旋振荡器，如腕式振动器。

6.7 移液枪及能精确移取 10~1000 μL 体积的枪头。

6.8 10 mL、40 mL 和 100 mL 移液管。

6.9 10 mL、25 mL 和 100 mL 容量瓶。

6.10 100 mL 和 200 mL 锥形瓶或合适的烧瓶。

6.11 二氧化硅固相萃取柱和滤芯（500 mg）：纯非晶态二氧化硅，如 Sep-pak Vac 硅胶滤芯（Waters）、Bond Elut Jr SI 滤芯（Agilent）、Supelclean LC-Si 滤芯（Supelco）、Strata SI-1 Silica（Phenomenex）等。

6.12 玻璃移液管。

6.13 带有合适烧瓶的旋转蒸发器，或带试管的 Turbovap。

6.14 配备计算机控制、数据采集和处理系统的 GC-MS。系统必须能够操作质谱仪，以便在单离子监测检测模式下获得色谱数据。所使用的色谱柱为毛细管色谱柱，模式为不分流进样。建议气相色谱仪配备自动进样器。

6.15 气相色谱柱：具有甲基苯基（5%）聚硅氧烷固定相的熔融石英毛细管色谱柱，具有 0.25 mm 内径和 0.25 μm 膜厚度的 30 m 柱适合本分析，如 DB-5ms 色谱柱（Agilent）。

6.16 2 mL 自动进样瓶。

7 试剂

除非另有说明，所有试剂应至少为分析纯。试剂尽可能以 CAS 号来识别。

7.1 环己烷 [110-82-7]。

7.2 B[*a*]P [50-32-8]。

7.3 B[*a*]P-d12 [63466-71-7]。

注意：B[*a*]P 和 B[*a*]P-d12 对人体有致癌作用。操作这些化合物或含有这些化合物的任何溶液时，应采取适当的安全预防措施。

8 玻璃器皿准备

清洁、干燥的玻璃器皿，避免残留物污染。

9 溶液配制

不适用。

10 标准溶液配制

下述标准溶液配制方法仅供参考，如有需要可进行调整。

10.1 同位素内标物（可根据以下两种方法中的任一种来配制）。

选项 A

  10.1.1 200 µg/mL 的初级 B[*a*]P-d12 储备液：

    10.1.1.1 用测量精度为 0.0001 g 的天平称取约 0.005 g B[*a*]P-d12 置于 25 mL 的容量瓶中，记录质量，精确到 0.0001 g。

    10.1.1.2 用环己烷溶解 B[*a*]P-d12，混合均匀。

    10.1.1.3 贴标签，转移到棕色瓶中并在–20℃±5℃下储存。

  10.1.2 2 µg/mL 的 B[*a*]P-d12 工作溶液：

    10.1.2.1 移取初级 B[*a*]P-d12 储备液[**10.1.1**]1 mL 置于 100 mL 容量瓶中。

    10.1.2.2 用环己烷稀释至刻度，摇匀。

    10.1.2.3 贴标签，在–20℃±5℃下储存。

选项 B

  10.1.3 从供应商处获得浓度为 1000 µg/mL 的初级 B[*a*]P-d12 储备液，在–20℃±5℃下储存。

  10.1.4 制备 2 µg/mL 的 B[*a*]P-d12 工作溶液：

    10.1.4.1 移取 200 µL 的 1000 µg/mL B[*a*]P-d12 储备液[**10.1.3**]置于 100 mL 容量瓶中。

    10.1.4.2 用环己烷稀释至刻度，摇匀。

    10.1.4.3 贴标签，在–20℃±5℃下储存。

10.2 B[a]P 标准溶液。

 10.2.1 200 μg/mL 的初级 B[a]P 储备液：

  10.2.1.1 用测量精度为 0.0001 g 的天平称取约 0.005 g 的 B[a]P 置于 25 mL 的容量瓶中，记录质量，精确至 0.0001 g。

  10.2.1.2 用 25 mL 环己烷溶解 B[a]P 标准品，混合均匀。

  10.2.1.3 贴标签，在–20℃±5℃下储存。

 10.2.2 1 μg/mL 的二级 B[a]P 储备液：

  10.2.2.1 移取 0.5 mL 的初级 B[a]P 储备液[10.2.1]置于 100 mL 容量瓶中。

  10.2.2.2 用环己烷稀释至刻度，摇匀。

  10.2.2.3 贴标签，在–20℃±5℃下储存。

如果储存在–20℃±5℃，标准溶液可稳定长达 6 个月。

注意：超声处理可用于制备初级储备液，以帮助 B[a]P 和 B[a]P-d12 标准品完全溶解。在这种情况下，首先将标准品溶解在约 15 mL 溶剂中。超声处理后，让溶液静置直至其冷却至工作温度（例如室温），然后用环己烷稀释至刻度。

10.3 B[a]P 工作溶液。

如表 1 所示准备工作溶液。

表 1 B[a]P 工作溶液

标准	二级 B[a]P 标准储备液的体积(1 μg/mL)(μL)	B[a]P-d12 注射液体积(1 μg/mL)[10.1.2 或 10.1.4](μL)	总体积(mL)	B[a]P 工作标准溶液的近似浓度(ng/mL)
1	20	100	10	2
2	40	100	10	4
3	80	100	10	8
4	200	100	10	20
5	400	100	10	40
6	600	100	10	60

注：所有溶剂和溶液在使用前调节至室温。

 10.3.1 将不同体积的二级 B[a]P 储备液[10.2.2]注入 10 mL 容量瓶中。

 10.3.2 加入 100 μL B[a]P-d12 内标溶液[10.1.2 或 10.1.4]。

 10.3.3 对于选项 A，B[a]P-d12 工作溶液[10.1.2]的浓度可以由下式确定：

$$最终浓度（μg/mL）= x \times 400$$

 式中：x ——标准品[10.1.1.1]的原始质量（g）。

 10.3.4 对于选项 B，B[a]P-d12 工作溶液[10.1.4]的浓度根据供应商提供的值计算。

10.3.5 添加环己烷至刻度，混合均匀。

10.3.6 标准溶液的最终浓度由下式确定：

$$最终浓度（ng/mL）$$
$$= \frac{x(g)}{25\ mL} \times \frac{0.5\ mL}{100\ mL} \times \frac{y(\mu L)}{10\ mL} \times \frac{1\ mL}{1000\ \mu L} \times \frac{1000000000\ ng}{1\ g}$$
$$= \frac{x \times 0.5 \times 1000 \times y(ng)}{25\ mL} = x \times y \times 20$$

式中：x ——标准品[**10.2.1.1**]的原始质量（g）；

y ——表 1 描述的二级 B[*a*]P 储备液[**10.2.2**]的体积（μL）。

10.3.7 标准溶液范围可根据使用的设备和待测样品进行调整，注意方法灵敏度可能对结果产生的影响。

11 抽样

11.1 根据 ISO 8243 [**2.6**]或替代方法，根据各实验室惯例或者要求的特殊规定或样品可用性，选取具有代表性的实验室卷烟样品。

11.2 试样的组成。

11.2.1 若可能，将实验室样品分成独立的单元（如包、条）。

11.2.2 从至少 $\sqrt{n}$ 个[**2.6**]单元中取出每种试样的等量产品。

12 卷烟准备

12.1 按照 ISO 3402 [**2.1**]调节所有待测卷烟。

12.2 按照 ISO 4387 [**2.2**]和 WHO TobLabNet SOP 01 [**2.3**]标记烟蒂长度。

12.3 按照 WHO TobLabNet SOP 01 [**2.3**]的深度抽吸方案或 ISO 方案准备待测样品。

13 吸烟机设置

13.1 环境条件。

按 ISO 3308[**2.4**]要求设定抽吸的环境条件。

13.2 吸烟机要求。

除深度抽吸方案按 WHO TobLabNet SOP 01 [**2.3**]进行准备外，其余按照 ISO 3308 [**2.4**]的要求设置吸烟机。

14 样品生成

在指定吸烟机上抽吸足够量的样品，样品含量应不足以使滤片发生穿透，且确保 B[a]P 的浓度落在分析的校准范围内。

14.1 抽吸卷烟试样并按照 ISO 4387 [2.2]或 WHO TobLabNet SOP 01[2.3]收集 TPM。

14.2 至少包含一个用于质量控制的参比试样。

14.3 当样品类型为第一次检测时，评估滤片穿透的可能性，调整卷烟的数量以避免发生穿透。当确定焦油、烟碱和一氧化碳时，对于 92 mm 滤片和 44 mm 滤片，穿透分别发生在总粒相物水平超过 600 mg 和 150 mg 时。一旦发生穿透，必须减少每个滤片上的卷烟数量。

14.4 表 2 显示了 ISO 和深度抽吸方案中直线型和转盘型吸烟机每次抽吸的数量。

表 2 直线型和转盘型吸烟机进行一次测量的卷烟数量

	ISO 方案		深度抽烟方案	
	直线型	转盘型	直线型	转盘型
每个滤片上的卷烟数量	5	20	3	10
每个结果包含的滤片数	4	1	7	2

注：应调整抽吸的卷烟数量，确保滤片不会发生穿透，且 B[a]P 的浓度落在分析的校准范围内。

14.5 记录每个滤片的卷烟数量和总抽吸量。

14.6 测量完成所需的样品量后，进行五次空吸，然后从吸烟机上取下滤片捕集器。

15 样品制备

15.1 滤片萃取。

15.1.1 从捕集器上取下滤片，将滤片对折，再对折，使滤片含有 TPM 的面向内。用不含 TPM 的另一面擦拭捕集器内表面，使残留在捕集器上的粒相物转移到滤片上。将每个滤片置于 60 mL 的棕色瓶中。

注意：在 ISO 方案中，用于直线型吸烟机的四个 44 mm 滤片或用于转盘型吸烟机的一个 92 mm 滤片被萃取到一个烧瓶中。在深度抽吸方案中，用于直线型吸烟机的三个 44 mm 滤片或用于转盘型吸烟机的一个 92 mm 滤片被萃取到一个烧瓶中。

15.1.2 向烧瓶中加入环己烷（44 mm 滤片加 40 mL，92 mm 滤片加 100 mL）。

15.1.3 用 B[a]P-d12 工作溶液[10.1.2 或 10.1.4]对每个烧瓶进行加标（44 mm 滤片为 40 μL，92 mm 滤片为 100 μL）。

15.1.4 使用扇形或圆形振动器以 250 r/min 振动 30 min，从滤片上萃取 B[a]P。

15.2 样品净化。

15.2.1 用 10 mL 环己烷活化固相萃取柱，并舍去洗脱液，且不要让萃取柱流干。

15.2.2 移取 10 mL 样品萃取物至萃取柱上，让其通过萃取柱，并收集洗脱液。

15.2.3 在最后一个试样通过后，另用两份 15 mL 的环己烷洗脱固相萃取柱。

15.2.4 蒸发环己烷溶液至几乎干燥。

注意：旋转蒸发器的条件应设定在 55℃，真空压力为 30 kPa（300 mbar）（约 10 min）。

15.2.5 移取 1 mL 环己烷到烧瓶中以溶解萃取物，盖上烧瓶，小心旋转，冲洗内表面，取等份试样进行 GC-MS 测试。

重要提示：样品中 B[a]P-d12 内标物的最终浓度约为 20 ng/mL。

16 样品分析

使用 GC-MS 定量分析卷烟主流烟气中的 B[a]P。分析物可以在 GC 柱上与其他潜在干扰物质分离开，比较未知物的峰面积比（分析物与同位素标记分析物）与已知浓度标准液的峰面积(分析物与同位素标记分析物)比，得到分析物的浓度。

16.1 GC-MS 操作条件举例。

GC 柱：具有甲基苯基（5%）聚硅氧烷固定相的熔融石英毛细管色谱柱，具有 0.25 mm 内径和 0.25 μm 膜厚度的 30 m 色谱柱（DB-5ms Agilent）；

进样口温度：280℃；

模式：恒流；

流速：1.2 mL/min；

进样模式：不分流，1 μL 或 2 μL；

升温程序：初始温度为 150℃，以 6℃/min 升高到 260℃，260℃保持 7 min，50℃/min 升高到 290℃，290℃保持 20 min；

传输线温度：280℃；

MS 源：230℃；

离子质荷比：B[*a*]P *m/z* 252，B[*a*]P-d12 *m/z* 264；

驻留时间：50 ms；

电离模式：电子电离。

注意：根据仪器和色谱柱条件以及色谱峰的分离度调整操作参数。

16.2 一般分析信息。

    16.2.1 对于此处描述的条件，预期的洗脱顺序是 B[*a*]P-d12、B[*a*]P。

    16.2.2 流速、流动相浓度和色谱柱的老化等差异可能会改变保留时间。

    16.2.3 根据每个实验室的做法来确定 B[*a*]P 的测定顺序。本节举例说明。

    16.2.4 注入空白溶液（不含内标物的萃取液）来检查系统或试剂中是否存在污染。

    16.2.5 注入标准空白溶液（带有内标物的空白溶液）来验证 GC-MS 系统的性能。

    16.2.6 注入标准液、质量控制溶液和样品。

    16.2.7 记录 B[*a*]P 和 B[*a*]P-d12 的峰面积。

    16.2.8 计算每种标准溶液（包括标准空白溶液）的 B[*a*]P 峰与 B[*a*]P-d12 峰的相对响应比（$A_{B[a]P}/A_{B[a]P\text{-}d12}$）。

    16.2.9 以 B[*a*]P 的浓度为 X 轴，以相应的峰面积比为 Y 轴作图。

    16.2.10 截距在统计上与零不应该有显著差异。

    16.2.11 标准曲线应在整个标准范围内呈线性。

    16.2.12 使用线性回归的斜率 b 和截距 a 来计算线性回归方程（$Y=a+bx$），如果线性回归方程的 R^2 小于 0.99，则应重新校准。如果单个校准点与预期值（通过线性回归估计）相差超过 10%，则应该舍弃该点。

    16.2.13 注入质量控制溶液和样品，用适当的软件确定峰面积。

    16.2.14 所有试料的峰值比必须落在标准曲线的工作范围内；否则，应调整标准溶液或试料的浓度。

代表性色谱图见附录 1。

17 数据分析和计算

    17.1 斜率和截距是由 B[*a*]P 和 B[*a*]P-d12 的相对响应与 B[*a*]P 的相对浓度确定的。

    17.2 针对每个标准溶液[10.3]计算峰面积的相对响应比：

$$RF = \frac{A_{B[a]P}}{A_{B[a]P\text{-}d12}}$$

式中：RF——相对响应比；

$A_{B[a]P}$——B[a]P（色谱图中 m/z 252 的离子）的峰面积；

$A_{B[a]P\text{-}d12}$ 是 B[a]P-d12（色谱图中 m/z 264 的离子）的峰面积。

17.3 绘制每种标准溶液的相对响应因子 $A_{B[a]P}/A_{B[a]P\text{-}d12}$（Y 轴）与浓度（X 轴）的关系图。使用线性回归的斜率 b 和截距 a 计算线性回归方程（Y=a+bX）。

17.4 标准曲线应在整个标准范围内呈线性。

17.5 B[a]P 的含量由计算的试样的相对响应比与标准曲线的斜率和截距来确定：

$$M = \frac{Y-a}{b} \times \frac{V \times V_e}{V_c \times n}$$

式中：M——B[a]P 的计算含量（ng/支）；

Y——相对响应比（$A_{B[a]P}/A_{B[a]P\text{-}d12}$）；

a——从标准曲线获得的线性回归方程的截距；

b——从标准曲线获得的线性回归方程的斜率；

V——样品溶液的体积（1 mL）；

V_e——萃取液的体积（对于 44 mm 滤片是 40 mL，对于 92 mm 滤片是 100 mL）；

V_c——用于净化萃取液的等份试样的体积（10 mL）；

n——抽吸卷烟的数量（表 2）。

如果适用，可以使用替代计算方程。

18 特别注意事项

安装新色谱柱后，按照制造商的规定，在指定的仪器条件下注入烟草样品萃取物进行调节。应重复进样，直到 B[a]P 和 B[a]P-d12 的峰面积（或峰高）可重复。

19 数据报告

19.1 报告每个评估样品的单独测量值。

19.2 以 ng/支或根据需要报告结果。

20 质量控制

20.1 控制参数。

如果控制测量值超出预期值的公差极限，则必须进行适当的调查并采取措施。如有必要，需要做额外的实验室质量保证规程，以符合各个实验室的做法。

20.2 实验室试剂空白样。

如 **16.3.5** 所述,要在样品制备和分析过程中检测潜在的污染,需包含实验室试剂空白样。空白样由分析测试样品中使用的所有试剂和材料组成,并像测试样品一样进行分析。应根据各个实验室的做法对空白样进行评估。

20.3 质量控制样品。

为了验证整个分析过程的一致性,根据各个实验室的做法分析参比卷烟(或适当的质量控制样品)。

20.4 每运行 20 次后制作新的标准曲线(20 个样品萃取物)。

21 方法性能

21.1 报告限。

报告限设置为所用标准曲线的最低浓度,重新计算为 ng/支卷烟。

21.2 实验室基质加标回收率。

加入到基质上的分析物的回收率用作准确度的替代量度。通过将已知含量的标准品(抽吸后)添加到捕集卷烟烟气的玻璃纤维滤片上,然后按照样品相同的萃取方法对玻璃纤维滤片进行萃取,未加标的玻璃纤维滤片也要进行分析。回收率由下式计算,如表 3 所示。

回收率(%)=100×(分析结果–未加标结果)/ 加标量

表 3　B[*a*]P 基质加标的平均值和回收率

加标量(ng/支)	平均值(ng/支)	回收率(%)
2.91	2.93	100.58
6.79	6.82	100.39
9.70	10.08	103.92

21.3 分析特异性。

GC-MS 具有分析特异性。保留时间和分子质荷比(*m/z*)用于验证未知样品的结果的特异性。

21.4 线性。

建立的 B[*a*]P 标准曲线在 2~60 ng/mL 的标准浓度范围内是线性的。

21.5 潜在干扰。

没有已知的组分具有与 B[*a*]P 或 B[*a*]P-d12 相似的保留时间和 *m/z*。

22 重复性和再现性

2012 年进行的一项国际合作研究[2.9]涉及 3 种参比卷烟(1R5F、3R4F 和 CM6)

以及8个实验室对2种商业品牌卷烟的测试,表4给出了所示方法的精密度限值。

常规正确操作该方法时,同一操作者使用同一设备在最短的操作时间内的两个单一结果之间的差异超过重复性 r 的情况,平均20个样品不超过1次。

常规正确操作该方法时,两个实验室报告的匹配卷烟样品的单一结果超过再现性 R 的情况,平均20次不超过1次。

根据 ISO 5725-1 [**2.7**]和 ISO 5725-2 [**2.8**]对测试结果进行统计分析,得到表4所示的准确数据。

表4 参比卷烟主流烟气中 B[*a*]P 的精密度限值（ng/支）

参比卷烟	ISO 方案				深度抽烟方案			
	N	m_{cig}	r_{limit}	R_{limit}	N	m_{cig}	r_{limit}	R_{limit}
1R5F	8	1.46	0.47	0.99	8	6.47	1.52	2.91
3R4F	8	5.99	1.03	2.09	8	14.51	2.52	3.15
CM6	7	13.70	1.43	5.83	7	25.86	2.81	8.45

注：1R5F、3R4F 和 CM6 是本研究中分析的三种参比卷烟;
N 为参与实验室数量;
m_{cig} 为每支卷烟 B[*a*]P 的平均值;
r_{limit} 为 B[*a*]P 的重复性限值;
R_{limit} 为 B[*a*]P 的再现性限值。

为了计算 r 和 R,在 ISO 条件下,测试结果被定义为来自直线型吸烟机的四个滤片（每个滤片上5支烟）和转盘型吸烟机的一个滤片（每个滤片上20支烟）平均释放量的7次重复的平均值。在深度抽吸方案下,测试结果被定义为来自直线型吸烟机的七个滤片（每个滤片上3支烟）和转盘型吸烟机的两个滤片（每个滤片上10支烟）平均释放量的7次重复的平均值。更多信息请参[**2.9**]。

23 检测报告

检测报告应包含以下内容：
（a）相关方法即 WHO TobLabNet SOP 05；
（b）收到样品的时间；
（c）结果及其单位。

附录1 卷烟主流烟气中苯并[a]芘的典型色谱图

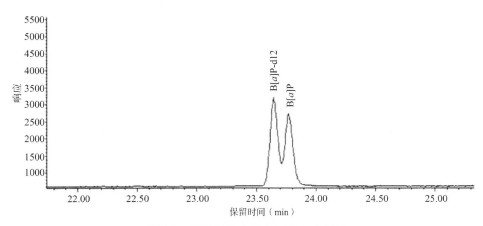

附图1 标准溶液中苯并[a]芘的色谱图

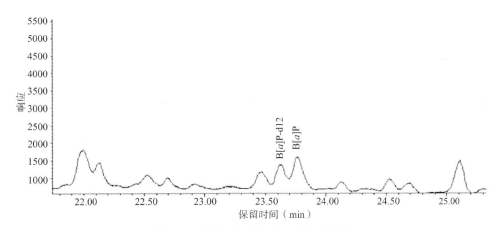

附图2 试样溶液中苯并[a]芘的色谱图

WHO TobLabNet SOP 06

卷烟烟丝中保润剂的测定标准操作规程

方　　法：卷烟烟丝中保润剂的测定
分 析 物：丙二醇（丙烷-1,2-二醇）（CAS 号：57-55-6）
丙三醇（丙烷-1,2,3-三醇）（CAS 号：56-81-5）
三甘醇（2,2′-二乙醚乙二醇）（CAS 号：112-27-6）
基　　质：卷烟烟丝
更新时间：2016 年 6 月

> 吸烟机的抽吸方案不能代表所有人的吸烟行为:吸烟机测试对用于卷烟设计及监管目的的卷烟释放物表征非常有用,但将吸烟机的测量结果披露给吸烟者则会导致对不同品牌的暴露和风险差异的误解。吸烟机测量的烟气释放物数据可以用于产品危害评估,但并不意味着这些数据等效于人体暴露或风险的测量值。将吸烟机测量结果的差异表示为暴露或风险的差异是对世界卫生组织烟草实验室网络标准的滥用。

提到的公司和制造商的产品,并不意味着 WHO 优先推荐这些产品。除错误和疏漏外,专有产品的名称以首字母大写表示。

本方法由世界卫生组织（WHO）烟草实验室网络（TobLabNet）制订，作为卷烟烟丝中保润剂的测定标准操作规程（SOP）。

引言

为了在全球范围内建立具有可比性的烟草制品检测方法，需要卷烟特定成分和释放物的一致性检测方法。2008 年 11 月在南非德班举行的世界卫生组织《烟草控制框架公约》（WHO FCTC）第三次缔约方会议回顾了 FCTC/COP1(15)和 FCTC/COP2(14)号决议关于制订 WHO FCTC 第 9 条（烟草制品成分管制）和第 10 条（烟草制品披露管制）实施指南的要求，并提到第三次缔约方会议工作报告中有关其工作进展的信息，提请公约秘书处授权 WHO 无烟草行动组在 5 年内对检测卷烟成分和释放物的分析化学方法进行验证（FCTC/COP/3/REC/1）。

根据 2006 年 10 月在加拿大渥太华举行的第三次会议上确定的优先次序标准，WHO FCTC 第 9 条和第 10 条工作组确定优先验证以下成分的分析化学检测方法：
- 烟碱；
- 氨；
- 保润剂[1,2-丙二醇、甘油（1,2,3-丙三醇）和三甘醇（2,2'-二乙醚乙二醇）]。

测量这些成分需要验证三个方法：一个适用于烟碱，一个适用于氨，一个适用于保润剂。

根据上述渥太华会议确定的优先次序标准，工作组确定了在卷烟主流烟气释放物中应该验证测试和测量方法（分析化学）的下列优先级物质清单：
- 4-(N-甲基亚硝胺基)-1,3-吡啶基-1-丁酮（NNK）；
- N-亚硝基去甲烟碱（NNN）；
- 乙醛；
- 丙烯醛；
- 苯；
- 苯并[a]芘；
- 1,3-丁二烯；
- 一氧化碳；
- 甲醛。

这些释放物使用下述两种抽吸方案测量，按五个方法进行验证：一个适用于烟草特有的亚硝胺（NNK 和 NNN），一个适用于苯并[a]芘，一个适用于醛类（乙醛、丙烯醛和甲醛），一个适用于挥发性有机物（苯和 1,3-丁二烯），一个适用于一氧化碳。

下表列出了两种用于验证上述测试方法的吸烟方案。

抽吸方案	抽吸容量（mL）	抽吸间隔（s）	滤嘴通风孔
ISO 方案：ISO 3308 常规分析用吸烟机 定义和标准条件	35	60	不封闭
深度抽吸方案：在 ISO 3308 基础上有所调整	55	30	所有通风孔必须如 SOP 01 **12.2** 所述 100%封闭

SOP 06 描述卷烟烟丝中保润剂的测定标准操作规程。对于卷烟以外的烟草制品，可根据方法给出的具体参数进行调整。

1 适用范围

本方法适用于气相色谱-质谱（GC-MS）法定量测定卷烟烟丝中的保润剂：甘油（1,2,3-丙三醇）、丙二醇和三甘醇（2,2′-二乙醚乙二醇，三乙二醇）。

2 参考标准

2.1 ISO 8243：卷烟 抽样。

2.2 联合国毒品和犯罪问题办公室（UNODC）《代表性药物取样指南》（http://www.unodc.org/documents/scientific/Drug_Sampling.pdf）。

2.3 WHO TobLabNet 官方方法：卷烟烟丝中保润剂的测定。

2.4 WHO TobLabNet SOP 02：烟草制品成分和释放物分析方法验证标准操作规程。

2.5 ISO 5725-1：检测方法和结果的准确度（正确度和精密度） 第 1 部分 一般原则和定义。

2.6 ISO 5725-2：检测方法和结果的准确度（正确度和精密度） 第 2 部分 确定标准检测方法重复性和再现性的基本方法。

3 术语和定义

3.1 保润剂含量：卷烟烟丝中的保润剂总量，以 mg/g 卷烟烟丝表示。

3.2 卷烟烟丝：卷烟中含烟草的部分，包括再造烟叶、烟梗、膨胀烟丝和添加物。

3.3 烟草制品：完全或部分以烟叶为原料制造，用于抽吸、吸吮、咀嚼或吸入的产品（WHO FCTC 第 1(f)条）。

3.4 实验室样品：用于进行实验室检测的样品，由一次性或在规定时间内送到实验室的单一类型产品组成。

3.5 试样：从实验室样品中随机抽取的待测样品。所取样品数量应能代表实验室样品。

3.6 试料：从试样中随机抽取的用于单次测试的样品。所取样品数量应能代表试样。

4 方法概述

4.1 使用甲醇和1,3-丁二醇溶液的混合物从卷烟烟丝中萃取保润剂。

4.2 用火焰离子化检测器或质谱仪检测器分析萃取物。

4.3 将每种保润剂与内标物的峰面积比对标准品中已知的每种保润剂浓度进行作图，建立标准曲线，从而确定试料的保润剂含量。

5 安全和环境预防措施

5.1 遵守所有化学实验活动的常规安全与环境预防措施。

5.2 使用本检测方法对特定产品进行检测和评估时所用材料或设备可能对环境有害。本检测方法无力解决所有与使用有关的安全问题。所有使用此方法的人有责任咨询相关机构并遵循现有监管要求，制定健康与安全规程以及环境预防措施。

5.3 应特别注意避免吸入或皮肤接触危险化学品。在制备或处理未稀释材料、标准溶液、萃取液和收集样品时，应在通风橱中进行，并穿戴合适的实验服、手套和护目镜。

6 仪器和设备

常规实验仪器，特别是：

6.1 测量精度为0.0001 g的分析天平。

6.2 萃取烧瓶：带有塞子的锥形瓶。

6.3 机械腕式振动器。

6.4 配有火焰离子化检测器的气相色谱仪。

6.5 能够明显分离溶剂、内标物、丙二醇、甘油、三甘醇和其他烟草成分色谱峰的毛细管气相色谱柱，如DB-Wax熔融石英色谱柱（30 m×0.32 mm×1 μm）。

7 试剂

除非另有说明，所有试剂应至少为分析纯。试剂尽可能以CAS号来识别。

7.1 甲醇[67-56-1]（色谱纯）。

7.2 甘油[56-81-5]。

7.3 三甘醇[112-27-6]。

7.4 丙二醇[57-55-6]。

7.5 1,3-丁二醇[107-88-0]（用作内标物）。

7.6 载气：高纯度（＞99.999%）的氦气[7440-59-7]。

7.7 辅助气体：用于火焰离子化检测器的高纯度(99.999%)空气和氢气[1333-74-0]。

8 玻璃器皿准备

清洁和干燥的玻璃器皿，避免残留物污染。

9 溶液配制

1,3-丁二醇初级储备液（约 2 mg/mL）：

9.1 称取约 20 g（±0.05g）的 1,3-丁二醇。

9.2 将称取的 1,3-丁二醇放入 100 mL 容量瓶中，并用甲醇定容。

9.3 用移液管将 20 mL 初级储备液移入 2 L 容量瓶中，并用甲醇定容。溶液的体积可以根据需要放大或缩小。混合均匀，储存于 4~8℃。

10 标准溶液配制

10.1 约 100 mg/mL 的甘油初级标准溶液：
准确称取 10 g ± 0.05 g 甘油到 100 mL 容量瓶中，用萃取液[**9.3**]定容至刻度。

10.2 约 50 mg/mL 的丙二醇初级标准溶液：
准确称取 5 g ± 0.05 g 丙二醇到 100 mL 容量瓶中,用萃取液[**9.3**]定容至刻度。

10.3 约 25 mg/mL 的三甘醇初级标准溶液：
准确称取 2.5 g ± 0.05 g 三甘醇到 100 mL 容量瓶中，并用萃取液[**9.3**]定容至刻度。

10.4 混合二级标准溶液（4 mg/mL 甘油，3 mg/mL 丙二醇，1.5 mg/mL 三甘醇）：
移取 4 mL 的甘油[**10.1**]、6 mL 的丙二醇[**10.2**]和 6 mL 的三甘醇[**10.3**]的初级储备液到 100 mL 容量瓶中，用萃取液[**9.3**]定容至刻度。

将所有标准溶液储存于 4~8℃。

10.5 工作溶液。

所有标准溶液均使用表 1 和表 2 所列稀释比例在容量瓶中配制。

表 1 标准工作溶液

标准	混合二级标准溶液体积（mL）	总体积（mL）	工作标准溶液中的近似甘油浓度（mg/mL）	工作标准溶液中的近似丙二醇浓度（mg/mL）	工作标准溶液中的近似三甘醇浓度（mg/mL）
1	0.2	10	0.4	0.005	0.08
2	0.4	10	0.8	0.01	0.16
3	0.8	10	1.6	0.02	0.32
4	1.6	10	3.2	0.04	0.64
5	3.2	10	6.4	0.08	1.28
6	4.8	10	9.6	0.12	1.92

表 2 低含量水平丙二醇的标准工作溶液

序号	混合二级标准溶液体积（mL）	总体积（mL）	标准工作溶液中丙二醇近似溶度（mg/mL）
1	0.017	10	0.005
2	0.050	10	0.015
3	0.100	10	0.030
4	0.250	10	0.075
5	0.500	10	0.150
6	1.000	10	0.300

注 1：用含有内标物的萃取溶液[9.3]在容量瓶中定容至刻度。

注 2：标准溶液范围可根据使用的设备和待测样品进行调整，注意可能对方法灵敏度产生的影响。

注 3：如有必要，可使用所需体积的混合二级储备液在 10 mL 容量瓶中加入萃取溶液制备定量下限或上限的低或高浓度的标准溶液。表 2 为丙二醇示例，必要时调整校准范围很重要。

所有溶剂和溶液在使用前调节至室温。

11 抽样

11.1 根据 ISO 8243 [2.1]或替代方法，根据各实验室惯例或者要求的特殊规定或样品可用性，选取具有代表性的实验室卷烟样品。

11.2 试样的组成。

11.2.1 若可能，将实验室样品分成独立的单元（如包、条）。

11.2.2 从至少 $\sqrt{n}$ 个[2.2]单元中取出每种试样的等量产品。

11.2.3 如果没有可用的单元，则将整个实验室样品合并为一个单元。

12 卷烟准备

12.1 从两盒卷烟和质量控制样品（若适用）中取出烟丝，或使用至少 10 g

加工过的卷烟烟丝。

12.2 将足够的卷烟烟丝进行混合至每个试料至少含 10 g 烟丝。

12.3 通过研磨使卷烟烟丝均匀化，并保存在密闭的棕色瓶或容器中。将样品分成几个试样。

12.4 按照 ISO 3402 [2.1]对磨碎的卷烟烟丝进行调制。

13 吸烟机设置

不适用。

14 样品生成

不适用。

15 样品制备

15.1 称取 4 g 经充分混合的研磨测试样品，精确至 0.001 g，放入合适的萃取瓶中。

15.2 向样品中加入 50 mL 萃取液[9.3]。

注：样品质量和稀释系数可根据保润剂的含量或样品的可用性进行调整。

15.3 将烧瓶塞住，并置于腕式振荡器或等效装置上以不低于 210 r/min 的速率振荡至少 60 min。

15.4 从振荡器中取出样品，旋转烧瓶使所有烟丝浸没在溶剂中。

15.5 让样品静置或使用离心机，直至上清液澄清。

15.6 将上清液转移至自动进样瓶里，并进行 GC 分析。

16 样品分析

使用配备火焰离子化检测器或与质谱检测器联用的气相色谱仪来定量测定卷烟烟丝中的保润剂。分析物可以从色谱柱上的其他潜在干扰物质中分离出来，比较未知物的峰面积比与已知浓度标准溶液的峰面积比，得到单个分析物的浓度。

16.1 气相色谱操作条件示例。

气相色谱柱：DB-WAX 熔融石英柱，30 m×0.32 mm id×25.0 μm；

固定相：DB-Wax 或等效物；

载气：氦气，流速为 1.8 mL/min；

进样量：1 μL；

进样模式：分流比 25∶1；

程序升温：进样口温度 220℃，检测器温度 260℃，初始温度 120℃保持 3 min，升温速率 10℃/min 至 180℃，保持 11 min；

总运行时间：20 min。

注：操作参数可能需要根据仪器和色谱柱条件以及色谱峰的分辨率进行调整。

16.2 预计保留时间。

 16.2.1 对于此处描述的条件，预期的洗脱顺序是丙二醇、1,3-丁二醇（内标物）、甘油、三甘醇。

 16.2.2 温度、气体流速和色谱柱的使用年限的差异可能会改变保留时间。

 16.2.3 在分析开始之前，必须验证洗脱顺序和保留时间。

16.3 保润剂测定。

测定步骤的顺序应符合实验室惯例。提供以下内容作为指南：

 16.3.1 在使用前通过注入两份 1 μL 的样品溶液作为预试样来调节系统。

 16.3.2 在与样品相同的条件下注入一个标准溶液（萃取溶液[9.3]），以验证气相色谱系统的性能。

 16.3.3 注入空白溶液（溶剂[7.1]）以检查系统或试剂中是否存在污染。

 16.3.4 将各标准溶液[10.5]注入气相色谱仪。

 16.3.5 评估标准溶液的保留时间和响应（峰面积）。如果保留时间与之前进样的保留时间接近（±0.2 min），并且响应值与之前进样的典型响应值的差异在±20%以内，则系统已准备就绪可进行分析。如果响应值超出规范范围，则应根据各实验室政策采取后续操作。

 16.3.6 记录每种保润剂和内标物的峰面积。

 16.3.7 计算保润剂峰面积与内标物峰面积比（相对响应因子，RF）。每种保润剂标准溶液（包括空白溶剂）的 $RF = Y = A_{humectants}/A_{IS}$

 16.3.8 根据峰面积比（RF，Y 轴）绘制添加保润剂的浓度（X 轴）曲线。

 16.3.9 截距在统计上不应与零有显著差异。

 16.3.10 标准曲线应在整个范围内呈线性。

 16.3.11 使用线性回归的斜率 b 和截距 a 来计算线性回归方程（$Y=a+bx$），如果线性回归方程的 R^2 小于 0.99，则应重新校准。如果单个校准点与预期值相差超过 10%（通过线性回归估计），则应忽略该点。注入质控品和样品，并用适当的仪器软件测定峰面积。

 16.3.12 注入 1 μL 的质量控制样品和试料，用适当的软件确定峰面积。

 16.3.13 所有测试部分获得的信号（峰面积）必须在校准曲线的工作范

围内，否则应对解决方案进行调整。

17 数据分析和计算

17.1 在相同条件下注射两份样品萃取液试样。

17.2 对于每个试料，计算每种保润剂（甘油、丙二醇、三甘醇）峰面积的响应对内标物响应的比值（Y_t）。

17.3 使用线性回归系数计算每种保润剂（甘油、丙二醇、三甘醇）在每个试料中的浓度（mg/mL）：

$$m_t = \frac{Y_t - a}{b}$$

17.4 根据下式计算每种保润剂（甘油、丙二醇、三甘醇）在烟草样品中的含量 m_h：

$$m_h = \frac{m_t \times V_e}{m_o}$$

其中：m_t——试样溶液中各保润剂（甘油、丙二醇、三甘醇）的浓度（mg/mL）；
V_e——所用萃取液的体积（mL）；
m_o——试料的质量（g）。

18 特别注意事项

18.1 安装新色谱柱后，在指定的仪器条件下注入烟草样品萃取物进行调节。应重复进样，直至丙二醇、甘油、三甘醇和内标物的峰面积（或峰高）可重复。这需要大约4次进样。

18.2 建议在每组样本（序列）运行时，通过将色谱柱温度升高至220℃ 30 min，从气相色谱柱中清除高沸点组分。

18.3 当内标物的峰面积（或峰高）明显高于预期时，建议用不加内标物的萃取溶液萃取烟草样品。可以确定是否有其他成分与内标物共洗脱，这会造成保润剂数值偏低的误差。

19 数据报告

19.1 报告每个评估样品的单独测量值。
19.2 按 mg/g 烟草或按要求报告结果。

20 质量控制

20.1 控制参数。

如果控制测量值超出预期值的公差极限,则必须进行适当的调查并采取措施。如有必要,需要做额外的实验室质量保证规程,以符合各个实验室的做法。

20.2 实验室试剂空白样。

如 16.3.2 所述,要在样品制备和分析过程中检测潜在的污染,需包含实验室试剂空白样。空白样由分析测试样品中使用的所有试剂和材料组成,并像测试样品一样进行分析。应根据各个实验室的做法对空白样进行评估。

20.3 质量控制样品。

为了验证整个分析的一致性,根据各实验室的实践分析参比卷烟。

20.4 可根据各实验室政策的要求添加额外的质量控制样品。

21 方法性能

21.1 报告限。

  21.1.1 报告限设置为校准曲线的最低浓度,重新计算为 mg/g(甘油为 0.08 mg/g,丙二醇为 0.06 mg/mL,三甘醇为 0.03 mg/mL)。因此,甘油的报告限为 1.0 mg/g,丙二醇为 0.75 mg/g,三甘醇为 0.375 mg/g。对于丙二醇工作标准溶液(表2),报告限设置为校准曲线的最低浓度,重新计算为 0.063 mg(在 0.005 mg 时,报告限为 0.063 mg)。

21.2 实验室基质加标回收率。

加入到基质上的分析物的回收率用作准确度的替代量度。通过将已知含量的标准品加到装有烟丝的锥形瓶中并通过用与样品的相同方法萃取烟丝来确定回收率,未加标的烟丝也要进行分析。回收率通过下式计算,如表3所示。

表 3 实验室基质加标平均回收率

丙二醇			甘油			三甘醇		
加标量(mg/g)	平均值(mg/g)	回收率(%)	加标量(mg/g)	平均值(mg/g)	回收率(%)	加标量(mg/g)	平均值(mg/g)	回收率(%)
5.00	4.94	99	5.00	5.47	109	5.00	5.70	114
10.00	9.87	99	10.01	10.04	100	10.00	11.05	110
20.00	19.64	98	20.02	19.74	99	20.01	20.76	104

回收率（%）= 100×（分析结果 – 未加标结果）/加标量

21.3 分析特异性。

相关分析物的保留时间用于验证分析特异性。质量控制卷烟烟丝的成分响应与内标成分响应的限定比范围用于验证未知样品的结果的特异性。

21.4 线性。

保润剂的校准曲线在标准浓度范围内呈线性：甘油为 0.08~3.2 mg/mL，丙二醇为 0.005~0.30 mg/mL，三甘醇为 0.03~1.2 mg/mL。

21.5 潜在干扰。

没有已知的组分具有与丙二醇、三甘醇、甘油和内标物（1,3-丁二醇）类似的保留时间。

22 重复性和再现性

2012~2014 年，13 个实验室根据 WHO TobLabNet SOP 02[2.4]进行了一项国际合作研究，对 7 个样品（5 种参比卷烟和 2 种商品卷烟）使用火焰离子化检测器方法[2.3]进行测试。

常规正确操作该方法时，同一操作者使用同一设备在最短的操作时间内两个单一结果之间的差异超过重复性 r 的情况，平均 20 个样品不超过 1 次。

常规正确操作该方法时，两个实验室报告的匹配卷烟样品的单一结果超过再现性 R 的情况，平均 20 次不超过 1 次。

2012~2014 年，7 个实验室对 7 个样本（5 种参比卷烟和 2 种商品卷烟）进行了一项使用质谱检测方法的共同研究，得出了等效值（附录 2）。

根据 ISO 5725-1[2.5]和 ISO 5725-2[2.6]对试验结果进行统计分析，以确定表4至表 6 中所示的气相色谱-火焰离子化检测器分析的重复性和再现性限值。平均标准偏差分析中气相色谱法的重复性和再现性限值见附录 2 的附表 2 至附表 4。

表 4　气相色谱-火焰离子化检测法测定参比卷烟和商品卷烟中
烟草中丙二醇的重复性和再现性限值

参比卷烟	n*	$\hat{m}$(mg/g)	重复性限（r）mg/g	再现性限（R）（mg/g）
1R5F	12	0.231	0.001	0.005
3R4F	12	0.180	0.000	0.001
CM6	12	0.140	0.000	0.004
商品卷烟 1	12	5.496	0.244	1.972
商品卷烟 2	12	0.078	0.000	0.001

*一个数据集的结果因异常而被删除。

表 5 气相色谱-火焰离子化检测法测定参比卷烟和商品卷烟中
烟草中甘油的重复性和再现性限值

参比卷烟	n^*	$\hat{m}$(mg/g)	重复性限（r）（mg/g）	再现性限（R）（mg/g）
1R5F	11	21.603	1.490	6.082
3R4F	11	21.203	1.299	7.638
CM6	11	1.160	0.006	0.063
商品卷烟 1	11	14.376	0.648	3.621
商品卷烟 2	11	1.222	0.015	0.049

*一个数据集的结果因异常而被删除。

表 6 气相色谱-火焰离子化检测法测定参比卷烟烟草中三甘醇的重复性和再现性限值

参比卷烟	n^*	$\hat{m}$(mg/g)	重复性限（r）（mg/g）	再现性限（R）（mg/g）
TG 样品 1	12	0.891	0.207	0.409
TG 样品 2	11	8.078	1.565	2.621

*TG 样品 1 的一个数据集因异常而被删除，TG 样品 2 的两个数据集因异常而被删除。

23 检测报告

检测报告应包含以下内容：
（a）相关方法即 WHO TobLabNet SOP 06；
（b）收到样品的时间；
（c）结果及其单位。

附录1 气相色谱-火焰离子化检测器方法示例

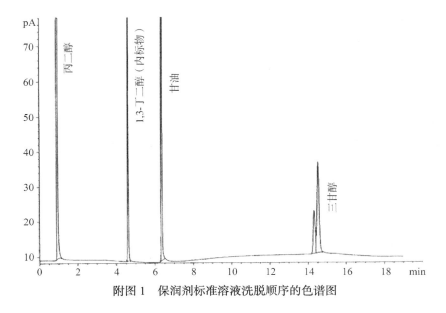

附图1 保润剂标准溶液洗脱顺序的色谱图

附录 2 使用质谱检测器代替火焰离子化检测器时应考虑的要点

A2.1 方法概要

该方法为气相色谱法,配置硅胶色谱柱和质谱检测器。称取 4 g 研磨的烟草烟丝加入 50 mL 甲醇萃取溶剂,在机械振动器上萃取 60 min。用移液管将 2 mL 样品萃取液转移到含有 150 mg 无水 $MgSO_4$ 和 25 mg N-丙基乙二胺的分散固相萃取小柱上。以 2000 r/min 的转速在涡旋振荡器上振荡萃取柱 2 min,必要时以 10000 r/min 的转速离心 10 min。将上清液转移到自动进样瓶中,并在气相色谱仪上进行分析。

A2.2 仪器和设备

- **A2.2.1** 分析天平,至少精确到四位小数。
- **A2.2.2** 125 mL 具塞锥形瓶。
- **A2.2.3** 10 mL、100 mL 和 2000 mL 容量瓶。
- **A2.2.4** 20 mL 吸管,用于制备萃取溶剂。
- **A2.2.5** 标准溶液制备用各种尺寸移液管。
- **A2.2.6** Brinkmann 分液器一次性枪头,10~50 mL。
- **A2.2.7** 2 mL 自动进样瓶和瓶盖,带聚四氟乙烯隔垫。
- **A2.2.8** 0.22 μm 过滤膜。
- **A2.2.9** 涡流振荡器。
- **A2.2.10** 离心机。
- **A2.2.11** 高速粉碎机。
- **A2.2.12** 0.45 mm 滤膜。
- **A2.2.13** 配备质谱检测器的气相色谱仪。
- **A2.2.14** 毛细管气相色谱柱能够明显分离溶剂、内标物、丙二醇、甘油、三甘醇和其他烟草成分的色谱峰(例如,DB-Wax 熔融石英色谱柱 30 m×0.32 mm×1 μm 或等效柱)。

A2.3 试剂和用品

如有必要,添加以下内容:

A2.3.1 无水硫酸镁[7487-88-9];

A2.3.2 正丙基乙二胺[111-39-7]。

注：**A2.3.1** 和 **A2.3.2** 可由含有 150 mg 无水硫酸镁[7487-88-9]和 25 mg 正丙基乙二胺[111-39-7]的商用分散固相萃取小柱代替。

A2.4 标准的制备

见气相色谱-火焰离子化检测方法[**10.1~10.5**]。

A2.5 样品制备

A2.5.1 如有必要，用移液管将 1~2 mL 样品萃取液转移到含有 150 mg 无水 $MgSO_4$ 和 25 mg 正丙基乙二胺的分散固相萃取小柱上。

A2.5.2 在 2000 r/min 的涡旋振荡器上振荡滤芯 2 min，并在 10000 r/min 的转速下离心 10 min。通过 0.22 μm 的注射器滤膜过滤上清液，直至其澄清。

A2.5.3 将上清液转移到自动进样瓶中，并在气相色谱仪上进行分析。

A2.6 样品分析

A2.6.1 气相色谱仪操作条件。

进样模式：分流比 100∶1；

色谱柱：DB-WAX，30 m×0.25 mm×1.0 μm；

检测器：质谱；

载气：氦气，流速为 1.0 mL/min；

升温程序：进样口温度 250℃，初始温度 90℃，以 15℃/min 升温至 180℃，保持 8 min，50℃/min 至 230℃，在 230℃下保持 5 min，在 250℃下运行 10 min；

总运行时间：17 min；

自动进样器条件：进样体积 1.0 μL。

A2.6.2 质谱工作条件。

传输线温度：280℃；

电离模式：电喷雾电离；

电离电压：70 eV；

离子源温度：250℃；

四极杆温度：150℃；

溶剂延迟时间:3 min。

注:应调整这些气相色谱-质谱操作条件,以获得甘油、三甘醇和丙二醇峰的正确分辨率。典型色谱图(总离子流图)如附录3所示。

附表 1　保润剂和内标物的定性和定量离子

保润剂	定量离子	第二定量离子	定性离子(丰度比)
甘油	61	43	61:43(100:79)
三甘醇	45	89	45:89(100:19)
丙二醇	45	43	45:43(100:20)
1,3-丁二醇	72	43	43:72(100:28)

附表 2　气相色谱-质谱法测定参比卷烟烟草中丙二醇的重复性和再现性限值

参比卷烟	n^*	$\hat{m}$(mg/g)	重复性限(r)(mg/g)	再现性限(R)(mg/g)
1R5F	6	0.215	0.001	0.003
3R4F	6	0.179	0.000	0.001
CM6	6	0.113	0.000	0.002
商品卷烟 1	6	6.310	0.128	0.668
商品卷烟 2	6	0.037	0.000	0.001

*一个数据集的结果因异常而被删除。

附表 3　气相色谱-质谱法测定参比卷烟和商品卷烟中烟草中甘油的重复性和再现性限值

参比卷烟	n^*	$\hat{m}$(mg/g)	重复性限(r)(mg/g)	再现性限(R)(mg/g)
1R5F	5	21.147	0.659	1.941
3R4F	5	21.510	0.715	1.331
CM6	5	1.466	0.006	0.106
商品卷烟 1	5	14.762	0.512	1.164
商品卷烟 2	5	1.457	0.005	0.110

*两个数据集的结果因异常而被删除。

附表 4　气相色谱-质谱法测定参比卷烟烟草中三甘醇的重复性和再现性限值

参比卷烟	n^*	$\hat{m}$(mg/g)	重复性限(r)(mg/g)	再现性限(R)(mg/g)
TG 样品 1	6	0.971	0.006	0.010
TG 样品 2	6	8.450	0.189	0.617

*一个数据集结果因异常而被删除。

附录 3 使用质谱法获得的色谱图示例

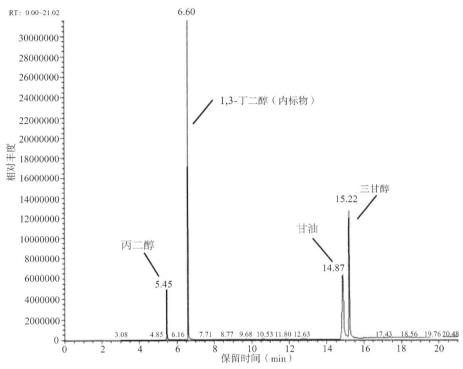

附图 1 标准溶液中保润剂的色谱图

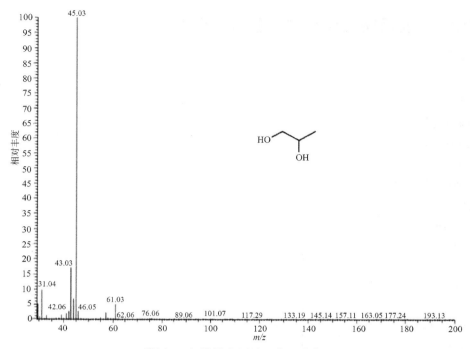

附图 2 扫描模式下丙二醇的质谱图

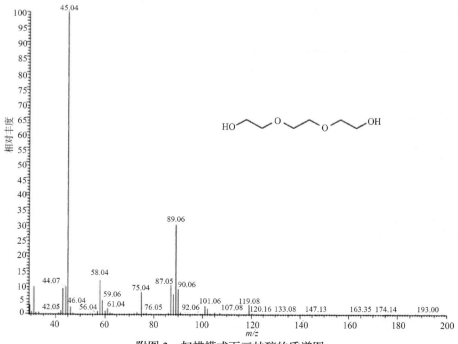

附图 3 扫描模式下三甘醇的质谱图

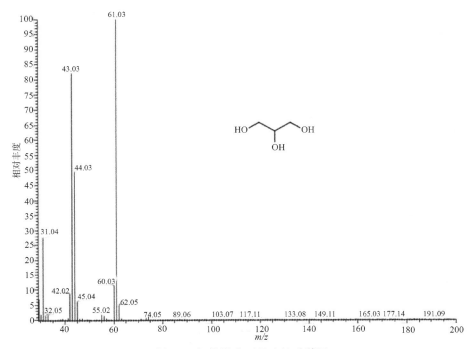

附图 4　扫描模式下甘油的质谱图

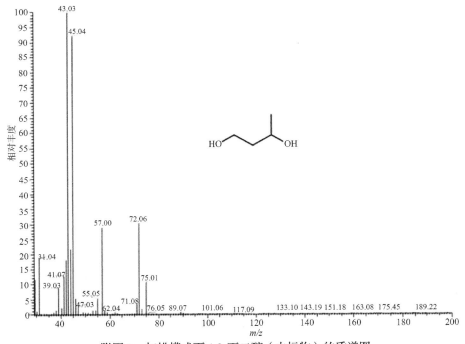

附图 5　扫描模式下 1,3-丁二醇（内标物）的质谱图

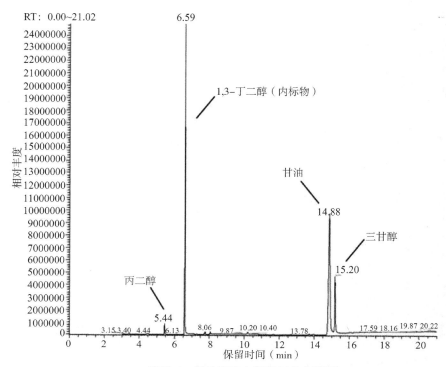

附图 6 样品溶液中保润剂的色谱图

WHO TobLabNet SOP 07

卷烟烟丝中氨的测定标准操作规程

方　　法：卷烟烟丝中氨的测定
分 析 物：氨（CAS 号：7664-41-7）
基　　质：卷烟烟丝
更新时间：2016 年 7 月

> 吸烟机的抽吸方案不能代表所有人的吸烟行为：吸烟机测试对用于卷烟设计及监管目的的卷烟释放物表征非常有用，但将吸烟机的测量结果披露给吸烟者则会导致对不同品牌的暴露和风险差异的误解。吸烟机测量的烟气释放物数据可以用于产品危害评估，但并不意味着这些数据等效于人体暴露或风险的测量值。将吸烟机测量结果的差异表示为暴露或风险的差异是对世界卫生组织烟草实验室网络标准的滥用。

提到的公司和制造商的产品，并不意味着 WHO 优先推荐这些产品。除错误和疏漏外，专有产品的名称以首字母大写表示。

本方法由世界卫生组织（WHO）烟草实验室网络（TobLabNet）制订，作为卷烟烟丝中氨的测定标准操作规程（SOP）。

引言

为了在全球范围内建立具有可比性的烟草制品检测方法，需要卷烟特定成分和释放物的一致性检测方法。2008年11月在南非德班举行的世界卫生组织《烟草控制框架公约》（WHO FCTC）第三次缔约方会议回顾了FCTC/COP1(15)和FCTC/COP2(14)号决议关于制订WHO FCTC第9条（烟草制品成分管制）和第10条（烟草制品披露管制）实施指南的要求，并提到第三次缔约方会议工作报告中有关其工作进展的信息，提请公约秘书处授权WHO无烟草行动组在5年内对检测卷烟成分和释放物的分析化学方法进行验证（FCTC/COP/3/REC/1）。

根据2006年10月在加拿大渥太华举行的第三次会议上确定的优先次序标准，WHO FCTC第9条和第10条工作组确定优先验证以下成分的分析化学检测方法：
- 烟碱；
- 氨；
- 保润剂[1,2-丙二醇、甘油(1,2,3-丙三醇)和三甘醇(2,2′-二乙醚乙二醇)]。

测量这些成分需要验证三个方法：一个适用于烟碱，一个适用于氨，一个适用于保润剂。

根据上述渥太华会议确定的优先次序标准，工作组确定了在卷烟主流烟气释放中应该验证测试和测量方法（分析化学）的下列优先级物质清单：
- 4-(*N*-甲基亚硝胺基)-1,3-吡啶基-1-丁酮（NNK）；
- *N*-亚硝基去甲烟碱（NNN）；
- 乙醛；
- 丙烯醛；
- 苯；
- 苯并[*a*]芘；
- 1,3-丁二烯；
- 一氧化碳；
- 甲醛。

这些释放物使用下述两种抽吸方案测量，按五个方法进行验证：一个适用于烟草特有的亚硝胺（NNK和NNN），一个适用于苯并[*a*]芘，一个适用于醛类（乙醛、丙烯醛和甲醛），一个适用于挥发性有机物（苯和1,3-丁二烯），一个适用于一氧化碳。

下表列出了两种用于验证上述测试方法的吸烟方案。

抽吸方案	抽吸容量（mL）	抽吸间隔（s）	滤嘴通风孔
ISO 方案：ISO 3308 常规分析用吸烟机 定义和标准条件	35	60	不封闭
深度抽吸方案：在 ISO 3308 基础上有所调整	55	30	所有通风孔必须如 SOP 01 **12.2** 所述 100%封闭

SOP 07 描述卷烟烟丝中氨的测定标准操作规程。

1 适用范围

本方法适用于离子色谱法测定卷烟烟丝中的氨。

2 参考标准

2.1 ISO 8243：卷烟 抽样。

2.2 联合国毒品和犯罪问题办公室（UNODC）《代表性药物取样指南》（http://www.unodc.org/documents/scientific/Drug_Sampling.pdf）。

2.3 ISO 5725-1：检测方法和结果的准确性（正确度和精密性） 第 1 部分 一般原则和定义。

2.4 ISO 5725-2：检测方法和结果的准确性（正确度和精密性） 第 2 部分 确定标准检测方法重复性和再现性的基本方法。

2.5 ISO 指南 34：标准物质生产商能力的一般要求。

2.6 AOAC 966.02 官方方法：烟草的干（湿）损失（2016 年第 20 版）。

3 术语和定义

3.1 氨含量：卷烟烟丝中的氨总量，以 mg/g 卷烟烟丝表示。

3.2 卷烟烟丝：卷烟中含烟草的部分，包括再造烟叶、烟梗、膨胀烟丝和添加剂。

3.3 烟草制品：完全或部分以烟叶为原料制造，用于抽吸、吸吮、咀嚼或吸入的产品（WHO FCTC 第 1(f)条）。

3.4 实验室样品：用于进行实验室检测的样品，由一次或在规定时间内送到实验室的单一类型产品组成。

3.5 试样：从实验室样品中随机抽取的待测样品。所取样品数量应能代表实验室样品。

3.6 试料：从试样中随机抽取的用于单次测试的样品。所取样品数量应能代表试样。

4 方法概述

4.1 使用稀硫酸萃取液从卷烟烟丝中萃取氨,然后离心(如果需要,则过滤)。
4.2 用离子色谱法结合电导检测分析萃取物。
4.3 采用外标法进行定量,通过比较样品中分析物与标准物电导率的响应来实现定量。

5 安全和环境预防措施

5.1 遵守所有化学实验活动的常规安全与环境预防措施。
5.2 使用本检测方法对特定产品进行检测和评估时所用材料或设备可能对环境有害。本检测方法无力解决所有与使用有关的安全问题。所有使用此方法的人有责任咨询相关机构并遵循现有监管要求,制定健康与安全规程以及环境预防措施。
5.3 应特别注意避免吸入或皮肤接触危险化学品。在制备或处理未稀释材料、标准溶液、萃取液和收集样品时,应在通风橱中进行,并穿戴合适的实验服、手套和护目镜。

6 仪器和设备

常规实验仪器,特别是:
6.1 测量精度为 0.0001 g 的分析天平。
6.2 250 mL 具塞锥形瓶。
6.3 各种尺寸的烧杯。
6.4 各种尺寸的容量瓶。
6.5 各种尺寸用于制备标准溶液的移液管。
6.6 自动进样器。
6.7 2 mL 的自动进样瓶。
6.8 机械腕式振荡器。
6.9 配有电导检测器的离子色谱仪。
6.10 阳离子交换柱(250×4 mm)。
6.11 阳离子交换保护柱(50×4 mm)。
6.12 可再生阳离子抑制剂(可选)
注 1:以上是此分析的典型仪器配置,也可使用能产生类似效果的替代配置

（如没有抑制器）。

注 2：样品应在分析前过滤或离心。

6.13 过滤设备。

  6.13.1 10 mL 注射器。

  6.13.2 0.45 mm 滤膜。

  6.13.3 15 mm×0.45 μm 水相过滤膜。

  6.13.4 无尘定量滤纸，如 Whatman 40 号、8 μm。

6.14 离心设备。

  6.14.1 Micro-17 离心机。

  6.14.2 微量离心瓶。

7 试剂

除非另有说明，所有试剂应至少为分析纯。试剂尽可能以 CAS 号来识别。

7.1 甲基磺酸[75-75-2]（色谱纯）。

7.2 质量分数 95%~98%的浓硫酸[7664-93-9]。

7.3 硫酸铵[7783-20-2]或由 ISO 指南 34 认可的 1000 mg/L 氨标准溶液。

7.4 一级水[7732-18-5]。

8 玻璃器皿准备

8.1 清洁、干燥的玻璃器皿，避免残留物污染。

注意：建议不要使用清洁剂进行清洁，以尽量减少干扰。

9 溶液配制

下述的溶液配制方法仅供参考，如有需要可进行调整。

9.1 萃取液（约 0.0125 mol/L 的稀硫酸）：

  9.1.1 为获得 0.0125 mol/L 的硫酸溶液，使用公式 $\rho = m/V$，其中 ρ 为硫酸的密度（1.84 g/mL），m 为硫酸的质量。

  注意：将 1.29 g 硫酸加入 1 L 的一级水[7.4]中可以得到 0.0125 mol/L 硫酸溶液。

  9.1.2 标记为"稀硫酸萃取液"，并存放在冰箱中。

9.2 流动相（0.02 mol/L 的甲基磺酸水溶液）：

准确地将 1.32 mL 甲基磺酸[7.1]转移到 1 L 容量瓶[6.4]中，并用一级水[7.4]定容至刻度。

10 标准溶液配制

下述标准溶液配制方法仅供参考，如有需要可进行调整。

10.1 约 100 μg/mL 的氨标准溶液：

    10.1.1 准确称取 0.092 g±0.002 g 硫酸铵[**7.3**]至 100 mL 烧杯中，并将其溶解在萃取溶液[**9.1**]中。所有溶剂和溶液必须冷却至室温。

    10.1.2 将硫酸铵溶液[**10.1.1**]转移到 250 mL 容量瓶中，用萃取液[**9.1**]定容至刻度。

10.2 工作溶液（表 1）。

表 1 氨工作溶液

标准	标准溶液[**10.1**]体积（mL）	最终体积（mL）	标准工作溶液中氨的近似浓度（μg/mL）
1	0.01	10	0.1
2	0.05	10	0.5
3	0.2	10	2
4	0.5	10	5
5	1	10	10

注 1：所有标准溶液均在容量瓶中配制，容量瓶用萃取液[**9.1**]定容至刻度。
注 2：标准溶液范围可根据使用的设备（带或不带抑制设备）和待测样品进行调整，注意方法灵敏度可能对结果产生的影响。
注 3：标准溶液的浓度在储存期间会发生显著变化，这将改变标准曲线的截距。因此其存放时间不应超过 2 周。
注 4：在计算浓度时必须考虑所用标准品中氨的纯度。

11 抽样

11.1 根据 ISO 8243 [**2.1**]或替代方法，根据各实验室惯例或者要求的特殊规定或样品可用性，选取具有代表性的实验室卷烟样品。

11.2 试样的组成。

    11.2.1 若可能，将实验室样品分成独立的单元（如包、条）。

    11.2.2 从至少 $\sqrt{n}$ 个[**2.2**]单元中取出每种试样的等量产品。

12 卷烟准备

12.1 从一包（例如含 20 支）卷烟和质量控制样品（若适用）中取出烟丝，或使用至少 7 g 加工过的卷烟烟丝。

12.2 将足够的卷烟烟丝混合至每个试料至少含 7 g 烟丝，制备至少三个重复的试料。

注意：如有必要，可以调整样品量。

13 吸烟机设置

不适用。

14 样品生成

不适用。

15 样品制备

15.1 对于每个试料，称取 0.7 g 充分混合的试料，精确至 0.0001 g，放入 250 mL 锥形瓶或其他合适的烧瓶中。

15.2 向样品中加入 50 mL 萃取液[9.1]。

注意：样品质量和稀释因子可根据氨的浓度或样品的可用性进行相应调整。

15.3 盖上瓶盖，并且将它们放置在振荡器上以至少 160 r/min 的转速振荡 30 min 或使用机械振动器以合适的速率振动。

15.4 从振荡器中取出样品，旋转烧瓶使所有烟丝浸没在溶剂中。

15.5 静置 30 min 直至上清液澄清。

15.6 采用 8 μm 无尘定量滤纸过滤萃取物。

15.7 将萃取物[15.6]过 0.45 μm 滤膜。

15.8 将滤液转移至自动进样瓶，并用离子色谱仪进行分析。

注意：如果样品信号（峰面积）不在标准曲线的工作范围内，则应相应调整溶液，即进一步稀释。所有最终稀释液应记录在样品报告中。

或者，可以遵循以下步骤：

15.5 取 1.5 mL 上清液在 13000 r/min 速率下离心 5 min。

15.6 将离心后的萃取物转移到自动进样瓶中，并用离子色谱仪进行分析。

16 样品分析

使用电导检测器、离子色谱法测定卷烟烟丝中的氨。分析物可以从色谱柱上的其他潜在干扰物质中分离出来，比较未知物的峰面积比与已知浓度标准液的峰面积比，得到单个分析物的浓度。

16.1 离子色谱仪操作条件。

流动相：0.02 mol/L 的甲基磺酸水溶液[9.2]；

流速：1.0 mL/min；

进样量：25 µL；

色谱柱温度：30℃；

这些仪器设置和其他规格（例如阳离子抑制器的电流设置）应根据每个设备和色谱柱配置进行调整，以获得良好的分离度和色谱性能。

注1：在没有抑制的情况下，建议在流动相中使用稀释度更高的甲基磺酸水溶液（稀释度高达10倍）。

注2：根据仪器和色谱柱条件以及色谱峰的分离度调整操作参数。

16.2 预计保留时间。

注意：柱的类型和年限差异将改变保留时间。

在上述条件下，预期的总分析时间约为15 min。

16.3 氨的测定。

  16.3.1 使用前调节系统，例如注入两份 25 µL 样品溶液作为预试样，也可以使用制造商建议的其他平衡规程。

  注意：根据固定相的不同，可能需要在校准标准后注入空白溶液，还应在序列结束时注入空白溶液，以减少交叉污染。

  16.3.2 调节后,注入 25 µL 空白溶液,检查系统或试剂中是否有污染。

  16.3.3 将每种氨标准溶液[10.2]的试样注入离子色谱仪。

  16.3.4 评估标准的保留时间和响应（面积计数）。如果保留时间与先前的进样保留时间相似（±0.2 min），计算的分辨率≥1.5 并且响应与先前进校中的典型响应差异在±20%范围内，则说明系统已准备好进行分析。

  16.3.5 记录氨的峰面积。

  16.3.6 根据氨浓度和对应的峰面积绘制标准曲线。

  16.3.7 应使用以下替代方案来计算氨浓度。

对于有抑制器的离子色谱：

从数据计算二次回归方程（$Y=ax^2+bx+c$）。

如果回归方程的 R^2 小于 0.99，则应重新校准。如果单个校准点与预期值（通过线性回归估计）相差超过 10%，则应该舍弃该点。

对于没有抑制器的离子色谱：

使用线性回归的斜率 b 和截距 a 来计算线性回归方程（$Y=a+bx$），如果线性回归方程的 R^2 小于 0.99，则应重新校准。如果单个校准点与预期值（通过线性回归估计）相差超过 10%，则应该舍弃该点。

  16.3.8 在这两种情况下，截距在统计上与零不应该有显著差异。

  16.3.9 注入质量控制样品和试料，用适当的软件确定峰面积。

16.3.10 所有试料的信号（峰面积）必须落在标准曲线的工作范围内。代表性标准曲线和色谱图见附录1。

17 数据分析和计算

17.1 对于每个试料，计算氨的峰面积。

17.2 根据二次或线性回归方程的系数计算每个试料中的铵离子（NH_4^+）的浓度（mg/mL）。

17.3 氨的量（以 mg/g）由下式确定：

$$M = \frac{C \times V \times 17.03}{m \times 18.04}$$

式中：M——卷烟烟丝中氨（NH_3）的浓度（mg/g）；

C——从标准曲线获得的样品溶液中 NH_4^+ 的浓度（μg/mL）；

V——样品溶液的体积（mL）；

m——烟丝的质量（mg）。

注1：NH_3 的相对分子质量为 17.03。

注2：NH_4^+ 的相对分子质量为 18.04。

18 特别注意事项

安装新色谱柱后，在指定的仪器条件下注入烟草样品萃取物进行调节。应重复进样，直至氨的峰面积（或峰高）可根据各个实验室的验收标准实现重复。

19 数据报告

19.1 报告每个评估样品的单独测量值。

19.2 以 mg/g 烟丝或根据需要报告结果。

19.3 结果可按原样或以干重报告。

注意：水分可用 AOAC 966.02 [2.6]或等效标准方法测量。

20 质量控制

20.1 控制参数。

如果控制测量值超出预期值的公差极限，则必须进行适当的调查和采取措施。如有必要，需要做额外的实验室质量保证规程，以符合各个实验室的做法。

20.2 实验室试剂空白样。

要在样品制备和分析过程中检测潜在的污染，需包含实验室试剂空白样。空白样由分析测试样品中使用的所有试剂和材料组成，并像测试样品一样进行分析。应根据各个实验室的做法对空白样进行评估。

20.3 质量控制样品。

为了验证整个分析过程的一致性，在每次分析运行时分析参比卷烟，例如肯塔基大学（美国肯塔基州列克星敦）研究卷烟或国际烟草科学研究合作中心（法国巴黎）质控样。

21 方法性能

21.1 报告限。

报告限设置为所用标样曲线的最低浓度，重新计算约为 0.1 μg/mL。

21.2 实验室基质加标回收率。

加入到基质上的分析物的回收率用作准确度的替代量度。通过将已知含量的标准品（萃取前）加到带有烟丝的锥形瓶中并通过用与样品相同的方法萃取烟丝来确定回收率，未加标的烟丝也要进行分析。回收率通过下式计算，如表 2 所示。

$$回收率（\%）=100×（分析结果-未加标结果）/加标量$$

表 2 烟丝中氨的样品回收率

加标量（mg）[a]	分析结果（mg）	未加标结果（mg）	回收率（%）
0.0799（低）	0.1751	0.0960	98.90
0.0999（中）	0.1944	0.0956	98.88
0.1199（高）	0.2141	0.0953	99.10
			平均：98.96

[a] "低、中、高"的加标量约为未加标量的 80%，100%和 120%。

21.3 检测限（LOD）和定量限（LOQ）。

LOD 可以通过在几天内分析最低标准至少 10 次获得结果的标准偏差的 3 倍来确定。LOQ 可以通过在几天内分析最低标准至少 10 次所获得结果的标准偏差的 10 倍来进行确定。

烟丝中氨的 LOD 为 0.0033 mg/g，LOQ 为 0.011 mg/g。

或者，使用的最低标准浓度可以作为 LOQ。

22 重复性和再现性

2014~2015 年进行的一项国际合作研究涉及 9 个实验室对 3 种参比卷烟和 2 种商品卷烟的测试,对该方法给出了以下精密度限制。

常规正确操作该方法时,同一操作者使用同一设备在最短的操作时间内两个单一结果之间的差异超过重复性 r 的情况,平均 20 个样品不超过 1 次。

常规正确操作该方法时,两个实验室报告的匹配卷烟样品的单一结果超过再现性 R 的情况,平均 20 次不超过 1 次。

根据 ISO 5725-1 [2.3]和 ISO 5725-2 [2.4]对测试结果进行统计分析,得到表 3 所示的准确数据。商品卷烟的氨浓度在参比卷烟的氨浓度范围内。

为了计算 r 和 R,将一个测试结果定义为 7 次重复的平均值。

表 3 参比试样中氨的精密度限值

参比卷烟	n	m	重复性限值 r(mg/g)	再现性限值 R(mg/g)
1R5F	9	1.214	0.002	0.153
3R4F	9	0.857	0.002	0.075
CM6	9	0.172	0.000	0.004
商品卷烟 1	9	0.383	0.001	0.010
商品卷烟 2	9	1.746	0.005	0.347

注:n 为参与实验室数量;
m 为每支卷烟的氨平均值。

23 检测报告

检测报告应包含以下内容:
(a)相关方法即 WHO TobLabNet SOP 07;
(b)收到样品的时间;
(c)结果及其单位。

附录1 卷烟烟丝中氨的标准曲线和色谱图

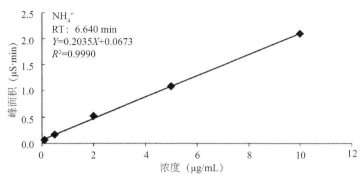

附图1a 氨的标准曲线（没有抑制器）

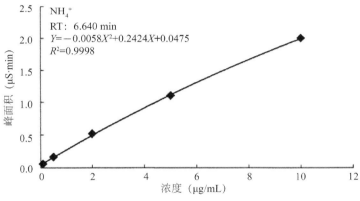

附图1b 氨的标准曲线（有抑制器）

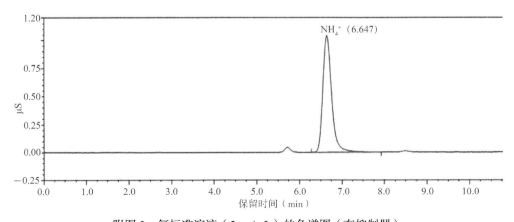

附图2 氨标准溶液（5 μg/mL）的色谱图（有抑制器）

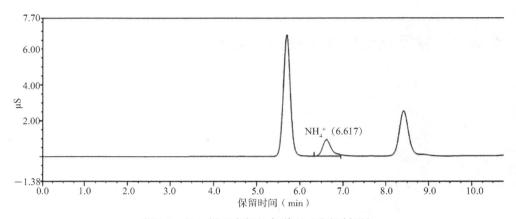

附图3　烟丝样品中氨的色谱图（有抑制器）

WHO TobLabNet SOP 08

ISO 和深度抽吸方案下卷烟主流烟气中醛类化合物的测定标准操作规程

方　　法：ISO 和深度抽吸方案下卷烟主流烟气中醛类化合物的测定
分 析 物：甲醛（CAS 号：50-00-0）
乙醛（CAS 号：75-07-0）
丙烯醛（CAS 号：107-02-8）
基　　质：卷烟主流烟气
更新时间：2018 年 8 月 31 日

吸烟机的抽吸方案不能代表所有人的吸烟行为：吸烟机测试对用于卷烟设计及监管目的的卷烟释放物表征非常有用，但将吸烟机的测量结果披露给吸烟者则会导致对不同品牌的暴露和风险差异的误解。吸烟机测量的烟气释放物数据可以用于产品危害评估，但并不意味着这些数据等效于人体暴露或风险的测量值。将吸烟机测量结果的差异表示为暴露或风险的差异是对世界卫生组织烟草实验室网络标准的滥用。

本方法由世界卫生组织（WHO）烟草实验室网络（TobLabNet）制订，作为国际标准化组织（ISO）和深度抽吸方案下卷烟主流烟气中醛类化合物的测定标准操作规程（SOP）。

引言

为了在全球范围内建立具有可比性的烟草制品检测方法，需要卷烟特定成分和释放物的一致性检测方法。2008年11月在南非德班举行的世界卫生组织《烟草控制框架公约》（WHO FCTC）第三次缔约方会议回顾了 FCTC/COP1(15)和FCTC/COP2(14)号决议关于制订 WHO FCTC 第9条（烟草制品成分管制）和第10条（烟草制品披露管制）实施指南的要求，并提到第三次缔约方会议工作报告中有关其工作进展的信息，提请公约秘书处授权 WHO 无烟草行动组在5年内对检测卷烟成分和释放物的分析化学方法进行验证（FCTC/COP/3/REC/1）。

根据2006年10月在加拿大渥太华举行的第三次会议上确定的优先次序标准，WHO FCTC 第9条和第10条工作组确定优先验证以下成分的分析化学检测方法：
- 烟碱；
- 氨；
- 保润剂[1,2-丙二醇、甘油（1,2,3-丙三醇）和三甘醇（2,2'-二乙醚乙二醇）]。

测量这些成分需要验证三个方法：一个适用于烟碱，一个适用于氨，一个适用于保润剂。

根据上述渥太华会议确定的优先次序标准，工作组确定了在卷烟主流烟气释放中应该验证测试和测量方法（分析化学）的下列优先级物质清单：
- 4-(N-甲基亚硝胺基)-1,3-吡啶基-1-丁酮（NNK）；
- N-亚硝基去甲烟碱（NNN）；
- 乙醛；
- 丙烯醛；
- 苯；
- 苯并[a]芘；
- 1,3-丁二烯；
- 一氧化碳；
- 甲醛。

这些释放物使用下述两种抽吸方案测量，按五个方法进行验证：一个适用于烟草特有的亚硝胺（NNK 和 NNN），一个适用于苯并[a]芘，一个适用于醛类（乙醛、丙烯醛和甲醛），一个适用于挥发性有机物（苯和1,3-丁二烯），一个适用于一氧化碳。

下表列出了用于验证上述测试方法的两种抽吸方案。

抽吸方案	抽吸容量（mL）	抽吸间隔（s）	滤嘴通风孔
ISO 方案：ISO 3308 常规分析用吸烟机 定义和标准条件	35	60	不封闭
深度抽吸方案：在 ISO 3308 基础上有所调整	55	30	所有通风孔必须如 SOP 01 **12.2** 所述 100%封闭

SOP 08 描述 ISO 和深度抽吸方案下卷烟主流烟气中醛类化合物的测定标准操作规程。对于卷烟以外的烟草制品，可根据方法给出的具体参数进行调整。

1 适用范围

本方法适用于液相色谱法定量测定卷烟主流烟气中的以下三种醛类化合物：甲醛、乙醛和丙烯醛。

请注意，要成功操作吸烟机或其他分析设备，须对操作人员进行培训。对操作吸烟机或其他分析设备没有经验的人，在检测烟草制品成分和释放物前，需要接受专门的培训。

2 参考标准

2.1 ISO 3308：常规分析用吸烟机 定义和标准条件。

2.2 ISO 4387：卷烟 常规分析用吸烟机测定总粒相物和去烟碱干粒相物。

2.3 ISO 3402：烟草和烟草制品 用于调节和测试的大气环境。

2.4 ISO 8243：卷烟 抽样。

2.5 ISO 5725-2：检测方法和结果的准确度（正确度和精密度） 第 2 部分 确定标准检测方法重复性和再现性的基本方法。

2.6 WHO TobLabNet SOP 01：卷烟深度抽吸标准操作规程。

2.7 WHO TobLabNet SOP 02：烟草制品成分和释放物分析方法验证标准操作规程。

2.8 WHO TobLabNet SOP 09： ISO 和深度抽吸方案下卷烟主流烟气中挥发性有机物的测定标准操作规程。

2.9 联合国毒品和犯罪问题办公室（UNODC）《代表性药物取样指南》（http://www.unodc.org/documents/scientific/Drug_Sampling.pdf）。

2.10 ISO 5725-1：检测方法和结果的准确度（正确度和精密度） 第 1 部分 一般原则和定义。

2.11 Uchiyama S, Tomizawa T, Inaba, Y, Kunugita, N. Simultaneous determination of volatile organic compounds and carbonyls in mainstream cigarette smoke using a sorbent cartridge followed by two-step elution. J Chromatogr A. 2013; 1314:31-7. doi: 10.1016/j.chroma.2013.09.019。

2.12 Uchiyama, S, Hayashida, H, Izu, R, Inaba, Y, Nakagome, H, Kunugita N. Determination of nicotine, tar, volatile organic compounds and carbonyls in mainstream cigarette smoke using a glass filter and a sorbent cartridge followed by the two-phase/one-pot elution method with carbon disulfide and methanol. J Chromatogr A. 2015; 1426:48-55. doi.org/10.1016/j. chroma.2015.11.058。

3 术语和定义

3.1 TPM：总粒相物。
3.2 烟气捕集器：为测定特定烟气成分从卷烟样品中收集部分烟气的装置。
3.3 醛类：甲醛、乙醛、丙烯醛。
3.4 CX-572：碳吸附剂 Carboxen 572。
3.5 DNPH-HCl：2,4-二硝基苯肼盐酸盐。
3.6 烟草制品：完全或部分以烟叶为原料制造，用于抽吸、吸吮、咀嚼或吸入的产品（WHO FCTC 第 1(f)条）。
3.7 深度抽吸方案：抽吸过程参数为抽吸容量 55 mL，抽吸间隔 30 s，每次抽吸持续时间 2 s，100%封闭滤嘴通风孔。
3.8 ISO 方案：抽吸过程参数为抽吸容量 35 mL，抽吸间隔 60 s，每次抽吸持续时间 2 s，不封闭滤嘴通风孔。
3.9 实验室样品：用于进行实验室检测的样品，由一次性或在规定时间内送到实验室的单一类型产品组成。
3.10 试样：从实验室样品中随机抽取的待测样品。所取样品数量应能代表实验室样品。
3.11 试料：从试样中随机抽取的用于单次测试的样品。所取样品数量应能代表试样。

4 方法概述

4.1 在深度抽吸方案或 ISO 方案下收集卷烟主流烟气。抽吸每支卷烟时每个烟气捕集器含有 300 mg CX-572 颗粒和一片玻璃纤维滤片。
4.2 卷烟试样的主流烟气总粒相物被捕集在含有 300 mg CX-572 颗粒和玻璃纤维滤片的烟气捕集器上。
4.3 CX-572 颗粒和玻璃纤维滤片上的醛类化合物用二硫化碳和甲醇的混合溶液来萃取。

4.4 用 2,4-二硝基苯肼（DNPH）进行醛的衍生化。

4.5 采用配备紫外/二极管阵列检测器（UV/DAD）的高效液相色谱（HPLC）测定醛类化合物。

4.6 本标准操作规程所述的方法可同时检测主流烟气中醛类化合物（SOP 08）和挥发性有机物（SOP 09）的含量。

4.7 将醛色谱峰面积与通过分析已知醛浓度的标准溶液建立的标准曲线进行比较，以确定试料的醛含量。

5 安全和环境预防措施

5.1 遵守所有化学实验活动的常规安全与环境预防措施。

5.2 使用本检测方法对特定产品进行检测和评估时所用材料或设备可能对环境有害。本检测方法无力解决所有与使用有关的安全问题。所有使用此方法的人有责任咨询相关机构并遵循现有监管要求，制定健康与安全规程以及环境预防措施。

5.3 应特别注意避免吸入或皮肤接触危险化学品。在制备或处理未稀释材料、标准溶液、萃取液和收集样品时，应在通风橱中进行，并穿戴合适的实验服、手套和护目镜。

6 仪器和设备

6.1 按照 ISO 3402 [**2.3**]规定调节卷烟所需设备。

6.2 按照 ISO 4387 [**2.2**]规定标记烟蒂长度所需设备。

6.3 按照 WHO TobLabNet SOP 01 [**2.6**]规定采用深度抽吸方案时封闭通风孔所需设备。

6.4 按照 ISO 3308 [**2.1**]规定抽吸烟草制品所需设备。

6.5 测量精度为 0.0001 g 的分析天平。

6.6 移液枪及能精确移取 40~5000 μL 体积的枪头。

6.7 5 mL 和 10 mL 容量瓶。

6.8 磁力搅拌装置。

6.9 能够转移 2~20 mL 有机溶剂的校准注射器或分液器。

6.10 容量至少为 20 mL 或 50 mL 的密封烧瓶，建议使用棕色瓶。

6.11 2 mL 的 LC 自动进样瓶。

6.12 配备紫外/二极管阵列检测器（UV/DAD）的 HPLC 系统，包含 360 nm 的检测波长。

6.13 能够明显分离烟气释放物中溶剂、DNPH 和醛类化合物峰的 HPLC 色谱柱，如 Ascentis® RP-Amide（150 mm×4.6 mm，3 μm）。

6.14 CX-572 碳吸附剂，20/45 目。

6.15 CX-572 滤芯：手动制备填充 300 mg CX-572 的 1 mL 固相萃取滤芯，或购买 SigmaAldrich®（No. 54294-U）预装的 CX-572 柱。

6.16 按照 ISO 4387 准备玻璃纤维滤片。

6.17 烟气捕集器由吸烟机支托器及 CX-572（带或不带滤芯）组成，支托器放置在玻璃纤维滤片前面，见图 1 至图 5。将 CX-572 柱朝着卷烟的开口侧放入支托架。当开口端朝相反方向放置时，柱中的成分可在抽吸时被吸烟机吸入。如有必要，可在柱中插入专用塞子，如图 6 所示。

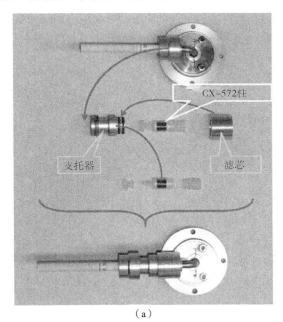

(a)

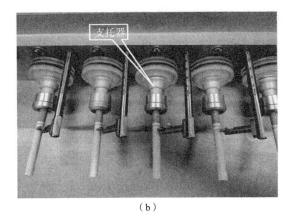

(b)

图 1 Borgwaldt 吸烟机

(a) 单通道；(b) 直线型

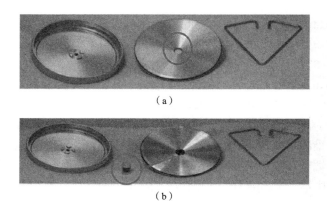

图2 适用于商品转盘型吸烟机（Cerulean 和/或 Borgwaldt 吸烟机）的支托器

（a）CX-572 颗粒直接压在支托器上；（b）将装有 CX-572 的滤芯插入支托器

卷烟　　CX-572柱子　剑桥滤片　吸烟机

图3 直线型商品吸烟机（Cerulean 或 Borgwaldt 吸烟机）的支托器

CX-572 可预先装到支托器中，也可使用商品 CX-572 柱

图4 直线型 Cerulean 吸烟机（也适用于直线型 Borgwaldt 或转盘型 Cerulean 吸烟机或者单通道吸烟机）的支托器

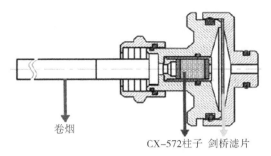

图 5 适用于直线型 Cerulean 吸烟机（也适用于直线型 Borgwaldt 吸烟机)的支托器

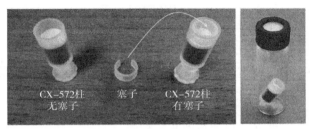

图 6 CX-572 柱中插入塞子，以防止内容物被吸入吸烟机

插入塞子后将 SPE 柱直接放入封闭容器（小瓶）中，以防污染 CX-572

7 试剂

除非另有说明，所有试剂应至少为分析纯。试剂尽可能以 CAS 号来识别。

7.1 2,4-二硝基苯肼盐酸盐（DNPH-HCl），>98%（东京化学工业株式会社，D0846）[55907-61-4]。

7.2 甲醛-DNPH 溶液，100 μg/mL（以醛计）（Sigma-Aldrich，CRM47177 Supelco）[1081-15-8]。

7.3 乙醛-DNPH 溶液，1000 μg/mL（以醛计）（Sigma-Aldrich，CRM4M7340 Supelco）[1019-57-4]。

7.4 丙烯醛-DNPH 溶液，1000 μg/mL（以醛计）（Sigma-Aldrich，CRM47342 Supelco）[888-54-0]。

7.5 乙腈[75-05-8]（色谱纯）。

7.6 85%的磷酸[7664-38-2]。

7.7 二硫化碳[75-15-0]。

7.8 无水甲醇[67-56-1]（分析纯）。

7.9 >98%的乙醇[64-17-5]。

7.10 碳吸附剂 CX-572 颗粒，20/45 目（Sigma-Aldrich，11072-U）。

8 玻璃器皿准备

清洁、干燥的玻璃器皿,避免残留物污染。

9 溶液配制

9.1 衍生化试剂(DNPH 试剂):
    9.1.1 称取约 1 g DNPH-HCL[**7.1**]置于 50 mL 容量瓶中。
    9.1.2 加入 10 mL 磷酸[**7.6**]。
    9.1.3 用乙腈[**7.5**]稀释至 50 mL。
    9.1.4 通过磁力搅拌装置连续搅拌溶液,直至获得澄清溶液。将溶液储存在 4~10℃,此溶液可稳定 1 个月。

9.2 萃取液(可根据以下两种方法中的任一种来配制):
    9.2.1 选项 A:二硫化碳和甲醇。
    9.2.2 选项 B:含有内标物的二硫化碳和甲醇的 1:4 混合溶液。
    9.2.3 参照 WHO TobLabNet SOP 09 分析苯和 1,3-丁二烯的方法,移取 20 mL 的二硫化碳和 500 μL 或者 250 μL 的内标物溶液(5mg/mL 的苯-d6)至 100 mL 的容量瓶,用甲醇定容到刻度。
    9.2.4 二硫化碳和甲醇的 1:4 混合溶液:移取 20 mL 二硫化碳至 100 mL 容量瓶,用甲醇定容至刻度。

注意:萃取液的量可以根据需要进行调整,以下仅为举例说明。

10 标准溶液配制

10.1 醛稀释混合标准溶液:
    10.1.1 将 1 mL 甲醛-DNPH [**7.2**],1 mL 乙醛-DNPH [**7.3**]和 0.1 mL 丙烯醛-DNPH [**7.4**]标准品移入 10 mL 容量瓶中。
    10.1.2 用乙醇[**7.9**]定容至刻度,并充分混合均匀。
    10.1.3 贴上标签,在 4℃±3℃ 的棕色样品瓶中保存。

10.2 醛最终混合标准溶液:
    10.2.1 根据表 1 配制醛最终标准溶液。
    10.2.2 将稀释的混合标准溶液[**10.1**]和 0.2 mL DNPH 溶液加入至 5 mL 容量瓶中(对于转盘型吸烟机,将 0.4 mLDNPH 溶液加入 10mL 容量瓶)。
    10.2.3 用乙醇[**7.9**]稀释至刻度,并充分混合。

10.2.4 贴上标签,在 4℃±3℃下保存。

10.2.5 注意：10 mL 是配制的混合标准溶液[**10.1**]的最终体积。工作溶液的最终体积为 5 mL（对于直线型吸烟机）或 10 mL（对于转盘型吸烟机），如表 1 所示。标准溶液中的最终醛浓度显示在表 1 中。

表 1a　直线型吸烟机：工作溶液中的醛浓度

标准	混合标准溶液的体积（mL），V_{mix}	总体积（mL），V_{tot}	最终混合标准溶液中的醛浓度（mg/L）					
			甲醛		乙醛		丙烯醛	
			mg/L	μg/cig	mg/L	μg/cig	mg/L	μg/cig
1	0.05	5	0.10	10	1.00	100	0.10	10
2	0.25	5	0.50	50	5.00	500	0.50	50
3	0.50	5	1.00	100	10.00	1000	1.00	100
4	1.00	5	2.00	200	20.00	2000	2.00	200
5	2.00	5	4.00	400	40.00	4000	4.00	400

表 1b　转盘型吸烟机：工作溶液中的醛浓度

标准	混合标准溶液的体积（mL），V_{mix}	总体积（mL），V_{tot}	最终混合标准溶液中的醛浓度（mg/L）					
			甲醛		乙醛		丙烯醛	
			mg/L	μg/cig	mg/L	μg/cig	mg/L	μg/cig
1	0.01	10	0.02	8	0.20	80	0.02	8
2	0.05	10	0.10	40	1.00	400	0.10	40
3	0.10	10	0.20	80	2.00	800	0.20	80
4	0.25	10	0.50	200	5.00	2000	0.50	200
5	0.50	10	1.00	400	10.00	4000	1.00	400

标准溶液范围可根据使用的设备和待测样品进行调整，注意方法灵敏度可能对结果产生的影响。

所有溶剂和溶液在使用前必须调节至室温。

11　抽样

11.1　根据 ISO 8243 [**2.4**]或替代方法，根据各实验室惯例或者要求的特殊规定或样品可用性，选取具有代表性的实验室卷烟样品。

11.2　试样的组成。

11.2.1　若可能，将实验室样品分成独立的单元（如包、条）。

11.2.2 从至少 $\sqrt{n}$ 个[2.9]单元中取出每种试样的等量产品。

注：采用随机抽样方法。$\sqrt{n}$ 是常用的方法之一，在许多情况下都很有效[2.9]。

12 卷烟准备

12.1 按照 ISO 3402 [2.3]调节所有待测卷烟。

12.2 按照 ISO 4387 [2.2]和 WHO TobLabNet SOP 01 [2.6]标记烟蒂长度。

12.3 按照 WHO TobLabNet SOP 01 [2.6]的深度抽吸方案准备要待测样品。

13 吸烟机设置

13.1 环境条件。

按 ISO 3308[2.1]要求设定抽吸的环境条件。

13.2 吸烟机要求。

除深度抽吸方案按 WHO TobLabNet SOP 01 [2.1]进行准备外，其余按照 ISO 3308 [2.6]的要求设置吸烟机。

13.3 如上所述[6.17]准备烟气捕集器，使其与图 1 至图 5 所示保持一致。

14 样品生成

14.1 抽吸卷烟试样并按照 ISO 4387 [2.2]或 WHO TobLabNet SOP 01 [2.6]收集主流烟气，并将醛收集到含有 CX-572 滤芯[6.15]和玻璃纤维滤片[6.16]的烟气捕集器上[6.17]。

14.2 至少包含一个用于质量控制的参比试样。

表 2 显示了 ISO 和深度抽吸方案中直线型和转盘型吸烟机每次抽吸的数量。

表 2 直线型和转盘型吸烟机进行一次测量的卷烟数量

有关适当的吸烟计划，请参阅附录 2

	ISO 方案		深度抽吸方案	
	直线型	转盘型	直线型	转盘型
每个烟气捕集器的卷烟数量	1	1	1	1
每个结果包含的烟气捕集器数量	3	3	3	3

14.3 记录每个烟气捕集器的卷烟数量和总抽吸量。

14.4 测量完成所需的样品量后，进行一次空吸，然后从吸烟机上取下烟气捕集器。

14.5 一个结果需要 3 次测量的平均值。

14.6 在抽吸卷烟样品之前，每天至少进行一次 10 口抽吸空白测试（图 7）。

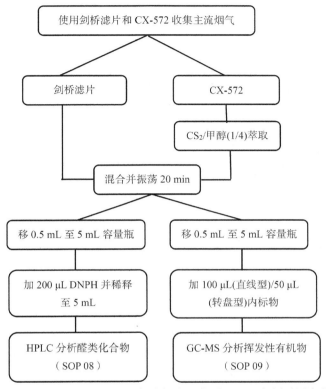

图 7 卷烟主流烟气中醛类（SOP 08）和挥发性有机物（SOP 09）同时测定分析方法的流程

如果需要可在吸烟前后称量烟气捕集器。质量差异为 TPM 信息提供参考

15 样品制备

先添加 CS_2，然后加入甲醇溶液（选项 A），或者直接用 CS_2 与甲醇混合溶液进行萃取（选项 B）。

15.1 选项 A：

15.1.1 向小瓶或烧瓶[6.9]中加入 2 mL（直线型吸烟机）或 4 mL（转盘型吸烟机）CS_2[7.4]，然后缓慢加入 8 mL（直线型吸烟机）或 16 mL（转盘型吸烟机）甲醇，不要将两种溶液混合。然后将 CX-572 迅速放入两相溶液中。

15.1.2 停止起泡后，轻轻旋转并将玻璃纤维滤片放入小瓶或烧瓶中。

15.1.3 旋紧瓶盖。

15.1.4 在旋转振荡器上以 120 r/min 的速度振荡 30 min。

15.1.5 静置 1 min（或粒相物沉淀）。

15.2 选项 B：

15.2.1 直接将玻璃纤维滤片和 CX-572 放入小瓶或烧瓶中。

15.2.2 向瓶子中慢慢加入 10 mL（直线型吸烟机）或 20 mL（转盘型吸烟机）的混合溶液[9.2.2]，旋紧瓶盖。

15.2.3 以 120 r/min 的速度振荡 30 min。

15.2.4 静置 1 min（或粒相物沉淀）。

15.3 衍生化：

15.3.1 静置样品至少 1 min（或直到颗粒沉降）。

15.3.2 使用注射器缓慢（不打开小瓶或烧瓶）将 0.5 mL 洗脱液转移到小瓶或烧瓶[6.10]中。

15.3.3 加入 0.2 mL DNPH 溶液[9.1]并轻轻摇动。

15.3.4 静置溶液 10 min，加入乙醇[7.9]稀释到 5 mL 并混合。

15.3.5 转移足够量的样品到 HPLC 自动进样瓶中。

16 样品分析

使用 HPLC 测定卷烟主流烟气中的甲醛、乙醛和丙烯醛。分析物可以从色谱柱上的其他潜在干扰物质中分离出来，比较未知物的峰面积比与已知浓度标准液的峰面积比，得到单个分析物的浓度。

16.1 HPLC 操作条件示例。

HPLC 色谱柱：Ascentis® RP-Amide（150 mm×4.6 mm，3 μm）；

进样量：10 μL；

柱温箱温度：(30±5)℃（根据所用的特定色谱柱进行调整）；

脱气装置：开；

二元泵流速：1.0 mL/min；

流动相 A：水；

流动相 B：乙腈；

梯度：如表 3 所示；

总分析时间：45 min；

UV 检测器：360 nm 或最大波长在 300~400 nm 处。

表3 醛分离所使用的梯度（A：水；B：乙腈）

时间（min）	A（%）	B（%）
0	55	45
10	55	45
25	0	100
30	0	100
31	55	45
40	55	45

注：根据仪器和色谱柱条件以及色谱峰的分离度调整操作参数。

16.2 一般分析信息。

16.2.1 对于此处描述的条件，预期的洗脱顺序为甲醛-DNPH、乙醛-DNPH和丙烯醛-DNPH。

16.2.2 流速、流动相浓度和色谱柱的老化等差异可能会改变保留时间。

16.2.3 根据每个实验室的做法来确定醛的测定顺序。本节举例说明。

16.2.4 注入空白溶液（萃取液）来检查系统或试剂中是否存在污染。

16.2.5 注入标准空白溶液来验证HPLC-DAD系统的性能。

16.2.6 注入标准液、质量控制溶液和样品。

16.2.7 记录甲醛-DNPH、乙醛-DNPH和丙烯醛-DNPH的峰面积。

注意： 醛-DNPH混合溶液中可能含有存在于样品溶液中的其他醛，应检查目标物质是否与干扰峰充分地分离，从而防止干扰，导致最终结果偏高。

16.2.8 根据峰面积（Y轴）和加入的甲醛、乙醛和丙烯醛浓度（X轴）来绘制标准曲线。

16.2.9 截距在统计上与零不应该有显著差异。

16.2.10 标准曲线应在整个标准范围内呈线性。

16.2.11 使用线性回归的斜率b和截距a来计算线性回归方程（$Y=a+bx$），如果线性回归方程的R^2小于0.99，则应重新校准。如果单个校准点与预期值（通过线性回归估计）相差超过10%，则应该舍弃该点。

16.2.12 所有试料的信号（峰面积）必须落在标准曲线的工作范围内；否则应根据需要调整溶液。

代表性色谱图见附录1。

17 数据分析和计算

17.1 确定每个试料甲醛、乙醛和丙烯醛的峰面积响应。

17.2 使用线性回归系数计算每个试料的甲醛、乙醛和丙烯醛浓度（mg/L）：

$$m_t = \frac{Y_t - a}{b}$$

使用下式计算卷烟样品中的甲醛、乙醛和丙烯醛含量 m_n：

$$M_n = (M_t - M_{bl}) \times \frac{V_t \times V_s}{V_e \times N_c}$$

其中：M_n——计算的醛含量（μg/支）；

M_t——测试溶液中醛的浓度（mg/L）；

M_{bl}——测试样品空白溶液的浓度（mg/L）；

V_t——测试溶液的体积（直线型吸烟机：5 mL；转盘型吸烟机：10 mL）；

V_s——溶剂（二硫化碳+甲醇）的体积（直线型吸烟机：10 mL；转盘型吸烟机：20 mL）；

V_e——用于衍生化的洗脱液的体积（0.5 mL）；

N_c——一个滤芯抽吸的卷烟数量。

注：如果适用，可以使用替代计算方程。

18 特别注意事项

安装新色谱柱后，在指定的仪器条件下注入样品来调节色谱柱。应重复进样，直至甲醛-DNPH、乙醛-DNPH 和丙烯醛-DNPH 的峰面积（或峰高）可重复。

19 数据报告

19.1 报告每个评估样品的单个测量值。

19.2 按方法规范指定的报告结果。

19.3 更多信息请参阅 WHO TobLabNet SOP 02 [2.7]。

20 质量控制

20.1 控制参数。

如果控制测量值超出预期值的公差极限，则必须进行适当的调查并采取措施。如有必要，需要做额外的实验室质量保证规程，以符合各个实验室的做法。

20.2 实验室试剂空白样。

如 **16.2.4** 所述，要在样品制备和分析过程中检测潜在的污染，需包含实验室试剂空白样（LRB）。空白样由分析测试样品中使用的所有试剂和材料组成，包括

CX-572 和玻璃纤维滤片的萃取,并作为试样进行分析。空白应小于检测限。

注:在某些实验室,由于环境条件的限制,低于检测限的空白无法实现。在这种情况下,应在每次试验运行中包括实验室试剂空白样,并应针对实验室试剂空白校正样品结果。

20.3 质量控制样品。

为了验证整个分析过程的一致性,根据各个实验室的做法分析参比卷烟。

20.4 附加质量控制。

可以根据实验室政策添加额外的质量控制样品。

21 方法性能

21.1 报告限。

报告限设置为所用标准曲线的最低浓度,重新计算为 µg/支卷烟(甲醛和丙烯醛为 8~10 µg/支,乙醛为 80~100 µg/支,相应地,0.04~0.10 mg/L 以及 0.410~1.0 mg/L 分别为最低标准浓度)。

21.2 实验室基质加标回收率。

将加入到基质上的分析物的回收率用作准确度的替代测量。通过添加已知含量的标准品(抽吸后)并用与样品相同的方法萃取来确定回收率,未加标的样品也要进行分析。回收率由下式计算,如表 4 所示。

回收率(%)=100 ×(分析结果 – 未加标结果)/加标量

表 4 基质加标的平均值和回收率

甲醛			乙醛			丙烯醛		
加标量(µg)	平均值(mg)	回收率(%)	加标量(µg)	平均值(mg)	回收率(%)	加标量(µg)	平均值(mg)	回收率(%)
12.7	14.4	113.4	385.0	416.5	108.2	21.9	22.3	101.8
25.4	27.6	108.7	770.0	722.2	93.8	43.9	41.4	94.3
50.7	52.1	102.8	1540.0	1407.8	91.4	87.7	86.0	98.1

21.3 分析特异性。

目标分析物的保留时间用于验证分析特异性。该成分对质量控制卷烟烟气样品的响应范围用于验证未知样品的结果的特异性。

21.4 线性。

建立的甲醛和丙烯醛的标准浓度范围为 0.04 ~ 4.0 mg/L,乙醛为 0.4 ~ 40 mg/L。

21.5 潜在干扰。

其他醛的存在可能引起干扰,其保留时间可能类似于待测定的醛的保留时间。

22 重复性和再现性

根据世界卫生组织的数据，2015~2017 年进行的一项国际合作研究涉及 10 个实验室和 5 个样品（3 种参比卷烟和 2 种商品卷烟）。验证 WHO TobLabNet SOP 02 [2.7]对该方法给出了以下精密度限制。

常规正确操作该方法时，同一操作者使用同一设备在最短的操作时间内两个单一结果之间的差异超过重复性 r 的情况，平均 20 个样品不超过 1 次。

常规正确操作该方法时，两个实验室报告的匹配卷烟样品的单一结果超过再现性 R 的情况，平均 20 次不超过 1 次。

根据 ISO 5725-1 [2.11]和 ISO 5725-2 [2.5]对测试结果进行统计分析，得到表 5 至表 7 所示的精密度数据。

为了计算 r 和 R，在 ISO 条件下，测验结果被定义为来自直线型吸烟机的三个捕集器（每个捕集器抽吸 1 支烟）和转盘型吸烟机的三个捕集器（每个捕集器抽吸 1 支烟）平均释放量的 7 次重复测试的平均值。在深度抽吸方案下，测验结果被定义为来自直线型吸烟机的三个捕集器和转盘型吸烟机的三个捕集器平均释放量的 7 次重复测试的平均值。更多信息请参考[2.6]。

表 5　参比试样的主流烟气中甲醛测定的精密度限值（μg/支）

参比卷烟	ISO 方案			
	N	m_{cig}	r_{limit}	R_{limit}
1R5F	8	5.92	1.43	2.08
3R4F	8	22.87	3.05	13.34
CM6	8	48.94	4.54	37.53
商品卷烟 1	9	31.58	6.27	31.82
商品卷烟 2	9	31.66	5.98	24.74
参比卷烟	深度抽吸方案			
	N	m_{cig}	r_{limit}	R_{limit}
1R5F	9	36.92	4.49	39.51
3R4F	6	72.99	5.38	7.62
CM6	6	84.83	6.76	12.17
商品卷烟 1	8	55.11	5.64	23.92
商品卷烟 2	9	58.05	7.67	30.90

表6 参比试样的主流烟气中乙醛测定的精密度限值（µg/支）

参比卷烟	ISO方案			
	N	m_{cig}	r_{limit}	R_{limit}
1R5F	9	156.21	26.00	36.80
3R4F	10	501.61	55.82	151.52
CM6	6	742.83	44.05	48.22
商品卷烟1	9	641.25	80.07	295.24
商品卷烟2	8	592.43	55.81	124.17
参比卷烟	深度抽吸方案			
	N	m_{cig}	r_{limit}	R_{limit}
1R5F	8	866.60	90.51	168.60
3R4F	9	1054.47	114.99	290.90
CM6	6	961.75	62.96	78.90
商品卷烟1	9	990.93	100.62	288.42
商品卷烟2	9	930.38	124.48	334.17

表7 参比试样的主流烟气中丙烯醛测定的精密度限值（µg/支）

参比卷烟	ISO方案			
	N	m_{cig}	r_{limit}	R_{limit}
1R5F	7	9.60	1.81	1.84
3R4F	7	48.12	4.77	19.29
CM6	9	67.77	9.70	37.30
商品卷烟1	7	45.22	7.31	16.19
商品卷烟2	9	57.31	7.45	54.40
参比卷烟	深度抽吸方案			
	N	m_{cig}	r_{limit}	R_{limit}
1R5F	8	107.40	17.37	120.55
3R4F	9	137.25	24.49	138.17
CM6	9	121.01	26.93	108.33
商品卷烟1	8	111.05	14.82	117.04
商品卷烟2	9	120.16	27.97	163.28

注：由于甲醛具有高的挥发性和亲水性，因此这个方法没有对甲醛的收集进行优化。

由于重复性（r）小于30%，且参与共同实验的实验室数目少于10个，因此精密度数据变异系数比较高，这种方法可以被认为是有条件的验证。

基于上述的结果，这个方法适合用来分析卷烟主流烟气中的醛类。

附录1 卷烟主流烟气中醛类化合物的典型色谱图

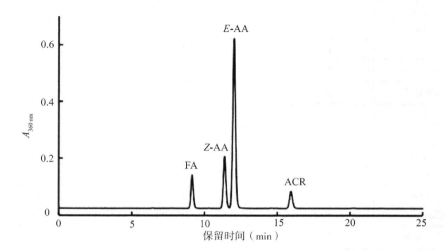

附图1 使用 Ascentis® RP-Amide（150 mm × 4.6 mm，3 μm）
分析醛类标准品的 HPLC 色谱图

FA：甲醛二苯腙；AA：乙醛二苯腙；ACR：丙烯醛二苯腙

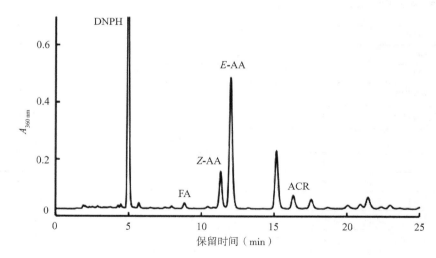

附图2 使用 Ascentis® RP-Amide（150 mm × 4.6 mm，3 μm）
分析主流烟气中醛类物质的 HPLC 色谱图

FA：甲醛二苯腙；AA：乙醛二苯腙；ACR：丙烯醛二苯腙

附录 2 吸烟计划表

所描述的吸烟计划基于 5 种不同的样品（卷烟品牌/类型）和每个样品的 7 次重复。如果需要不同数量的样品或重复样品，则应相应调整计划。

20 通道直线吸烟机吸烟计划（每通道 1 支烟）

天	1	2	3	4	5	6	7	8	9	10	11	12	13	14	15	16	17	18	19	20
1	A				B				C				D				E			
		A				B				C				D				E		
			A				B				C				D				E	
2		D				E				A				B				C		
			D				E				A				B				C	
				D				E				A				B				C
3	B				C				D				E				A			
		B				C				D				E				A		
			B				C				D				E				A	
4	E				A				B				C				D			
		E				A				B				C				D		
			E				A				B				C				D	
5	C				D				E				A				B			
		C				D				E				A				B		
			C				D				E				A				B	
6	D				C				B				A				E			
		D				C				B				A				E		
			D				C				B				A				E	
7	C				B				A				E				D			
		C				B				A				E				D		
			C				B				A				E				D	

转盘型吸烟机抽吸计划（每通道1支烟）

天	编号	样品	天	编号	样品	天	编号	样品	天	编号	样品
1	1	A	3	31	B	5	61	C	7	91	C
1	2	A	3	32	B	5	62	C	7	92	C
1	3	A	3	33	B	5	63	C	7	93	C
1	4	B	3	34	C	5	64	D	7	94	B
1	5	B	3	35	C	5	65	D	7	95	B
1	6	B	3	36	C	5	66	D	7	96	B
1	7	C	3	37	D	5	67	E	7	97	A
1	8	C	3	38	D	5	68	E	7	98	A
1	9	C	3	39	D	5	69	E	7	99	A
1	10	D	3	40	E	5	70	A	7	100	E
1	11	D	3	41	E	5	71	A	7	101	E
1	12	D	3	42	E	5	72	A	7	102	E
1	13	E	3	43	A	5	73	B	7	103	D
1	14	E	3	44	A	5	74	B	7	104	D
1	15	E	3	45	A	5	75	B	7	105	D
2	16	D	4	46	E	6	76	D			
2	17	D	4	47	E	6	77	D			
2	18	D	4	48	E	6	78	D			
2	19	E	4	49	A	6	79	C			
2	20	E	4	50	A	6	80	C			
2	21	E	4	51	A	6	81	C			
2	22	A	4	52	B	6	82	B			
2	23	A	4	53	B	6	83	B			
2	24	A	4	54	B	6	84	B			
2	25	B	4	55	C	6	85	A			
2	26	B	4	56	C	6	86	A			
2	27	B	4	57	C	6	87	A			
2	28	C	4	58	D	6	88	E			
2	29	C	4	59	D	6	89	E			
2	30	C	4	60	D	6	90	E			

WHO TobLabNet SOP 09

ISO 和深度抽吸方案下卷烟主流烟气中挥发性有机物的测定标准操作规程

方　　法：ISO 和深度抽吸方案下卷烟主流烟气中挥发性有机物的测定
分 析 物：苯（CAS 号：71-43-2）
1,3-丁二烯（CAS 号：106-99-0）
基　　质：卷烟主流烟气粒相物
更新时间：2018 年 8 月 31 日

吸烟机的抽吸方案不能代表所有人的吸烟行为：吸烟机测试对用于卷烟设计及监管目的的卷烟释放物表征非常有用，但将吸烟机的测量结果披露给吸烟者则会导致对不同品牌的暴露和风险差异的误解。吸烟机测量的烟气释放物数据可以用于产品危害评估，但并不意味着这些数据等效于人体暴露或风险的测量值。将吸烟机测量结果的差异表示为暴露或风险的差异是对世界卫生组织烟草实验室网络标准的滥用。

本方法由世界卫生组织（WHO）烟草实验室网络（TobLabNet）制订，作为国际标准化组织（ISO）和深度抽吸方案下卷烟主流烟气中挥发性有机物的测定标准操作规程（SOP）。

引言

为了在全球范围内建立具有可比性的烟草制品检测方法，需要卷烟特定成分和释放物的一致性检测方法。2008年11月在南非德班举行的世界卫生组织《烟草控制框架公约》（WHO FCTC）第三次缔约方会议回顾了 FCTC/COP1(15)和FCTC/COP2(14)号决议关于制订 WHO FCTC 第9条（烟草制品成分管制）和第10条（烟草制品披露管制）实施指南的要求，并提到第三次缔约方会议工作报告中有关其工作进展的信息，提请公约秘书处授权 WHO 无烟草行动组在5年内对检测卷烟成分和释放物的分析化学方法进行验证（FCTC/COP/3/REC/1）。

根据2006年10月在加拿大渥太华举行的第三次会议上确定的优先次序标准，WHO FCTC 第9条和第10条工作组确定优先验证以下成分的分析化学检测方法：

- 烟碱；
- 氨；
- 保润剂[1,2-丙二醇、甘油（1,2,3-丙三醇）和三甘醇（2,2′-二乙醚乙二醇）]。

测量这些成分需要验证三个方法：一个适用于烟碱，一个适用于氨，一个适用于保润剂。

根据上述渥太华会议确定的优先次序标准，工作组确定了在卷烟主流烟气释放中应该验证测试和测量方法（分析化学）的下列优先级物质清单：

- 4-(N-甲基亚硝胺基)-1,3-吡啶基-1-丁酮（NNK）；
- N-亚硝基去甲烟碱（NNN）；
- 乙醛；
- 丙烯醛；
- 苯；
- 苯并[a]芘；
- 1,3-丁二烯；
- 一氧化碳；
- 甲醛。

这些释放物使用下述两种抽吸方案测量，按五个方法进行验证：一个适用于烟草特有亚硝胺（NNK 和 NNN），一个适用于苯并[a]芘，一个适用于醛类（乙醛、丙烯醛和甲醛），一个适用于挥发性有机物（苯和 1,3-丁二烯），一个适用于一氧化碳。

下表列出了用于验证上述测试方法的两种抽吸方案。

抽吸方案	抽吸容量（mL）	抽吸间隔（s）	滤嘴通风孔
ISO 方案：ISO 3308 常规分析用吸烟机 定义和标准条件	35	60	不封闭
深度抽吸方案：在 ISO 3308 基础上有所调整	55	30	所有通风孔必须如 SOP 01 **12.2** 所述 100%封闭

SOP 09 描述 ISO 和深度抽吸方案下卷烟主流烟气中挥发性有机物的测定标准操作规程。对于卷烟以外的烟草制品，可根据方法给出的具体参数进行调整。

1 适用范围

本方法适用于气相色谱-质谱（GC-MS）法测定卷烟主流烟气中的苯和1,3-丁二烯两种挥发性有机物。

苯和 1,3-丁二烯是强致癌物质，并不存在于烟叶中，而是在卷烟抽吸过程中由有机物质的燃烧形成。

请注意，要成功操作吸烟机或其他分析设备，须对操作人员进行培训。对操作吸烟机或其他分析设备没有经验的人，在检测烟草制品成分和释放物前，需要接受专门的培训。

2 参考标准

2.1 ISO 3308：常规分析用吸烟机 定义和标准条件。

2.2 ISO 4387：卷烟 常规分析用吸烟机测定总粒相物和去烟碱干粒相物。

2.3 ISO 3402：烟草和烟草制品 用于调节和测试的大气环境。

2.4 ISO 8243：卷烟 抽样。

2.5 ISO 5725-2：检测方法和结果的准确度（正确度和精密度） 第 2 部分 确定标准检测方法重复性和再现性的基本方法。

2.6 WHO TobLabNet SOP 01：卷烟深度抽吸标准操作规程。

2.7 WHO TobLabNet SOP 02：烟草制品成分和释放物分析方法验证标准操作规程。

2.8 WHO TobLabNet SOP 08：ISO 和深度抽吸方案下卷烟主流烟气中醛类化合物的测定标准操作规程。

2.9 联合国毒品和犯罪问题办公室（UNODC）《代表性药物取样指南》（http://www.unodc.org/documents/scientific/Drug_Sampling.pdf）。

2.10 ISO 5725-1：检测方法和结果的准确度（正确度和精密度） 第 1 部分 一般原则和定义。

2.11 Uchiyama S, Tomizawa T, Inaba, Y, Kunugita, N. Simultaneous determination of volatile organic compounds and carbonyls in mainstream cigarette smoke using a sorbent cartridge followed by two-step elution. J Chromatogr A. 2013; 1314:31-7. doi: 10.1016/j.chroma.2013.09.019。

2.12 Uchiyama, S, Hayashida, H, Izu, R, Inaba, Y, Nakagome, H, Kunugita N. Determination of nicotine, tar, volatile organic compounds and carbonyls in mainstream cigarette smoke using a glass filter and a sorbent cartridge followed by the two-phase/one-pot elution method with carbon disulfide and methanol. J Chromatogr A. 2015; 1426:48-55. doi.org/10.1016/j. chroma.2015.11.058。

3 术语和定义

3.1 CX-572：碳吸附剂 Carboxen 572。

3.2 深度抽吸方案：抽吸过程参数为抽吸容量 55 mL，抽吸间隔 30 s，每次抽吸持续时间 2 s，100%封闭滤嘴通风孔。

3.3 ISO 方案：抽吸过程参数为抽吸容量 35 mL，抽吸间隔 60 s，每次抽吸持续时间 2 s，不封闭滤嘴通风孔。

3.4 实验室样品：用于进行实验室检测的样品，由一次性或在规定时间内送到实验室的单一类型产品组成。

3.5 烟草制品：完全或部分以烟叶为原料制造，用于抽吸、吸吮、咀嚼或吸入的产品（WHO FCTC 第 1(f)条）。

3.6 试样：从实验室样品中随机抽取的待测样品。所取样品数量应能代表实验室样品。

3.7 试料：从试样中随机抽取的用于单次测试的样品。所取样品数量应能代表试样。

3.8 烟气捕集器：为测定特定烟气成分从卷烟样品中收集部分烟气的装置。

3.9 TPM：总粒相物

4 方法概述

4.1 在深度抽吸方案或 ISO 方案下收集卷烟主流烟气。抽吸每支卷烟时每个烟气捕集器含有 300 mg CX-572 颗粒和一片玻璃纤维滤片。

4.2 卷烟试样的主流烟气总粒相物被捕集在含有 300 mg 的 CX-572 颗粒和玻璃纤维滤片的烟气捕集器上。

4.3 CX-572 颗粒和玻璃纤维滤片上的挥发性有机物通过二硫化碳和甲醇的混合溶液来进行萃取。

4.4 采用配备电子轰击离子源的气相色谱-质谱（GC-MS）测定挥发性有机物（1,3-丁二烯和苯）。

4.5 通过将分析物与同位素标记内标物的重建离子色谱图峰面积比进行作图，建立已知浓度苯和 1,3-丁二烯的校准曲线，从而确定试料的苯和 1,3-丁二烯含量。

4.6 本标准操作规程所述的方法可同时检测主流烟气中醛类化合物（SOP 08）和挥发性有机物（SOP 09）的含量。

5 安全和环境预防措施

5.1 遵守所有化学实验活动的常规安全与环境预防措施。

5.2 使用本检测方法对特定产品进行检测和评估时所用材料或设备可能对环境有害。本检测方法无力解决所有与使用有关的安全问题。所有使用此方法的人有责任咨询相关机构并遵循现有监管要求，制定健康与安全规程以及环境预防措施。

5.3 应特别注意避免吸入或皮肤接触危险化学品。在制备或处理未稀释材料、标准溶液、萃取液和收集样品时，应在通风橱中进行，并穿戴合适的实验服、手套和护目镜。

6 仪器和设备

6.1 按照 ISO 3402 [**2.3**]规定调节卷烟所需设备。

6.2 按照 ISO 4387 [**2.2**]规定标记烟蒂长度所需设备。

6.3 按照 WHO TobLabNet SOP 01 [**2.6**]规定采用深度抽吸方案时封闭通风孔所需设备。

6.4 按照 ISO 3308 [**2.1**]规定抽吸烟草制品所需设备。

6.5 测量精度为 0.0001 g 的分析天平。

6.6 移液枪及能精确移取 100~5000 μL 的枪头。

6.7 聚四氟乙烯内衬隔垫。

6.8 5 mL 和 10 mL 容量瓶。

6.9 容量至少为 10 mL 的密封烧瓶。

6.10 用于转移有机溶剂的 1 mL 和 5 mL 注射器。

6.11 2 mL GC 自动进样瓶。

6.12 配有质谱检测器的气相色谱仪。

6.13 能够明显分离烟气释放物中苯、1,3-丁二烯和同位素标记的苯峰的气相色谱柱，如 InertCap AQUATIC-2（60 m × 0.25 mm i.d., d=1.4 μm）。

6.14 CX-572 碳吸附剂，20/45 目。

6.15 CX-572 滤芯：手动制备填充 300 mg CX-572 的固相萃取滤芯，或购买 SigmaAldrich®（No. 54294-U）预装的 CX-572 柱。

6.16 按照 ISO 4387 准备玻璃纤维滤片。

6.17 烟气捕集器由吸烟机支托器和 CX572（带或不带滤芯）组成，支托器放置在玻璃纤维滤片前面，见图 1 至图 5。将 CX-572 柱朝着卷烟的开口侧放入支托器。当开口端朝相反方向放置时，柱中的成分会在抽吸时被吸烟机吸入。如有必要，可在柱中插入专用塞子，如图 6 所示。

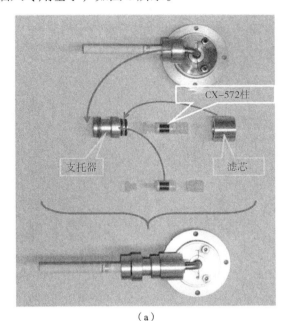

(a)

(b)

图 1 Borgwaldt 吸烟机

(a) 单通道；(b) 直线型

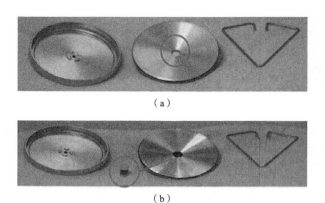

(a)

(b)

图 2 适用于商品转盘型吸烟机（Cerulean 和/或 Borgwaldt 吸烟机）的支托器

（a）CX-572 颗粒直接压在支托器上；（b）将装有 CX-572 的滤芯插入支托器

卷烟　　　CX-572 柱　剑桥滤片　　吸烟机

图 3 直线型商品吸烟机（Cerulean 或 Borgwaldt 吸烟机）的支托器

CX-572 可预先装到支托器中，也可使用商品 CX-572 柱

图 4 直线型 Cerulean 吸烟机（也适用于直线型 Borgwaldt 或
转盘型 Cerulean 吸烟机或者单通道吸烟机）的支托器

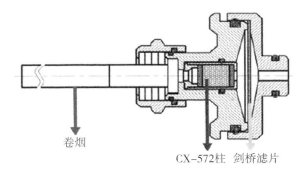

图 5 适用于直线型 Cerulean 吸烟机（也适用于直线型 Borgwaldt 吸烟机）的支托器

图 6 CX-572 柱中插入塞子，以防止内容物被吸入吸烟机

插入塞子后将 SPE 柱直接放入封闭容器（小瓶）中，以防污染 CX-572

7 试剂

除非另有说明，所有试剂应至少为分析纯。试剂尽可能以 CAS 号来识别。

7.1 2 mg/mL 苯标准溶液[71-43-2]。

7.2 纯度>99%的同位素标记的苯（苯-d6）[1076-43-3]。

7.3 2 mg/mL 1,3-丁二烯标准溶液(AccuStandard, S-406A-10X) [106-99-0]。

7.4 二硫化碳[75-15-0]。

7.5 无水甲醇 [67-56-1]。

7.6 碳吸附剂 CX-572 颗粒，20/45 目（Sigma-Aldrich，11072-U）。

8 玻璃器皿准备

清洁、干燥玻璃器皿，避免残留物污染。

9 溶液配制

9.1 同位素标记的内标物（5 mg/mL 的苯-d6 标准溶液）：
　9.1.1 准确称取 500 mg 苯-d6 到已加入约 80 mL 甲醇的 100 mL 容量瓶中，用甲醇定容至 100 mL。
　9.1.2 标记并储存–20℃±5℃。
9.2 萃取液（可根据以下两种方法中的任一种来配制）：
　9.2.1 选项 A：二硫化碳和甲醇。
　9.2.2 选项 B：含有内标物的二硫化碳和甲醇的 1：4 混合溶液。
　　移取 20 mL 的二硫化碳和 500 μL（直线型吸烟机）或 250 μL（转盘型吸烟机）同位素标记的内标物溶液（5 mg/mL 的苯-d6）至 100 mL 的容量瓶中，用甲醇定容至刻度。
　　工作溶液中内标物（苯-d6）的浓度为 25 μg/mL（直线型吸烟机）或 12.5 μg/mL（转盘型吸烟机）
　9.2.3 二硫化碳和甲醇的 1：4 混合溶液：移取 20 mL 二硫化碳至 100 mL 容量瓶，用甲醇定容至刻度。
　　注意：萃取液的量可以根据需要进行调整，以下仅为举例说明。

10 标准溶液配制

注意：可以根据标准范围和内标浓度保持不变的条件进行相应调整。
10.1 VOC 标准储备溶液（苯：80 μg/mL；1,3-丁二烯：200 μg/mL）。
　10.1.1 2.0 mg/mL 的苯标准溶液：
　　准确称取 200 mg 的苯[7.1]标准品到已加入 80 mL 甲醇的 100 mL 容量瓶中，并用甲醇定容到刻度。
　10.1.2 移取 0.8 mL 苯标准储备液[10.1]和 2.0 mL 1,3-丁二烯标准储备液 [7.3]到 20 mL 容量瓶中。
　10.1.3 加入 4 mL 二硫化碳[7.4]，用甲醇定容至 20 mL [7.5]。
　10.1.4 标记，并在 4℃±3℃的棕色瓶中储存。
10.2 VOC 标准工作溶液：
　10.2.1 根据所使用的吸烟机的不同，按表 1a（直线型吸烟机）和表 1b（转盘型吸烟机）移取不同体积的储备液[10.1]和混合溶剂[9.2.3]至 GC 自动进样瓶中。
　10.2.2 根据表 1a（直线型吸烟机）和表 1b（转盘型吸烟机）添加不同体积的同位素标记内标物至 GC 自动进样瓶中，混合均匀。

10.2.3 贴标签,在 4℃±3℃储存。

表 1a （选项 A）直线型吸烟机:标准溶液中 VOC 的浓度
每 5 mL 标准溶液加入的内标物

标准	VOC 储备液体积（mL）[10.1]	总体积（mL）	内标物体积（μL）[9.1]	VOC 浓度（mg/L）	
				苯	1,3-丁二烯
1	0	5.0	100	0	0
2	0.50	5.0	100	8	20
3	1.00	5.0	100	16	40
4	1.50	5.0	100	24	60
5	2.00	5.0	100	32	80
6	4.00	5.0	100	80	200

表 1b （选项 A）转盘型吸烟机:标准溶液中 VOC 的浓度
每 5 mL 标准溶液加入的内标物

标准	VOC 储备液体积（mL）[10.1]	总体积（mL）	内标物体积（μL）[9.1]	VOC 浓度（mg/L）	
				苯	1,3-丁二烯
1	0	5.0	50	0	0
2	0.25	5.0	50	4	10
3	0.50	5.0	50	8	20
4	0.75	5.0	50	12	30
5	1.00	5.0	50	16	40
6	2.00	5.0	50	40	80

表 1c （选项 B）直线型吸烟机:标准溶液中 VOC 的浓度
每 5 mL 标准溶液加入的内标物

标准	VOC 储备液体积（mL）[10.1]	内标物体积（μL）[9.1]	总体积（mL）	VOC 浓度（mg/L）	
				苯	1,3-丁二烯
1	0.00	20	5.0	0	0
2	0.50	20	5.0	8	20
3	1.00	20	5.0	16	40
4	1.50	20	5.0	24	60
5	2.00	20	5.0	32	80
6	4.00	20	5.0	64	160

表1d （选项B）转盘型吸烟机：标准溶液中VOC的浓度
每5 mL标准溶液加入的内标物

标准	VOC储备液体积（mL）[10.1]	内标物体积（μL）[9.1]	总体积（mL）	VOC浓度（mg/L）	
				苯	1,3-丁二烯
1	0.0	10	5	0	0
2	0.25	10	5	4	10
3	0.50	10	5	8	20
4	0.75	10	5	12	30
5	1.00	10	5	16	40
6	2.50	10	5	40	100

标准溶液范围可根据使用的设备和待测样品进行调整，注意方法灵敏度可能对结果产生的影响。

所有溶剂和溶液在使用前调节至室温。

11 抽样

11.1 根据ISO 8243 [**2.4**]或替代方法，根据各实验室惯例或者要求的特殊规定或样品可用性，选取具有代表性的实验室卷烟样品。

11.2 试样的组成。

11.2.1 若可能，将实验室样品分成独立的单元（如包、条）。

11.2.2 从至少 $\sqrt{n}$ 个[**2.9**]单元中取出每种试样的等量产品。

12 卷烟准备

12.1 按照ISO 3402 [**2.3**]调制所有卷烟。

12.2 按照ISO 4387 [**2.2**]和WHO TobLabNet SOP 01 [**2.6**]标记烟蒂长度。

12.3 按照WHO TobLabNet SOP 01 [**2.6**]的深度抽吸方案准备待测样品。

13 吸烟机设置

13.1 环境条件。

按ISO 3308 [**2.1**]要求设定抽吸的环境条件。

13.2 吸烟机要求。

除深度抽吸方案按WHO TobLabNet SO P01 [**2.1**]进行准备外，其余按照ISO 3308 [**2.6**]的要求设置吸烟机。

如上所述[**6.17**]准备烟气捕集器，使其与图1至图5所示保持一致。

14 样品生成

14.1 抽吸卷烟试样并按照 ISO 4387 [2.2]或 WHO TobLabNet SOP 01 [2.6]将主流烟气 VOC 收集到含有 CX-572 滤芯[6.15]和玻璃纤维滤片[6.16]的烟气捕集器上[6.17]。

14.2 至少包含一个用于质量控制的参比试样。

14.3 表2显示了 ISO 和深度抽吸方案中直线型和转盘型吸烟机每次抽吸的数量（图7）。

表2 直线型和转盘型吸烟机进行一次测量的卷烟数量
适当的吸烟计划，请参阅附录4

	ISO 方案		深度抽吸方案	
	直线型	转盘型	直线型	转盘型
每个烟气捕集器的卷烟数量	1	1	1	1
每个结果包含的烟气捕集器数量	3	3	3	3

注：在指定的吸烟机上抽吸1支卷烟，样品含量应不足以发生穿透，且确保苯和1,3-丁二烯的浓度均落在分析的校准范围内。

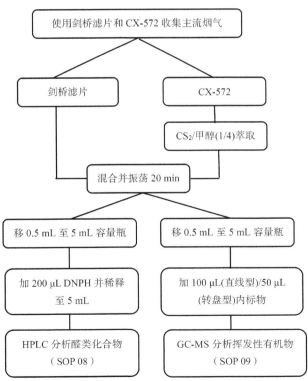

图7 卷烟主流烟气中醛类化合物（SOP 08）和挥发性有机物（SOP 09）同时测定方法流程

如果需要，可在抽吸前后称量烟气捕集器，质量差异为 TPM 的信息提供参考

14.4 记录每个烟气捕集器的卷烟数量和总抽吸口数。

14.5 测量完成所需的样品量后，进行一次空吸，然后从吸烟机上取下烟气捕集器。

14.6 一个结果需要 3 次测量的平均值。

14.7 在抽吸卷烟样品之前，每天至少进行一次 10 口抽吸空白测试。

15 样品制备

先添加 CS_2、然后加入甲醇溶液（选项 A），或者直接用 CS_2 与甲醇混合溶液进行萃取（选项 B）。

15.1 选项 A：

15.1.1 向烧瓶中加入 2 mL（直线型吸烟机）或 4 mL（转盘型吸烟机）CS_2[7.4]，然后缓慢加入 8 mL（直线型吸烟机）或 16 mL（转盘型吸烟机）甲醇，两种溶液不需要进行混合，然后将 CX-572 迅速放入两相溶液中。

15.1.2 停止起泡后，轻轻旋转并将玻璃纤维滤片放入小瓶或烧瓶中。

15.1.3 旋紧瓶盖。

15.1.4 以 120 r/min 的速率振荡 30 min。

15.1.5 静置 1 min（或直到粒相物沉淀）。

15.1.6 移取 5 mL 上清液放入烧瓶中，并加入 100 μL（直线型吸烟机）或 50 μL（转盘型吸烟机）内标物溶液。

15.1.7 混合均匀后，移取 1 mL 到 1.5 mL 自动进样瓶[6.10]，然后进行 GC-MS 分析。

15.2 选项 B：

15.2.1 直接将玻璃纤维滤片和 CX-572 放入小瓶或烧瓶中。

15.2.2 向烧瓶中加入 10 mL（直线型吸烟机）或 20 mL（转盘型吸烟机）的混合溶液[9.2.2]，旋紧瓶盖。

15.2.3 以 120 r/min 的速率振荡 30 min。

15.2.4 静置 1 min（或直到粒相物沉淀）。

15.2.5 混合均匀后，移取 1 mL 到 1.5 mL 自动进样瓶[6.10]，然后进行 GC-MS 分析。

16 样品分析

使用气相色谱-质谱联用（GC-MS）技术测定卷烟主流烟气中的苯和 1,3-丁二烯。分析物可以从色谱柱上的其他潜在干扰物质中分离出来，比较未知物的峰面积比(分析物峰面积与同位素标记的分析物峰面积)与已知浓度标准液的峰面积,得到单个分析物浓度。

16.1 气相色谱条件。

16.1.1 气相色谱条件见表3。

表3 GC 操作条件示例

气相色谱柱	InertCap AQUATIC-2 60 m × 0.25 mm i.d., d=1.4 µm, GL Sciences
进样口温度	240℃（离子阱200℃）
载气控制模式	恒流
流速	1 mL/min（离子阱1.0 mL/min）
进样量	1 µL 或 2 µL, 分流比为1:10（离子阱1:5, 1.5 min 后1:120）
柱温	40℃保持6 min, 以6℃/min升至250℃
总运行时间	45 min
接口/传输线温度	180℃（离子阱200℃）

- 对于此处描述的条件，预期的洗脱顺序为 1,3-丁二烯、二硫化碳、苯-d6 和苯。
- 流速、流动相浓度和色谱柱的老化等差异可能会改变保留时间。
- 洗脱顺序和保留时间在样品分析之前应进行确认。
- 使用上述条件，预期的总分析时间约为 60 min。

注意：根据仪器和色谱柱条件以及色谱峰的分离度调整操作参数。

16.1.2 质谱条件见表4。

表4 所使用的质谱条件
在全扫描模式下采集 m/z 信号

离子源	200℃（离子阱180℃）
驻留时间	50 ms
电离模式	电子轰击电离（电离电压为70 eV）
检测	全扫描：m/z 30~300
特征离子峰	苯：m/z 78；苯-d6：m/z 84；1,3-丁二烯：m/z 54

16.2 VOC 测定。

16.2.1 在使用前，通过注射两份 1 µL 或 2 µL 的样品溶液模拟进样来调节系统。

16.2.2 在与样品相同的条件下注入检查标准品（带内标物的空白溶液）以验证气相色谱-MS 系统的性能。

16.2.3 注入空白溶液（萃取液）以检查系统或萃取溶液或试剂中是否存在污染物。

16.2.4 将每种分析物标准液的试样注入 GC-MS。

16.2.5 评估同位素标记的标准品的保留时间和相对响应比（基于峰面积）。如果保留时间与先前的进样的保留时间相似（±0.2 min），且响应与先前进样中的典型响应差异在±20%以内，则说明系统已准备好进行分析。

16.2.6 注入标准溶液、质量控制溶液和样品，用适当的软件确定峰面积和相对响应比。

16.2.7 记录苯、1,3-丁二烯和苯-d6 的峰面积。

16.2.8 计算每个苯和 1,3-丁二烯标准溶液的苯与内标峰面积的比值（RF=$A_{Benzene}/A_{Benzene-d6}$），包括标准空白溶液。

16.2.9 根据添加的苯浓度（X轴）和峰面积比（Y轴）来绘制标准曲线。

16.2.10 截距在统计上与零不应该有显著差异。

16.2.11 标准曲线在整个标准范围内应该呈线性。

16.2.12 使用线性回归的斜率 b 和截距 a 来计算线性回归方程（$Y=a+bx$）。

注意： 如果线性回归方程的 R^2 小于 0.99，则应重新校准。如果单个校准点与预期值（通过线性回归估计）相差超过 10%，则应该舍弃该点。

16.2.13 注入质量控制样品和试料，用适当的软件确定峰面积。

16.2.14 所有试料的相对响应因子（峰面积比）必须落在标准曲线的工作范围内。

代表性色谱图见附录 1。

17 数据分析和计算

斜率和截距是由分析物和内标物的相对响应与分析物和内标物的相对浓度确定的。

17.1 根据每个标准曲线[10.2.3]计算每种挥发性有机物峰面积的相对响应比：

$$RF = \frac{A_a}{A_{IS}}$$

式中：RF——相对响应比；

A_a——目标分析物的峰面积；

A_{IS}——相应内标物的峰面积。

绘制每种挥发性有机物的标准曲线，相对响应因子 A_a/A_{IS}（Y轴）对浓度（X

轴),并使用线性回归的斜率(k)和截距(m)计算线性回归方程($RF=kX+m$)。

17.2 标准曲线应在整个标准范围内呈线性。

17.3 目标分析物中挥发性有机物的含量 M_{VOC} 由试料的相对响应比、标准曲线的斜率和截距计算得出:

$$M_{VOC} = \frac{RF - m}{k} \times \frac{V}{N}$$

式中:M_{VOC}——计算的含量(μg/支);

V——萃取液的体积(10 mL 或 20 mL);

$N=1$(每个滤片上卷烟的数量);

RF——相对响应比;

m——标准曲线的截距;

k——标准曲线的斜率。

如果适用,可以使用替代计算方程。

18 特别注意事项

安装新色谱柱后,在指定的仪器条件下注入样品来调节色谱柱。应重复进样,直至目标分析物的峰面积(或峰高)可重复。

19 数据报告

19.1 报告每个评估样品的单个测量值。

19.2 按方法规范指定的报告结果。

19.3 更多信息请参阅 WHO TobLabNet SOP 02 [2.7]。

20 质量控制

20.1 控制参数。

如果控制测量值超出预期值的公差极限,则必须进行适当的调查并采取措施。如有必要,需要做额外的实验室质量保证规程,以符合各个实验室的做法。

20.2 实验室试剂空白样。

如 16.2.3 所述,要在样品制备和分析过程中检测潜在的污染,需包含实验室试剂空白样(LRB)。空白样由分析测试样品中使用的所有试剂和材料组成,包括 CX-572 和玻璃纤维滤片的萃取,并作为试样进行分析。空白应小于检测限。

注:在某些实验室,由于环境条件的限制,低于检测限的空白无法实现。在

这种情况下，应在每次试验运行中包括实验室试剂空白样，并应针对实验室试剂空白校正样品结果。

20.3 质量控制样品。

为了验证整个分析过程的一致性，根据各个实验室的做法分析参比卷烟。

20.4 附加质量控制。

可以根据实验室政策添加额外的质量控制样品。

21 方法性能

21.1 报告限。

报告限设置为所用标准曲线的最低浓度，重新计算为 µg/支卷烟（以 0.05 µmol/mL 作为最低标准浓度）。

21.2 实验室基质加标回收率。

加入到基质上的分析物的回收率用作准确度的替代量度。通过添加已知量的标准品（抽吸后）并用与样品相同的方法萃取来测定回收率。未加标的样品也要进行分析。回收率由下式计算，如表 5 所示。

回收率（%）=100 ×（分析结果 – 未加标结果）/加标量

表 5　加标到基质的平均值和回收率（待验证完成后添加）

苯			1,3-丁二烯		
加标量（µg）	平均值（mg）	回收率（%）	加标量（µg）	平均值（mg）	回收率（%）
16.9	18.4	108.9	10	9.4	94.0
33.8	37.0	109.5	20	18.6	93.0
67.6	75.2	111.2	40	41.9	104.8

21.3 分析特异性。

GC-MS 具有良好的分析特异性。利用保留时间、确认离子比、定量离子与确认离子响应比的范围以及质量控制样品来验证未知样品结果的特异性。

21.4 线性。

建立的挥发性有机物标准曲线在 0.5~200 ng/mL 的标准浓度范围内是线性的。

21.5 潜在干扰。

没有已知的组分具有与挥发性有机物或内标物相似的保留时间和确认离子对比引起干扰。

22 重复性和再现性

2015~2017 年进行的一项国际合作研究涉及 10 个实验室对 3 种参比卷烟和 2 种商品卷烟的测试,根据 WHO TobLabNet SOP 02 [2.7]进行的实验对该方法给出了以下精密度限值。

常规正确操作该方法时,同一操作者使用同一设备在最短的操作时间内两个单一结果之间的差异超过重复性 r 的情况,平均 20 个样品不超过 1 次。

常规正确操作该方法时,两个实验室报告的匹配卷烟样品的单一结果超过再现性 R 的情况,平均 20 次不超过 1 次。

根据 ISO 5725-1 [2.10]和 ISO 5725-2 [2.5]对测试结果进行统计分析,得到表 6 和表 7 所示的准确数据。

为了计算 r 和 R,在 ISO 条件下,测验结果被定义为来自直线型吸烟机的三个捕集器(每个捕集器上 1 支卷烟)和转盘型吸烟机的三个捕集器(每个捕集器上 1 支卷烟)平均释放量的 7 次重复的平均值。在深度抽吸方案下,测验结果被定义为来自直线型吸烟机的三个捕集器(每个捕集器上 1 支卷烟)和转盘型吸烟机的三个捕集器(每个捕集器上 1 支卷烟)平均释放量的 7 次重复的平均值。更多信息请参考 [2.6]。

表 6 参比试样主流烟气中 1,3-丁二烯的精密度限值(μg/支)

参比卷烟	ISO 方案			
	N	m_{cig}	r_{limit}	R_{limit}
1R5F	8	11.06	1.96	4.27
3R4F	9	39.23	7.20	16.34
CM6	6	60.86	1.74	2.18
商业卷烟 1	8	72.71	8.03	42.79
商业卷烟 2	8	61.58	9.97	16.80
参比卷烟	深度抽吸方案			
	N	m_{cig}	r_{limit}	R_{limit}
1R5F	7	82.26	8.11	8.63
3R4F	8	92.22	11.51	13.18
CM6	7	106.20	12.59	12.67
商业卷烟 1	8	153.96	17.20	70.04
商业卷烟 2	6	141.52	11.00	14.55

表 7 参比试样主流烟气中苯的精密度限值（μg/支）

参比卷烟	ISO 方案			
	N	m_{cig}	r_{limit}	R_{limit}
1R5F	9	13.47	2.54	6.46
3R4F	9	44.50	6.94	10.15
CM6	8	72.86	5.94	11.40
商业卷烟 1	9	58.78	6.54	15.18
商业卷烟 2	9	49.94	8.83	15.37
参比卷烟	深度抽吸方案			
	N	m_{cig}	r_{limit}	R_{limit}
1R5F	8	85.59	8.17	16.68
3R4F	9	95.95	8.49	40.92
CM6	8	108.35	6.93	23.52
商业卷烟 1	8	122.00	7.21	48.76
商业卷烟 2	8	111.77	6.60	36.32

附录1 卷烟主流烟气中挥发性有机化合物的典型色谱图

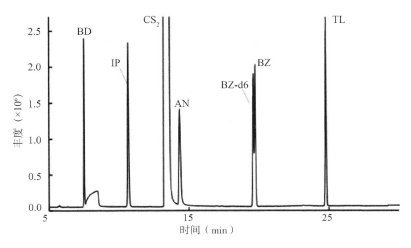

附图1 标准品2①的代表性色谱图

混合标准溶液的总离子流色谱图。BD：1,3-丁二烯；BZ-d6：苯-d6；BZ：苯

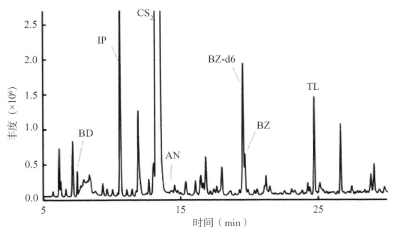

附图2 卷烟主流烟气样品中VOC的代表性色谱图②

样品溶液总离子流色谱图。BD：1,3-丁二烯；BZ-d6：苯-d6；BZ：苯

---

① 源自2012年进行的一项国际合作研究，由日本国立公共卫生研究所和中国疾病预防控制中心共同提供。仅供参考。

② 源自2012年进行的一项国际合作研究，由日本国立公共卫生研究所和中国疾病预防控制中心共同提供。仅供参考。

附录 2 吸烟计划

所描述的吸烟计划基于 5 种不同的样品（卷烟品牌/类型）和每个样品的 7 次重复。如果需要不同数量的样品或重复样品，则应相应调整计划。

20 通道直线型吸烟机吸烟计划（每个滤芯一支卷烟）

天	通道编号																			
	1	2	3	4	5	6	7	8	9	10	11	12	13	14	15	16	17	18	19	20
1	A			B			C				D				E					
		A			B			C				D				E				
			A			B			C				D				E			
2		D			E			A				B				C				
			D			E			A				B				C			
				D			E			A				B				C		
3	B			C					D			E					A			
			B			C				D			E					A		
				B			D				E				A					
4	E			A				B				C				D				
		E			A				B				C				D			
			E			A				B				C				D		
5	C			D				E			A					B				
		C			D				E			A					B			
			C			D				E				A				B		
6				D		C						B					A			E
	D			C			B						A							E
		D			C			B					A					E		
7			C			B			A					E				D		
		C			B			A					E				D			
				C			B			A				E						D

转盘型吸烟机抽吸计划（每个滤芯1支卷烟）

天	编号	样品	天	编号	样品	天	编号	样品	天	编号	样品
1	1	A	3	31	B	5	61	C	7	91	C
	2	A		32	B		62	C		92	C
	3	A		33	B		63	C		93	C
	4	B		34	C		64	D		94	B
	5	B		35	C		65	D		95	B
	6	B		36	C		66	D		96	B
	7	C		37	D		67	E		97	A
	8	C		38	D		68	E		98	A
	9	C		39	D		69	E		99	A
	10	D		40	E		70	A		100	E
	11	D		41	E		71	A		101	E
	12	D		42	E		72	A		102	E
	13	E		43	A		73	B		103	D
	14	E		44	A		74	B		104	D
	15	E		45	A		75	B		105	D
2	16	D	4	46	E	6	76	D			
	17	D		47	E		77	D			
	18	D		48	E		78	D			
	19	E		49	A		79	C			
	20	E		50	A		80	C			
	21	E		51	A		81	C			
	22	A		52	B		82	B			
	23	A		53	B		83	B			
	24	A		54	B		84	B			
	25	B		55	C		85	A			
	26	B		56	C		86	A			
	27	B		57	C		87	A			
	28	C		58	D		88	E			
	29	C		59	D		89	E			
	30	C		60	D		90	E			

WHO TobLabNet SOP 10

深度抽吸方案下卷烟主流烟气中烟碱和一氧化碳的测定标准操作规程

方　　法：深度抽吸方案下卷烟主流烟气中烟碱和一氧化碳的测定
分 析 物：烟碱（CAS 号：54-11-5）
一氧化碳（CAS 号：630-08-0）
基　　质：卷烟主流烟气
更新时间：2016 年 9 月 21 日

吸烟机的抽吸方案不能代表所有人的吸烟行为：吸烟机测试对用于卷烟设计及监管目的的卷烟释放物表征非常有用，但将吸烟机的测量结果披露给吸烟者则会导致对不同品牌的暴露和风险差异的误解。吸烟机测量的烟气释放物数据可以用于产品危害评估，但并不意味着这些数据等效于人体暴露或风险的测量值。将吸烟机测量结果的差异表示为暴露或风险的差异是对世界卫生组织烟草实验室网络标准的滥用。

本方法由世界卫生组织（WHO）烟草实验室网络（TobLabNet）制订，作为深度抽吸方案下卷烟主流烟气中的烟碱和一氧化碳的测定标准操作规程（SOP）。

引言

为了在全球范围内建立具有可比性的烟草制品检测方法，需要卷烟特定成分和释放物的一致性检测方法。2008年11月在南非德班举行的世界卫生组织《烟草控制框架公约》（WHO FCTC）第三次缔约方会议回顾了 FCTC/COP1(15)和FCTC/COP2(14)号决议关于制订 WHO FCTC 第 9 条（烟草制品成分管制）和第10条（烟草制品披露管制）实施指南的要求，并提到第三次缔约方会议工作报告中有关其工作进展的信息，提请公约秘书处授权 WHO 无烟草行动组在 5 年内对检测卷烟成分和释放物的分析化学方法进行验证（FCTC/COP/3/REC/1）。

根据2006年10月在加拿大渥太华举行的第三次会议上确定的优先次序标准，WHO FCTC 第 9 条和第 10 条工作组确定优先验证以下成分的分析化学检测方法：

- 烟碱；
- 氨；
- 保润剂[1,2-丙二醇、甘油（1,2,3-丙三醇）和三甘醇（2,2′-二乙醚乙二醇）]。

测量这些成分需要验证三个方法：一个适用于烟碱，一个适用于氨，一个适用于保润剂。

根据上述渥太华会议确定的优先次序标准，工作组确定了在卷烟主流烟气释放中应该验证测试和测量方法（分析化学）的下列优先级物质清单：

- 4-(N-甲基亚硝胺基)-1,3-吡啶基-1-丁酮（NNK）；
- N-亚硝基去甲烟碱（NNN）；
- 乙醛；
- 丙烯醛；
- 苯；
- 苯并[a]芘；
- 1,3-丁二烯；
- 一氧化碳；
- 甲醛。

这些释放物使用下述两种抽吸方案测量，按五个方法进行验证：一个适用于烟草特有的亚硝胺（NNK 和 NNN），一个适用于苯并[a]芘，一个适用于醛类（乙醛、丙烯醛和甲醛），一个适用于挥发性有机物（苯和1,3-丁二烯），一个适用于一氧化碳。

下表列出了用于验证上述测试方法的两种抽吸方案。

抽吸方案	抽吸容量（mL）	抽吸间隔（s）	滤嘴通风孔
ISO 方案：ISO 3308 常规分析用吸烟机 定义和标准条件	35	60	不封闭
深度抽吸方案：在 ISO 3308 基础上有所调整	55	30	所有通风孔必须如 SOP 01 **12.2** 所述 100%封闭

SOP 10 描述深度抽吸方案下卷烟主流烟气中烟碱和一氧化碳的测定标准操作规程。

1 适用范围

本方法适用于定量测定卷烟主流烟气中烟碱和一氧化碳。分别采用气相色谱（GC）法和非散射红外光谱（NDIR）法对烟碱和一氧化碳进行分析。

请注意，要成功操作吸烟机或其他分析设备，须对操作人员进行培训。对操作吸烟机或其他分析设备没有经验的人，在检测烟草制品成分和释放物前，需要接受专门的培训。

2 参考标准

2.1 ISO 3308：常规分析用用吸烟机 定义和标准条件。

2.2 ISO 4387：卷烟 常规分析用吸烟机测定总粒相物和去烟碱干粒相物。

2.3 ISO 3402：烟草和烟草制品 用于调节和测试的大气环境。

2.4 ISO 10315：卷烟 气相色谱法测定烟气冷凝物中的烟碱。

2.5 ISO 8454：卷烟 NDIR 法测定卷烟烟气气相中的一氧化碳。

2.6 ISO 8243：卷烟 抽样。

2.7 ISO 5725-1：检测方法和结果的准确度（正确度和精密度） 第 1 部分 一般原则和定义。

2.8 ISO 5725-2：检测方法和结果的准确度（正确度和精密度） 第 2 部分 确定标准检测方法重复性和再现性的基本方法。

2.9 WHO TobLabNet SOP 01：卷烟深度抽吸标准操作规程。

2.10 WHO TobLabNet SOP 02：烟草制品成分和释放物分析方法验证标准操作规程。

2.11 联合国毒品和犯罪问题办公室（UNODC）《代表性药物取样指南》（http://www.unodc.org/documents/scientific/Drug_Sampling.pdf）。

2.12 WHO TobLabNet 官方方法：深度抽吸方案下卷烟主流烟气中的焦油、烟碱和一氧化碳的测定。

3 术语和定义

3.1 TPM：总粒相物。

3.2 烟气捕集器：为测定特定烟气成分从卷烟样品中收集部分烟气的装置。

3.3 气相：根据 ISO 4387，使用符合 ISO 3308 标准的吸烟机在吸烟过程中穿过粒相捕集器的部分烟气。

3.4 清除抽吸：卷烟熄灭或从捕集器中取出后进行的任何抽吸。

3.5 烟草制品：完全或部分以烟叶为原料制造，用于抽吸、吸吮、咀嚼或吸入的产品（WHO FCTC 第 1(f)条）。

3.6 深度抽吸方案：抽吸过程参数为抽吸容量 55 mL，抽吸间隔 30 s，每次抽吸持续时间 2 s，100%封闭滤嘴通风孔。

3.7 ISO 方案：抽吸过程参数为抽吸容量 35 mL，抽吸间隔 60 s，每次抽吸持续时间 2 s，不封闭滤嘴通风孔。

3.8 实验室样品：用于进行实验室检测的样品，由一次性或在规定时间内送到实验室的单一类型产品组成。

3.9 试样：从实验室样品中随机抽取的待测样品。所取样品数量应能代表实验室样品。

3.10 试料：从试样中随机抽取的用于单次测试的样品。所取样品数量应能代表试样。

4 方法概述

4.1 所有样品按照 ISO 标准规程进行处理和标记。

4.2 100%封闭通风孔。

4.3 除抽吸容量和抽吸频率外，按照 ISO 标准规程抽吸卷烟。

4.4 收集卷烟烟气的气相，并使用非散射红外分析仪（NDIR）检测一氧化碳。

4.5 将卷烟烟气的粒相物收集在捕集器上用于进一步分析。

4.6 从捕集器中萃取烟碱后，使用气相色谱分析烟碱含量。

5 安全和环境预防措施

5.1 遵守所有化学实验活动的常规安全与环境预防措施。

5.2 使用本检测方法对特定产品进行检测和评估时所用材料或设备可能对环境有害。本检测方法无力解决所有与使用有关的安全问题。所有使用此方法的人有责任咨询相关机构并遵循现有监管要求，制定健康与安全规程以及环境预防措施。

5.3 应特别注意避免吸入或皮肤接触危险化学品。在制备或处理未稀释材料、标准溶液、萃取液和收集样品时，应在通风橱中进行，并穿戴合适的实验服、手套和护目镜。

6 仪器和设备

常用实验仪器，特别是：

6.1 按照 ISO 3402 **[2.3]**规定调节卷烟所需设备。

6.2 按照 ISO 4387 **[2.2]**规定标记烟蒂长度所需设备。

6.3 按照 WHO TobLabNet SOP 01 **[2.10]**规定采用深度抽吸方案时封闭通风孔所需设备。

6.4 按照 ISO 3308 **[2.1]**规定抽吸烟草制品所需设备。

6.5 按照 ISO 8454 **[2.5]**规定收集气相所需设备。

6.6 按照 ISO 8454 **[2.5]**规定检测一氧化碳所需设备。

6.7 剑桥滤片：ISO 4387 **[2.2]**中描述的玻璃纤维滤片（直线型吸烟机：滤片直径 44 mm；转盘型吸烟机：滤片直径 92 mm）。

6.8 萃取瓶：250 mL 锥形瓶、带塞子的 100 mL 三角瓶或其他合适的烧瓶。

6.9 腕式或环形振荡器，将萃取瓶保持在适当位置。

6.10 配备火焰离子化检测器（FID）的气相色谱，用于烟碱测定。

6.11 能够明显分离烟气释放物中溶剂、内标物和烟碱峰的气相色谱柱，如 Varian WCOT 熔融石英柱（涂层 CP-WAX 51，25 m×0.25 mm×0.20 μm；DB-Wax，30 m×0.53 mm×1 μm）。

6.12 非散射红外分析仪（NDIR），用于测量 ISO 8454 规定的气相中的一氧化碳。

6.13 测量精度为 0.1 kPa±0.1 kPa 的气压计。

6.14 精确精度为 0.1℃± 0.1℃的温度计。

7 试剂

除非另有说明，所有试剂应至少为分析纯。试剂尽可能以 CAS 号来识别。

7.1 载气：高纯度（＞99.999%）的氦气[7440-59-7]或氮气（CAS:7727-37-9）。

7.2 辅助气体：用于火焰离子化检测器的高纯度（＞99.999%）的空气和氢气[1333-74-0]。

7.3 最大含水量为 1.0 g/L 的异丙醇[67-63-0]。

7.4 已知纯度不低于 98%的烟碱[54-11-5]，也可以使用已知纯度不低于 98%

的水杨酸烟碱盐[29790-52-1]。如有必要，实验室可以验证烟碱纯度。

7.5 用于 GC 烟碱分析的内标物：

纯度不低于 99%的正十七烷[629-78-7]或喹哪啶 [91-63-4]。

纯度不低于 98%的二十烷[112-95-8]、异喹啉[119-65-3]、喹啉[91-22-5]或其他合适的替代品。

7.6 标准一氧化碳（CO）气体混合物：

至少三种已知浓度的标准气体混合物，相对误差为 2%，覆盖到预计标准测试范围以避免标准曲线的外推。

一氧化碳在氮气中的体积分数通常为 1%、3%和 5%。对于低水平的一氧化碳，建议延长标准曲线，例如 0.25%。

8 玻璃器皿准备

清洁、干燥的玻璃器皿，避免残留物污染。

9 溶液配制

9.1 萃取液。

含有适当浓度内标物的异丙醇[7.3]。

准确称取 0.10 g 正十七烷[7.5]至 1 L 容量瓶中。

混合均匀并将溶液转移到具有防止污染的储存容器中。

注意：可以调整内标物的浓度或类型，同时考虑内标物对灵敏度和选择性的可能影响以及方法的线性范围。

10 标准溶液配制

10.1 烟碱测定。

10.1.1 10 mg/mL 的烟碱储备液：

称取约 1000 mg 烟碱（精确至 0.1 mg）置于 100 mL 容量瓶中，用萃取液稀释至刻度。

混合均匀并在 0～4℃条件下避光储存。

在低温下储存的溶剂和溶液在使用前应调至室温（22℃±2℃）。

10.1.2 烟碱标准工作溶液：

如表 1 所示，分别移取不同体积的烟碱储备液[**10.1.1**]至 100 mL 容量瓶中，并用萃取液稀释至刻度。

在 0～4℃条件下避光储存。在低温下储存的溶剂和溶液在使用前应调至室温（22℃±2℃）。

注意：除烟碱之外，可以同时在 GC 上进行水分的测定，可以将烟碱和水的工作标准溶液配成混合标准溶液。

10.1.3 烟碱标准工作溶液的最终浓度（mg/L）由下式确定：

$$最终浓度（mg/L）=\frac{x \times y}{10} \times 标准品纯度$$

式中：x——称取的烟碱[**10.1.1**]的原始质量（mg）；

y——移取的烟碱储备液[**10.1.2**]体积。

标准工作溶液中烟碱浓度如表 1 所示。

表 1 标准工作溶液中烟碱的浓度

标准	烟碱储备液体积（10 mg/mL）	内标物体积（mL）	总体积（mL）	标准工作溶液中的烟碱最终浓度（mg/L）
1	0.2	不适用，包括萃取液	100	20
2	2		100	200
3	4		100	400
4	6		100	600
5	8		100	800

标准溶液范围可根据使用的设备和待测样品进行调整，注意方法灵敏度可能对结果产生的影响。

11 抽样

11.1 根据 ISO 8243 [**2.4**]或替代方法，根据各实验室惯例或者要求的特殊规定或样品可用性，选取具有代表性的实验室卷烟样品。

11.2 试样的组成。

11.2.1 若可能，将实验室样品分成独立的单元（如包、条）。

11.2.2 从至少 $\sqrt{n}$ 个[**2.11**]单元中取出每种试样的等量产品。

12 卷烟准备

12.1 按照 ISO 3402 [**2.3**]调制所有卷烟。

12.2 按照 ISO 4387 [**2.2**]和 WHO TobLabNet SOP 01 [**2.9**]标记烟蒂的长度。

12.3 按照 WHO TobLabNet SOP 01）[**2.9**]的深度抽吸方案准备待测样品。

13 吸烟机设置

13.1 环境条件。

按 ISO 3308 [**2.1**]要求设定抽吸的环境条件。

13.2 吸烟机要求。

除深度抽吸方案按 WHO TobLabNet SOP 01 [**2.9**]进行准备外，其余按照 ISO 3308 [**2.1**]的要求设置吸烟机。

14 样品生成

在指定吸烟机上抽吸足够量的样品，样品含量应不足以使滤片发生穿透，且烟碱的浓度落在分析的标准曲线范围内。

14.1 抽吸卷烟试样并按照 ISO 4387 [**2.2**]或 WHO TobLabNet SOP 01 [**2.9**]收集 TPM。

14.2 按照 ISO 8554 [**2.5**]，在合适的收集系统[**6.5**]中收集气相。

14.3 至少包含一个用于质量控制的参比试样。

14.4 当样品类型为第一次检测时，评估滤片穿透的可能性，调整卷烟的数量以避免穿透[**6.7**]。对于 92 mm 滤片和 44 mm 滤片，穿透分别发生在总粒相物水平超过 600 mg 和 150 mg 时。一旦发生穿透，必须减少每个滤片上的卷烟数量。

14.5 表 2 显示了 ISO 和深度抽吸方案中直线型和转盘型吸烟机每次抽吸的数量。

表 2 在直线型和转盘型吸烟机进行一次测量的卷烟数量

	ISO 方案		深度抽吸方案	
	直线型	转盘型	直线型	转盘型
每个烟气捕集器的卷烟数量	5	20	3	10
每个结果包含的烟气捕集器数量	4	1	7	2

14.6 抽吸 1 根卷烟后，取下烟蒂并对每个烟气捕集器进行一次空吸。测量完成所需的样品量后，对每个烟气捕集器进行五次空吸，然后从吸烟机上取下烟气捕集器。

14.7 记录每个烟气捕集器的卷烟数量和总抽吸量，包含空吸。

14.8 每天至少完成一次空白测试，方法是在吸烟机区域放置一个剑桥滤片[**6.7**]。

15 样品制备

15.1 收集 TPM 后，将剑桥滤片（CF）[**6.7**]（直线型吸烟机：44 mm 滤片收集 3 支卷烟的粒相物；转盘吸烟机：92 mm 滤片收集 10 支卷烟的粒相物）从捕集器上转移至一个干净、干燥的锥形瓶（44 mm 滤片使用 100 mL 的锥形瓶；92 mm 滤片使用 250 mL 的锥形瓶）[**6.8**]中。

15.2 用 1/4 个干净的 44 mm 剑桥滤片擦拭捕集器的前半部分，并使用干净的 44 mm CF 擦拭转盘烟机两次，并将它们放入相应的烧瓶中。

15.3 在每个烧瓶中加入 20 mL（44 mm 滤片）或者 50 mL（92 mm 滤片）的萃取溶剂[**9.1**]，确保滤片完全被浸没。

15.4 在振荡器[**6.9**]上以约 200 r/min 的速率振荡约 30 min，调整时间或速率以防止滤片解体。

16 样品分析

16.1 TPM 质量。
ISO 4387 [**2.2**]描述了测量 TPM 质量的方法。
16.2 烟碱含量。
ISO 10315 [**2.4**]描述了测量卷烟主流烟气中烟碱的方法。
合适的 GC-FID 操作条件示例：
进样口温度：250℃；
检测器温度：250℃；
载气：流速为 25 mL/min 的氦气或氮气；
进样量：1 μL 或者 2 μL；
柱温：170℃（等温）。
注意：根据仪器和色谱柱条件以及色谱峰的分离度调整操作参数。
16.3 一氧化碳（CO）含量。
ISO 8454 [**2.5**]描述了测定烟气气相中一氧化碳的方法。

17 数据分析和计算

17.1 烟碱含量。
    17.1.1 对于每个样品萃取物，计算烟碱响应与内标物响应的峰面积比。
    17.1.2 利用线性回归系数计算每个样品提取物中的烟碱浓度：

$$m_t = \frac{(Y_t - a)}{b}$$

式中：m_t——样品萃取物中烟碱的浓度（mg/L）；
Y_t——烟碱响应与内标物响应的峰面积比；
a——从标准曲线获得的线性回归方程的截距；
b——从标准曲线获得的线性回归方程的斜率。

17.1.3 每个样品萃取物的烟碱含量由以下公式计算：

$$烟碱含量（mg/支）= \frac{m_t \times V}{1000 \times N}$$

式中：m_t——样品萃取物中烟碱的浓度（mg/L）；
N——每个烟气捕集器所抽吸的卷烟数（直线型吸烟机：3 支；转盘型吸烟机：10 支）；
V——萃取液的体积（直线型吸烟机：20 mL；转盘型吸烟机：50 mL）。

17.2 一氧化碳（CO）含量。

17.2.1 每支卷烟的平均一氧化碳含量由下式计算得出：

$$V_{as} = \frac{C \times V \times N \times p \times T}{S \times 100 \times p_0 \times (t + T_0)}$$

式中：V_{as}——每支卷烟一氧化碳平均体积（mL）；
C——观察到的一氧化碳体积分数；
V——抽吸容量（mL）；
N——测量样品中的抽吸口数（包括空吸）；
p——大气压（kPa）；
p_0——标准大气压（kPa）；
S——抽吸卷烟的数量；
T——环境温度（℃）；
T_0——水的三相点的温度（K）。

17.2.2 每支卷烟的平均一氧化碳质量由下式给出：

$$m_{cig} = V_{as} \times \frac{M_{CO}}{V_m}$$

式中：V_{as}——每支卷烟一氧化碳平均体积（mL）；
m_{cig}——每支卷烟一氧化碳平均质量（mg）；
M_{CO}——一氧化碳的摩尔质量（g/mol）；
V_m——理想气体的摩尔体积（L/mol）。

18 特别注意事项

18.1 安装新色谱柱后,在指定的仪器条件下注入烟草样品萃取物进行调节。应重复进样,直至分析物和内标物的峰面积(或峰高)可重现。

18.2 当内标物的峰面积(或峰高)显著高于预期时,建议使用不含内标物的萃取剂提取烟草样品。

这可以确定是否有任何组分与内标物共同洗脱,从而导致烟碱含量偏低。

19 数据报告

19.1 报告每个评估样品的单个测量值。
19.2 按方法规范指定的报告结果。
19.3 更多信息请参阅 TobLabNet SOP 02 [2.10]。

20 质量控制

20.1 控制参数。
如果控制测量值超出预期值的公差极限,则必须进行适当的调查并采取措施。如有必要,需要做另外的实验室质量保证规程,以符合各个实验室的做法。

20.2 实验室试剂空白样。
要在样品制备和分析过程中检测潜在的污染,需包含实验室试剂空白样。空白样由分析测试样品中使用的所有试剂和材料组成,并像测试样品一样进行分析。应根据各个实验室的做法对空白样进行评估。

20.3 质量控制样品。
为了验证整个分析过程的一致性,根据各个实验室的做法分析参比卷烟(或适当的质量控制样品)。

21 方法性能

21.1 报告限。
报告限设置为所用标准曲线的最低浓度,重新计算为 mg/支卷烟(例如,对于直线型吸烟机,最低烟碱标准浓度为 20 mg/L,相当于 0.1 mg/支)。

21.2 分析特异性。
对于气相色谱测定烟碱,目标分析物的保留时间用于验证分析特异性。使用质量控制卷烟该组分的响应与内标组分的响应比范围来验证未知样品结果的特异性。

21.3 线性。

建立的烟碱标准曲线在 20~800 mg/L 的标准浓度范围内是线性的。

21.4 潜在干扰。

丁香酚的存在会引起干扰,因为其保留时间与烟碱的保留时间相似。含有丁香的样品最容易发生这种干扰。实验室可能需要通过调整分析仪器参数来解决这个问题。

22 重复性和再现性

2007 年进行的一项国际合作研究涉及 14 个实验室和 3 种参比卷烟(1R5F、3R4F 和 CM6)以及 2 种商品卷烟,根据 ISO 标准使用深度抽吸方案对卷烟主流烟气中 TNCO 的方法[**2.12**]进行了验证。表 3 和表 4 列出了该方法的精密度限值。

常规正确操作该方法时,同一操作者使用同一设备在最短的操作时间内两个单一结果之间的差异超过重复性 r 的情况,平均 20 个样品不超过 1 次。

常规正确操作该方法时,两个实验室报告的匹配卷烟样品的单一结果超过再现性 R 的情况,平均 20 次不超过 1 次。

根据 ISO 5725-1 [**2.7**]和 ISO 5725-2 [**2.8**]对测试结果进行统计分析,得到表 3 和表 4 所示的准确数据。

表 3 卷烟中一氧化碳(CO)的精密度限值(mg/支)

参比卷烟	N	m_{cig}	r_{limit}	R_{limit}
1R5F	13	27.39	2.53	4.62
3R4F	13	32.19	2.24	2.97
CM6	13	26.87	1.40	2.91

表 4 测定卷烟中烟碱的精密度限值(mg/支)

参比卷烟	N	m_{cig}	r_{limit}	R_{limit}
1R5F	13	1.02	0.08	0.17
3R4F	13	1.91	0.14	0.31
CM6	13	2.73	0.20	0.42

附录1 GC法分析烟碱的典型色谱图

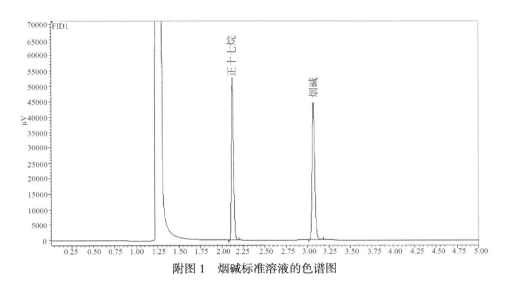

附图1 烟碱标准溶液的色谱图

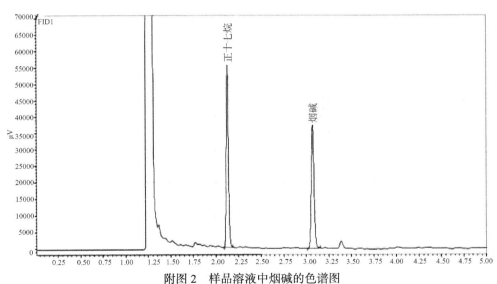

附图2 样品溶液中烟碱的色谱图

WHO TobLabNet
Official Method
SOP 01

Standard Operating Procedure for Intense Smoking of Cigarettes

Method:	Intense smoking of cigarettes
Analytes:	Not applicable
Matrix:	Cigarettes
Last update:	April 2012

No machine smoking regimen can represent all human smoking behaviour: machine smoking testing is useful for characterizing cigarette emissions for design and regulatory purposes, but communication of machine measurements to smokers can result in misunderstanding about differences between brands in exposure and risk. Data on smoke emissions from machine measurements may be used as inputs for product hazard assessment, but they are not intended to be nor are they valid as measures of human exposure or risks. Representing differences in machine measurements as differences in exposure or risk is a misuse of testing with WHO TobLabNet standards.

FOREWORD

This document was prepared by members of the World Health Organization (WHO) Tobacco Laboratory Network (TobLabNet) as a standard operating procedure (SOP) for intense smoking of cigarettes.

INTRODUCTION

In order to establish comparable measurements for testing tobacco products globally, consensus methods are required for measuring specific contents and emissions of cigarettes. The Conference of the Parties to the WHO Framework Convention on Tobacco Control (FCTC) at its third session in Durban, South Africa, in November 2008, recalling its decisions FCTC/COP1(15) and FCTC/COP2(14) on the elaboration of guidelines for implementation of Articles 9 (*Regulation of the contents of tobacco products*) and 10 (*Regulation of tobacco product disclosures*) of the WHO FCTC, noting the information contained in the report of the working group to the third session of the Conference of the Parties on the progress of its work … requested the Convention Secretariat to invite WHO's Tobacco Free Initiative to … validate, within five years, the analytical chemical methods for testing and measuring cigarette contents and emissions (FCTC/COP/3/REC/1).

Using the criteria for prioritization set at its third meeting in Ottawa, Canada, in October 2006, the working group on Articles 9 and 10 identified the following contents for which methods for testing and measurement (analytical chemistry) should be validated as a priority:

- nicotine
- ammonia
- humectants (propane-1,2-diol, glycerol (propane-1,2,3-triol) and triethylene glycol (2,2-ethylenedioxybis(ethanol)).

Measurement of these contents will require validation of three methods: one for nicotine, one for ammonia and one for humectants.

Using the criteria for prioritization set at the meeting in Ottawa mentioned above, the working group identified the following emissions in mainstream smoke for which methods for testing and measurement (analytical chemistry) should be validated as a priority:

- 4-(methylnitrosamino)-1-(3-pyridyl)-1-butanone (NNK)
- *N*-nitrosonornicotine (NNN)
- acetaldehyde
- acrylaldehyde (acrolein)
- benzene
- benzo[a]pyrene
- 1,3-butadiene
- carbon monoxide
- formaldehyde

Measurement of these emissions with the two smoking regimens described below will require validation of five methods: one for tobacco-specific nitrosamines (NNK and NNN), one for benzo[a]pyrene, one for aldehydes (acetaldehyde, acrolein and formaldehyde), one for volatile organic compounds (benzene and 1,3-butadiene), and one for carbon monoxide.

The table below sets out the two smoking regimens for validation of the test methods referred to above.

Smoking regimen	Puff volume (ml)	Puff frequency	Filter ventilation holes
ISO regimen: ISO 3308; *Routine analytical cigarette smoking machine—definitions and standard conditions*	35	Once every 60 s	No modifications
Intense regimen: Same as ISO 3308, but modified as indicated	55	Once every 30 s	All ventilation holes must be blocked 100% as described in **12.2**.

This SOP was prepared to describe the procedure for intense smoking of cigarettes.

1 SCOPE

This SOP describes the overall procedures for machine smoking of cigarettes under intense conditions.

Note: Training in use of the smoking machine and other analytical equipment is important for successful operation. People not experienced in operating smoking machines or in using the analytical methods for measuring tobacco product emissions and contents should be trained.

2 REFERENCES

2.1 ISO 3308: *Routine analytical cigarette-smoking machine—Definitions and standard conditions.*

2.2 ISO 4387: *Cigarettes—Determination of total and nicotine-free dry particulate matter using a routine analytical smoking machine.*

2.3 ISO 3402: *Tobacco and tobacco products—Atmosphere for conditioning and testing.*

3 TERMS AND DEFINITIONS

3.1 TPM: Total particulate matter

3.2 ISO regimen: Parameters used to smoke tobacco products that include a 35-ml puff volume, a 60-s puff interval, 2-s puff duration and no blocking of the filter ventilation holes

3.3 Intense Regimen – Parameters used to smoke tobacco products which include 55-ml puff volume, 30-s puff interval, 2-s puff duration and 100% blocking of the filter ventilation holes.

3.4 *Tobacco products*: Products entirely or partly made of leaf tobacco as the raw material that are manufactured to be used for smoking, sucking, chewing or snuffing (Article 1(f) of the WHO FCTC)

3.5 *Laboratory sample*: Sample intended for testing in a laboratory, consisting of a single type of product delivered to the laboratory at one time or within a specified period

3.6 *Test sample*: Product to be tested, taken at random from the laboratory sample. The number of products taken shall be representative of the laboratory sample.

3.7 *Test portion*: Random sample from the test sample to be used for a single determination. The number of products taken shall be representative of the test sample.

4   METHOD SUMMARY

4.1 All samples are conditioned and marked according to ISO standard procedures.

4.2 Ventilation holes are blocked 100%.

4.3 Cigarettes are smoked according to ISO standard procedures with the exception of puff volume and puff frequency.

5   SAFETY AND ENVIRONMENTAL PRECAUTIONS

5.1 Follow routine safety and environmental precautions, as in any chemical laboratory activity.

5.2 The testing and evaluation of certain products with this test method may require the use of materials or equipment that could be hazardous or harmful to the environment. This document does not purport to address all the safety aspects associated with its use. All persons using this method have the responsibility to consult the appropriate authorities and to establish health and safety practices as well as environmental precautions in conjunction with any existing, applicable regulatory requirements prior to its use.

5.3 Special care should be taken to avoid inhalation or dermal exposure to harmful chemicals. Use a chemical fume hood, and wear an appropriate laboratory coat, gloves and safety goggles when preparing or handling undiluted materials, standard solutions, extraction solutions or collected samples.

6   APPARATUS AND EQUIPMENT

Usual laboratory apparatus, in particular:

6.1 Equipment needed to condition cigarettes as specified in ISO 3402

6.2 Equipment needed to mark butt length as specified in ISO 4387

6.3 Equipment needed to cover ventilation holes for the intense regimen as specified in section 12.2

6.4 Equipment needed to perform smoking of tobacco products as specified in ISO 3308

6.5 Cellophane tape, 20 mm (¾") wide, such as Scotch® tape, (3M, Maplewood, Minnesota, USA)

6.6 Cigarette holders for blocking 100% ventilation holes

7   REAGENTS AND SUPPLIES

All reagents shall be of at least analytical reagent grade unless otherwise noted. When possible, reagents are identified by their Chemical Abstracts Service (CAS) registry numbers.

8 PREPARATION OF GLASSWARE

Clean and dry glassware in a manner to ensure no contamination from residues.

9 PREPARATION OF SOLUTIONS

Not applicable

10 PREPARATION OF STANDARDS

Not applicable

11 SAMPLING

Sampling should be done as described in the specific method SOP.

12 CIGARETTE PREPARATION

12.1 Mark cigarettes at a butt length in accordance with ISO 4387.

12.2 Block all ventilation holes, as specified below.

- **12.2.1** For the intense regimen, block filter ventilation holes completely by applying 20 mm (¾") wide cellophane tape [**6.5**] around the entire circumference of the cigarette.
- **12.2.2** Measure out a length of cellophane tape of 50–55 mm.
- **12.2.3** Attach the cut end of the tape parallel to the long axis of the cigarette with the side of the tape within 1 mm of the mouth end of the filter (see Figure 1).
- **12.2.4** Carefully wrap the tape around the filter to ensure complete bonding to the paper with no wrinkles or air holes. If wrinkles or air holes appear, reject the sample and do not include it in the analysis.
- **12.2.5** The tape should circle the cigarette twice, with a small overlap of less than 5 mm (see Figure 2).
- **12.2.6** The tape should not extend beyond the mouth end of the filter.

As an alternative to tape, special holders [6.6] for blocking 100% of ventilation holes can be used.

12.3 Condition all cigarettes to be smoked in accordance with ISO 3402.

13 PREPARATION OF THE SMOKING MACHINE

13.1 Ambient conditions

The ambient conditions for smoking are specified in ISO 3308.

Figure 1

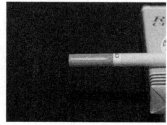

Figure 2

13.2 Machine specifications

Follow ISO 3308 machine specifications, except the following:

13.2.1 For the intense regimen, set the smoking machine to draw a puff volume of 55 ± 0.1 ml.

13.2.2 For the intense regimen, set the smoking machine to take puffs at a frequency of 30 s.

13.2.3 Programme each smoking run to end when the cigarette has burnt down to the mark placed earlier [**12.1**].

14 SAMPLE GENERATION

Smoke a sufficient amount of product on the specified smoking machine such that breakthrough does not occur.

14.1 Smoke the test samples as specified in ISO 4387, and collect the analyte of interest as described in the specific SOP.

14.2 Include at least one reference test sample for quality control.

14.3 When testing sample types for the first time, evaluate breakthrough. The number of cigarettes might have to be adjusted to prevent breakthrough. During determination of tar, nicotine and carbon monoxide, breakthrough of cigarette smoke occurs at TPM levels exceeding 600 mg for a 92-mm filter pad or 150 mg for a 44-mm filter pad. If breakthrough occurs, the number of cigarettes smoked onto each pad must be decreased. The breakthrough of the filter pads or other collection devices might differ, however, depending on the analyte of interest.

15 SAMPLE PREPARATION

Not applicable.

16 SAMPLE ANALYSIS

Not applicable.

17 DATA ANALYSIS AND CALCULATIONS

Not applicable.

18 SPECIAL PRECAUTIONS

None.

19 DATA REPORTING

Data will be reported as described in the specific SOP.

20 QUALITY CONTROL

Quality control will be performed as described in the specific SOP.

21 METHOD PERFORMANCE SPECIFICATIONS

Not applicable.

22 REPEATABILITY AND REPRODUCIBILITY

Not applicable.

23 BIBLIOGRAPHY

23.1 *ISO 10185: Tobacco and tobacco products—Vocabulary.*

**WHO TobLabNet
Official Method
SOP 02**

Standard Operating Procedure for Validation of Analytical Methods of Tobacco Product Contents and Emissions

Method:	Standard operating procedures for validation of analytical methods of tobacco contents and emissions
Analytes:	Not applicable
Matrix:	Tobacco cigarette mainstream smoke and tobacco filler contents
Last update:	November 2016

> No machine smoking regimen can represent all human smoking behaviour: machine smoking testing is useful for characterizing cigarette emissions for design and regulatory purposes, but communication of machine measurements to smokers can result in misunderstanding about differences between brands in exposure and risk. Data on smoke emissions from machine measurements may be used as inputs for product hazard assessment, but they are not intended to be nor are they valid as measures of human exposure or risks. Representing differences in machine measurements as differences in exposure or risk is a misuse of testing with WHO TobLabNet standards.

FOREWORD

This document was prepared by members of the World Health Organization (WHO) Tobacco Laboratory Network (TobLabNet) as a standard operating procedure (SOP) for the validation of analytical methods of tobacco cigarette mainstream smoke contents and cigarette tobacco filler contents.

INTRODUCTION

In order to establish comparable measurements for testing tobacco products globally, consensus methods are required for measuring specific contents and emissions of cigarettes. The Conference of the Parties (COP) to the WHO Framework Convention on Tobacco Control (WHO FCTC) at its third session in Durban, South Africa, in November 2008, recalling decisions FCTC/COP1(15) and FCTC/COP2(14) on the elaboration of guidelines for implementation of Articles 9 (*Regulation of the contents of tobacco products*) and 10 (*Regulation of tobacco product disclosures*) of the WHO FCTC, noting the information contained in the report of the working group to the third session of the Conference of the Parties on the progress of its work ... requested the Convention Secretariat to invite WHO's Tobacco Free Initiative to ... validate, within five years, the analytical chemical methods for testing and measuring cigarette contents and emissions (FCTC/COP/3/REC/1).

Using the criteria for prioritization set at its third meeting in Ottawa, Canada, in October 2006, the working group on Articles 9 and 10 identified the following contents for which methods for testing and measurement (analytical chemistry) should be validated as a priority:

- nicotine
- ammonia
- humectants (propane-1,2-diol, glycerol (propane-1,2,3-triol) and triethylene glycol (2,2-ethylenedioxybis(ethanol)).

Measurement of these contents will require validation of three methods: one for nicotine, one for ammonia and one for humectants.

Using the criteria for prioritization set at the meeting in Ottawa mentioned above, the working group identified the following emissions in mainstream smoke for which methods for testing and measurement (analytical chemistry) should be validated as a priority:

- 4-(methylnitrosamino)-1-(3-pyridyl)-1-butanone (NNK)
- *N*-nitrosonornicotine (NNN)

- acetaldehyde
- acrylaldehyde (acrolein)
- benzene
- benzo[a]pyrene
- 1,3-butadiene
- carbon monoxide
- formaldehyde

Measurement of these emissions with the two smoking regimens described below will require validation of five methods: one for tobacco-specific nitrosamines (NNK and NNN), one for benzo[a]pyrene, one for aldehydes (acetaldehyde, acrolein and formaldehyde), one for volatile organic compounds (benzene, 1,3-butadiene) and one for carbon monoxide.

The table below sets out the two smoking regimens for validation of the test methods referred to above.

Smoking regimen	Puff volume (mL)	Puff frequency	Filter ventilation holes
ISO regimen: ISO 3308; Routine analytical cigarette smoking machine — definitions and standard conditions	35	Once every 60 s	No modifications
Intense regimen: Same as ISO 3308, but modified as indicated	55	Once every 30 s	All ventilation holes must be blocked 100% as described in WHO TobLabNet SOP 01.

This method SOP was prepared to describe the procedure for the determination of nicotine and carbon monoxide in mainstream cigarette smoke under intense smoking conditions.

1. SCOPE

This document describes the standard procedures for validating analytical methods used for analysing tobacco cigarette mainstream smoke contents and cigarette tobacco filler contents.

2. REFERENCES

2.1 *ISO 5725-2: Accuracy (trueness and precision) or measurement methods and results – Part 2: Basic method for the determination of repeatability and reproducibility of a standard measurement method*

3. TERMS AND DEFINITIONS

3.1 *Tobacco products*: Products entirely or partly made of leaf tobacco as the raw material that are manufactured to be used for smoking, sucking, chewing or snuffing (Article 1(f) of the WHO FCTC).

3.2 *Cigarette tobacco filler*: Tobacco-containing part of a cigarette, including reconstituted tobacco, stems, expanded tobacco and additives.

3.3 *ISO Regimen:* Parameters used to smoke tobacco products which include 35-mL puff volume, 60-second puff interval, 2-second puff duration and no blocking of the filter ventilation holes.

3.4 *Laboratory sample*: Sample intended for testing in a laboratory, consisting of a single type of product delivered to the laboratory at one time or within a specified period

3.5 *Test sample*: Product to be tested, taken at random from the laboratory sample. The number of products taken shall be representative of the laboratory sample.

3.6 *Test portion*: Random portion from the test sample to be used for a single determination. The number of products taken shall be representative of the test sample.

3.7 *Repeatability:* All measurements carried out on each individual test sample shall be carried out under repeatability conditions, i.e. within a short interval, using the same instrument and by the same operator. Measurements of different test samples may be carried out on different days, if needed.

3.8 *Reproducibility:* All measurements are carried out under reproducibility conditions, i.e. allowing for differences between days, operators, laboratories, instruments, reagents and environmental conditions. Measurements are carried out on the same or identical materials using the same methodology.

4. METHOD SUMMARY

4.1 For each individual analytical method to be validated, a leading laboratory will be appointed by WHO's TobLabNet.

4.2 A Standard Operating Procedure and validation study protocol are written for each analytical method to be validated and sent to all participating laboratories.

4.3 For the purposes of validation of analytical methods, products are chosen whenever possible to cover the range of design properties to be found worldwide, while using specific product varieties that are intended to be consistent from pack to pack. In general, products to be analysed include three reference cigarettes and two commercially available brand varieties. The specific commercial brand varieties change with the specific test(s) to be performed.

4.4 Testing of analytical methods for validation purposes consists of an initial evaluation of a single cigarette variety, followed by a full evaluation of four additional varieties. After confirmation of the initial evaluation results, the other samples are sent for analysis. The leading laboratory has the discretion to include, or not to include, an initial evaluation.

4.5 Data analysis and quality control are carried out according to the Standard Operating Procedure to be validated, or the validation study protocol, taking into account procedures as defined at each laboratory.

4.6 Individual sample results are reported to WHO's Tobacco Free Initiative. After assigning a code to each laboratory and removing all information that can lead to identification of a laboratory, the results are sent to at least two entities. These entities will evaluate the data for outliers and calculate repeatability and reproducibility independently from each other, according to ISO procedures. After completing the evaluation of the data, the results from the entities will be compared.

4.7 A report of the validation study will be written and the results will be included in the analytical Standard Operating Procedure.

5. SAFETY AND ENVIRONMENTAL PRECAUTIONS

5.1 Follow routine safety and environmental precautions, as in any chemical laboratory activity, taking into account the specific safety precautions as described in the analytical SOP to be validated.

5.2 The testing and evaluation of certain products with the test method to be validated may require the use of materials or equipment that could be hazardous or harmful to the environment and this document does not purport to address all the safety aspects associated with its use. All persons using this method, or the analytical method to be validated, have a responsibility to consult with the appropriate authorities and to establish health and safety practices, as well as environmental precautions in conjunction with any existing applicable regulatory requirements prior to its use.

5.3 Special care should be taken to avoid inhalation or dermal exposure to harmful chemicals. Use a chemical fume hood, and wear an appropriate laboratory coat, gloves and safety goggles when preparing or handling undiluted materials, standard solutions, extraction solutions or collected samples.

6. LEADING LABORATORY

For each analytical method to be validated, WHO's TobLabNet will appoint a laboratory from its members to act as leading laboratory for that specific validation study.

The leading laboratory will:

- Develop or select the appropriate analytical procedure, write the SOP and validation protocol
- Lead the validation study and track progress
- Provide support to participating laboratories if needed
- Generate a validation study report

7. STANDARD OPERATING PROCEDURE

A detailed description of all operations, including reagents, supplies and any equipment needed are documented in the SOP. The SOP also contains any special precautions that are required and a summary of method performance specifications, including repeatability and reproducibility.
The lay-out of the SOP will be based on a standardized template – see Annex 1.

Training in the use of the smoking machine and other analytical equipment is important for successful performance of the analytical SOP. For those not experienced in operating smoking machines or analysis using the analytical method(s) for measuring tobacco product emissions and contents, training is available through the WHO TobLabNet.

8. VALIDATION STUDY PROTOCOL

A study protocol is sent to all participants of the validation study of an analytical method. The study protocol should always contain details of the following:

- Timeline of the validation study
- Number and type of samples to be analysed
- Number of replicates per sample
- Instruction on how to report the results, preferably using an Excel template
- Detailed schedule of smoke plans (if applicable)

Examples of study protocols are given in Annex 2 (cigarette tobacco filler contents) and Annex 3 (mainstream cigarette smoke contents).

9. SAMPLES

9.1 Description

9.1.1 Cigarettes are sampled randomly from the products available, mixed, and repackaged before shipping for analysis.

9.1.2 For the purposes of method validation only, products are checked for physical anomalies. All products having visual physical anomalies are excluded from despatch to participants.

9.2 Shipment and Receipt

9.2.1 TobLabNet will ship cigarettes of each cigarette variety to each participating laboratory. Sufficient samples will be sent to meet the measurement requirements plus additional spare samples.

9.2.2 Additional samples, beyond those required for the testing, are intended to provide extras in case of problems encountered during the testing.

9.2.3 TobLabNet will inform all participating laboratories that samples have been sent.

9.2.4 Upon receipt, each laboratory will notify TobLabNet by e-mail of the receipt of samples and any problems with or damage to the materials.

9.2.5 If problems were encountered during shipment, TobLabNet will send replacement samples as soon as practicable after notification.

9.2.6 After receipt of samples, each laboratory will complete analysis and report the data within the study protocol specified time limit.

9.2.7 Cigarettes should be carefully handled to avoid causing damage.

9.2.8 Cigarettes that are obviously bent, torn, crushed, or missing significant amounts of tobacco should be discarded.

9.3 Storage

Samples can be stored at room temperature for up to three months. Beyond this period, samples are to be stored in the original packaging or in airtight containers in a freezer at −20 °C or lower.

10. DATA REPORTING

10.1 Report results as specified in the SOP or study protocol to WHO's Prevention of Noncommunicable Diseases. For consistency and efficient statistical evaluation of the results, a Microsoft Excel template is recommended for reporting. If Microsoft Excel or compatible spreadsheet software is not available, laboratories can also report data electronically in an ASCII format, or if this is not possible, by hard copy. If available, an Excel file template will be sent to each laboratory by e-mail.

10.2 To maintain confidentiality, a laboratory code will be issued to each laboratory and all information that can lead to identification of a participating laboratory will be deleted by WHO's Tobacco Free Initiative (TFI). The assigned laboratory code should accompany each data record when results are reported.

10.3 Results below the limit of reporting (LOR) should be reported as < LOR. Do not leave data fields blank when results are below the LOR.

10.4 Each data record will contain the information as required in the study protocol, with a minimum of:

10.4.1 Laboratory code

10.4.2 Sample ID

10.4.3 Analyte 1 result

10.4.4 Analyte 2 result

10.4.5 Analyte n result

10.4.6 Specific analytical method

10.4.7 Comments

Do not report data with quality control or quality assurance parameters out of control.

11. DATA ANALYSIS

The coded results are sent to at least two entities by WHO's TFI for statistical evaluation.
The entities will independently evaluate the data according to ISO procedures.

11.1 Data analysis will follow procedures as described in ISO 5725-2 [**2.1**].

11.2 Identification of outliers will be performed as described in ISO 5725-2. Results identified as outliers will not be included in the final statistical determination of method repeatability and reproducibility.

11.3 If there is any inconsistency in data, or if data has been identified as an outlier, the laboratory reporting this data will be requested to check for irregularities via WHO's TFI to maintain confidentiality.

11.4 Repeatability and reproducibility will be calculated independently by the entities according to ISO procedures ISO 5725-1 [**13.2**] and ISO 5725-2 [**13.3**] to give the precision data. After completing the evaluation of the data, the results of the entities will be compared.

11.5 The results of the statistical evaluation will be sent to WHO and the lead laboratory.

12. VALIDATION REPORT

An extended report will be made of the validation study, including:

- Outline of the study
- Overview of participants and equipment used
- Description of data analysis methods used
- Summary of data analysis;
    - anomalous data
    - missing data
    - mean values and standard deviations
    - consistency data (Mandel's h and k statistics)
    - outlier tests (Cochran's and Grubbs' test)
    - exploratory data analysis
    - precision analysis (repeatability/reproducibility variance and precision limits)

The report will be submitted to WHO's TFI for publication.

13. BIBLIOGRAPHY

13.1 ISO Standards – Products by TC (Website for ordering methods: ISO/TC 126).

13.2 *ISO 5725-1. Accuracy (trueness and precision) of measurement methods and results — Part 1: General principles and definitions.*

13.3 *ISO 5725-2. Accuracy (trueness and precision) of measurement methods and results — Part 2: Basic method for the determination of repeatability and reproducibility of a standard measurement method.*

Annex 1. Example / Template Standard Operating Procedure

WHO TobLabNet
Official Method
SOP 0#

Standard operating procedure for method

Determination of NAME CONTENT(S) in mainstream cigarette smoke under ISO and intense smoking conditions / cigarette tobacco filler

Method:	Determination of NAME CONTENT(S) in mainstream cigarette smoke using ISO and intense smoking conditions / cigarette tobacco filler
Analytes:	CONTENT (CAS #) CONTENT (CAS #)
Matrix:	Tobacco cigarette mainstream smoke particulate matter / cigarette tobacco filler
Last update:	Month yyyy

> No machine smoking regimen can represent all human smoking behaviour: machine smoking testing is useful for characterizing cigarette emissions for design and regulatory purposes, but communication of machine measurements to smokers can result in misunderstanding about differences between brands in exposure and risk. Data on smoke emissions from machine measurements may be used as inputs for product hazard assessment, but they are not intended to be nor are they valid as measures of human exposure or risks. Representing differences in machine measurements as differences in exposure or risk is a misuse of testing with WHO TobLabNet standards.

FOREWORD

This document was prepared by members of the World Health Organization (WHO) Tobacco Laboratory Network (TobLabNet) as an analytical method standard operating procedure (SOP) for measuring CONTENTS in mainstream cigarette smoke under ISO and intense smoking conditions / cigarette tobacco filler.

INTRODUCTION

In order to establish comparable measurements for testing tobacco products globally, consensus methods are required for measuring specific contents and emissions of cigarettes. The Conference of the Parties (COP) to the WHO Framework Convention on Tobacco Control (WHO FCTC) at its third session in Durban, South Africa, in November 2008, recalling its decisions FCTC/COP1(15) and FCTC/COP2(14) on the elaboration of guidelines for implementation of Articles 9 *(Regulation of the contents of tobacco products)* and 10 *(Regulation of tobacco product disclosures)* of the WHO FCTC, noting the information contained in the report of the working group to the third session of the Conference of the Parties on the progress of its work ... requested the Convention Secretariat to invite WHO's Tobacco Free Initiative to ... validate, within five years, the analytical chemical methods for testing and measuring cigarette contents and emissions (FCTC/COP/3/REC/1).

Using the criteria for prioritization set at its third meeting in Ottawa, Canada, in October 2006, the working group on Articles 9 and 10 identified the following contents for which methods for testing and measurement (analytical chemistry) should be validated as a priority:

- nicotine
- ammonia
- humectants (propane-1,2-diol, glycerol (propane-1,2,3-triol) and triethylene glycol (2,2-ethylenedioxybis(ethanol)).

Measurement of these contents will require validation of three methods: one for nicotine, one for ammonia and one for humectants.

Using the criteria for prioritization set at the meeting in Ottawa mentioned above, the working group identified the following emissions in mainstream smoke for which methods for testing and measurement (analytical chemistry) should be validated as a priority:

- 4-(methylnitrosamino)-1-(3-pyridyl)-1-butanone (NNK)
- *N*-nitrosonornicotine (NNN)
- acetaldehyde
- acrylaldehyde (acrolein)

- benzene
- benzo[a]pyrene
- 1,3-butadiene
- carbon monoxide
- formaldehyde

Measurement of these emissions with the two smoking regimens described below will require validation of five methods: one for tobacco-specific nitrosamines (NNK and NNN), one for benzo[a]pyrene, one for aldehydes (acetaldehyde, acrolein and formaldehyde), one for volatile organic compounds (benzene, 1,3-butadiene) and one for carbon monoxide.

The table below sets out the two smoking regimens for validation of the test methods referred above.

Smoking regimen	Puff volume (mL)	Puff frequency	Filter ventilation holes
ISO regimen: ISO 3308; Routine analytical cigarette smoking machine — definitions and standard conditions	35	Once every 60 s	No modifications
Intense regimen: Same as ISO 3308, but modified as indicated	55	Once every 30 s	All ventilation holes must be blocked 100% as described in WHO TobLabNet SOP 01.

This SOP was prepared to describe the procedure for the determination of NAME CONTENT(S) in mainstream cigarette smoke / cigarette tobacco filler.

1. **SCOPE**

    This Standard Operating Procedure is suitable for the quantitative determination of NAME CONTENT(S) in mainstream (MS) cigarette smoke / cigarette tobacco filler: NAME ALL CONTENT(S) TO BE ANALYSED SEPERATELY by SPECIFY TECHNIQUE TO BE USED. When more contents (as requested by COP) are determined using this method: ADD "The COP has recommended measurements for …. and …. only …. The inclusion of information on …. and …. in this method is intended for those who choose to include these measurements. IF APPLICABLE

2. REFERENCES

2.1 *ISO 3308: Routine analytical cigarette-smoking machine – Definitions and standard conditions*

2.2 *ISO 4387: Cigarettes – Determination of total and nicotine-free dry particulate matter using a routine analytical smoking machine*

2.3 *ISO 3402: Tobacco and tobacco products – Atmosphere for conditioning and testing*

2.4 *ISO 5725-2: Accuracy (trueness and precision) or measurement methods and results – Part 2: Basic method for the determination of repeatability and reproducibility of a standard measurement method*

2.5 *World Health Organization Tobacco Laboratory Network, Standard Operating Procedure for Intense Smoking of cigarettes (WHO TobLabNet SOP_01).* IF APPLICABLE

2.6 *World Health Organization Tobacco Laboratory Network, Standard Operating Procedure for validation of analytical methods of tobacco product contents and emissions (WHO TobLabNet SOP_02).*

3. TERMS AND DEFINITIONS

3.1 AS NEEDED

3.2

3.3

3.4

3.5 *Tobacco products*: Products entirely or partly made of leaf tobacco as the raw material that are manufactured for smoking, sucking, chewing or snuffing (Article 1(f) of the WHO FCTC).

3.6 *Intense Regimen*: Parameters used to smoke tobacco products which include 55-mL puff volume, 30-second puff interval, 2-second puff duration and 100% blocking of the filter ventilation holes. IF APPLICABLE

3.7 *ISO Regimen*: Parameters used to smoke tobacco products which include 35-mL puff volume, 60-second puff interval, 2-second puff duration and no blocking of the filter ventilation holes. IF APPLICABLE

3.8 *Laboratory sample*: Sample intended for testing in a laboratory, consisting of a single type of product delivered to the laboratory at one time or within a specified period.

3.9 *Test sample*: Product to be tested, taken at random from the laboratory sample. The number of products taken shall be representative of the laboratory sample.

3.10 *Test portion:* Random portion from the test sample to be used for a single determination. The number of products taken shall be representative of the test sample.

4. METHOD SUMMARY

4.1 SUMMARY OF SELECTED ANALYTICAL APPROACH

4.2

4.3

4.4

5. SAFETY AND ENVIRONMENTAL PRECAUTIONS

5.1 Follow routine safety and environmental precautions, as in any chemical laboratory activity.

5.2 The testing and evaluation of certain products with this test method may require the use of materials or equipment that could be hazardous or harmful to the environment; this document does not purport to address all the safety aspects associated with its use. All persons using this method have a responsibility to consult with the appropriate authorities and to establish health and safety practices as well as environmental precautions in conjunction with any existing applicable regulatory requirements prior to its use.

5.3 Special care should be taken to avoid inhalation or dermal exposure to harmful chemicals. Use a chemical fume hood, and wear an appropriate laboratory coat, gloves and safety goggles when preparing or handling undiluted materials, standard solutions, extraction solutions or collected samples.

5.4 Include specific safety and / or environmental precautions if applicable.

6. APPARATUS AND EQUIPMENT

6.1 Equipment needed to condition cigarette tobacco filler / cigarettes as specified in ISO 3402 [2.3].

6.2 Equipment needed to perform marking for butt length as specified in ISO 4387 [**2.2**]. ONLY FOR SMOKING METHOD

6.3 Equipment needed to cover ventilation holes for the intense regimen as specified in World Health Organization Tobacco Laboratory Network, *Standard Operating Procedure for Intense Smoking of cigarettes (WHO TobLabNet SOP_01)* [**2.5**]. ONLY FOR SMOKING METHOD

6.4 Equipment needed to perform smoking of tobacco products as specified in ISO 3308 [**2.1**]

6.5 Calibrated analytical balance capable of measuring to at least X decimal places

6.6

6.7

6.8 Capillary Gas Chromatograph (GC), equipped with a flame ionization detector (FID)
= EXAMPLE

6.9 Capillary Gas Chromatography column capable of distinct separation of solvent peaks, the peaks for the internal standard, nicotine and other tobacco contents (e.g. Varian WCOT FUSED SILICA, 25 m x 0.25 mm ID, Coating: CP-WAX 51).
= EXAMPLE

7. **REAGENTS AND SUPPLIES**

All reagents shall be of at least analytical reagent grade unless otherwise noted. When possible, reagents are identified by their Chemical Abstract Service (CAS) registry numbers.

7.1

7.2

7.3

7.4

7.5

8. **PREPARATION OF GLASSWARE**

Clean and dry glassware so as to avoid contamination.

9. **PREPARATION OF SOLUTIONS**

9.1 NAME OF SOLUTION

    9.1.1

    9.1.2

    9.1.3

    9.1.4

9.2   NAME OF SOLUTION

    9.2.1

    9.2.2

    9.2.3

    9.2.4

9.3   NAME OF SOLUTION

    9.3.1

    9.3.2

    9.3.3

    9.3.4

10. PREPARATION OF STANDARDS

Preparation of the standard solutions as described below is for reference purposes. The preparation of the standard solutions can be adjusted if necessary.

10.1   COMPONENT internal standard solution (if applicable)

    10.1.1 Component internal standard stock solution (X.X mg/mL)

        10.1.1.1

        10.1.1.2

        10.1.1.3

    10.1.2 COMPONENT(S) internal standard (mixed) solution (X.X mg/mL)

        10.1.2.1

        10.1.2.2

        10.1.2.3

10.2   **COMPONENT(S) standard solution (x g/L)**

    10.2.1 COMPONENT(S) stock solution (X.X mg/mL)

        10.2.1.1

        10.2.1.2

    10.2.2 COMPONENT(s) diluted (mixed) standard solution (X.X mg/mL)

        10.2.2.1

        10.2.2.2

10.2.3 COMPONENT(s) final (mixed) standard solutions

10.2.3.1 Prepare final standard solutions according to Table X.

10.2.3.2

Final concentrations of the internal standards can be determined using the following equation:

Final concentration (IS) (ng/mL) = x * XXXX

Where x is the original weight (in g) of standard as determined in ....

Final concentrations of standards are determined using the following equation:

Final concentration (ng/mL) = x * y / XXX

Where x is the original weight (in g) of standard as determined in ... y is the volume (in μl) of the spiking solution as described in ...

The final COMPONENT concentrations in the standard solutions are shown in Table X.

Table X. Concentrations of COMPONENT(s) in standard solutions

Standard	Volume of Component(s) in Standard Solution (X g/L) (mL)	Volume of IS Solution (μl)	Total Volume (mL)	Approximate Component(s) concentration in Final Mixed Standard Solution (mg/L)	Approximate Level Equivalent to Unknown Levels in Mainstream Cigarette Smoke (mg/cig) or Cigarette Tobacco Filler (mg/g)
1					
2					
3					
4					
5					

The range of the standard solutions may be adjusted depending on the equipment used and samples to be tested, keeping in mind the possible influence on the sensitivity of the applied method.

All solvents and solutions must be equilibrated to room temperature before use.

11. SAMPLING

11.1 Sample cigarettes according to ISO 8243 [23.4] or as required for specific application of the method. Alternative methods may be used to obtain a representative laboratory sample when required by specific regulation or availability of samples.

11.2 Constitution of test sample

11.2.1 Divide the laboratory sample into separate sales units, if applicable.

11.2.2 Take an equal amount of product for each test sample from at least $\sqrt{n}$ [**23.5**] of the individual sales units.

11.2.3 If no individual sales units are available, combine the entire laboratory sample into one unit.

11.2.4 Take the amount of required product randomly from the combined unit.

12. CIGARETTE PREPARATION

12.1 Condition all cigarettes to be smoked in accordance to ISO 3402 [**2.3**]

12.2 Mark cigarettes at a butt length in accordance with ISO 4387 [**2.2**] and World Health Organization Tobacco Laboratory Network, *Standard Operating Procedure for Intense Smoking of cigarettes (WHO TobLabNet SOP_01)* [**2.5**].

12.3 Prepare test samples to be smoked under intense smoking conditions as specified in World Health Organization Tobacco Laboratory Network, *Standard Operating Procedure for Intense Smoking of cigarettes (WHO TobLabNet SOP_01)* [**2.5**].

OR

12.1 Remove the cigarette tobacco filler from the cigarettes and QC samples (where applicable) of at least one packet (e.g. containing 20 cigarettes) or use at least XX grams of processed cigarette tobacco filler.

12.2 Combine and mix sufficient cigarette tobacco filler to constitute at least XX g for each test sample.

12.3 Mix and grind the cigarette tobacco filler until the cigarette tobacco filler is sufficiently reduced to pass through a 4 mm screen.

12.4 Condition the ground cigarette tobacco filler (for non-volatile constituents) as required for the tobacco product according to ISO 3402 [**2.3**]. Note to editor: Conditioning is recommended unless detrimental to the analytical method.

13. PREPARATION OF THE SMOKING MACHINE

(IF NOT APPLICABLE KEEP PARAGRAPH TITLE AND NOTE "Not applicable for this method")

13.1 Ambient conditions

The ambient conditions for smoking are specified in ISO 3308 [**2.1**].

13.2 Machine specifications

Follow ISO 3308 [**2.1**] machine specifications, except for intense smoking as described in World Health Organization Tobacco Laboratory Network, *Standard Operating Procedure for Intense Smoking of cigarettes (WHO TobLabNet SOP_01)* [**2.5**].

14. SAMPLE GENERATION

(IF NOT APPLICABLE KEEP PARAGRAPH TITLE AND NOTE "Not applicable for this method")

Note: Smoke a sufficient number of cigarettes on the specified smoking machine such that breakthrough does not occur and the concentrations of the NAME CONTENT(S) fall within the calibration range prepared for the analysis.

14.1 Smoke the cigarette test samples and collect the TPM as specified in ISO 4387 [**2.2**] or in WHO TobLabNet SOP_01 [**2.5**].

14.2 Include one reference test sample to be used for quality control, if applicable.

14.3 When testing sample types for the first time, an evaluation of breakthrough should be carried out. The number of cigarettes may need to be adjusted to prevent breakthrough of the filter pad. If the TPM exceeds 600 mg for a 92-mm filter pad or 150 mg for a 44-mm filter pad, the number of cigarettes smoked onto each pad must be decreased.

14.4 Number of cigarettes per result for linear and rotary smoking machines at ISO and intense smoking regimens are shown in Table X.

Table X. The number of cigarettes to be smoked for one measurement in linear and rotary smoking machines.

	ISO smoking regimen		Intense smoking regimen	
	Linear	Rotary	Linear	Rotary
No. of cigarettes per pad	5	20	3	10
No. of pads per result	4	1	7	2

14.5 Record the number of cigarettes and total puffs for each filter pad.

14.6 After smoking the required number of test samples, perform five clearing puffs and remove the pad holder from the smoking machine.

15. SAMPLE PREPARATION

FOR METHOD INCLUDING SMOKING:

15.1 Extraction of Filter Pads

Remove the pads from the holders. Fold each pad loosely in half, and then in half again, with the TPM on the inside. Use the side opposite where TPM is collected to wipe the inner surface of the pad holder, thus including any particulate matter that may have been left in the holder. Place each filter pad ....

15.2

15.3

15.4

FOR METHOD WITHOUT SMOKING:

15.1 Take X.X g of the well mixed, ground and conditioned test sample and weigh it to X.XXX g accuracy into an extraction vessel.

15.2

15.3

15.4

16. SAMPLE ANALYSIS

SPECIFY TECHNIQUE is used to quantify NAME GROUP CONTENTS in MAINSTREAM CIGARETTE SMOKE / CIGARETTE TOBACCO FILLER. The analytes are resolved from other potential interferences on the column used. Comparison of the area ratio of the unknowns with the area ratio of known standard concentrations yields the concentration of individual analytes.

16.1 NAME INSTRUMENT TYPE Operating Conditions: example

This section illustrates an example of HPLC or GC equipment and operating conditions. Sufficient information should be provided so results could be duplicated by a different laboratory.

HPLC Column:	NAME BRAND AND TYPE OF COLUMN (XX x XXX mm, X.X μm particle size) or equivalent
Injection Volume:	X μl
Column oven temperature:	XX ± X°C (Alter as appropriate for the specific column used)
Degasser:	On/Off
Binary pump flow:	X.X mL/min
Mobile phase A:	
Mobile phase B:	
Gradient:	Shown in Table X below
Total Analysis Time:	XX min

Table X. HPLC gradient for NAME CONTENT(S) separation

Time (min)	Flow (µl/min)	Mobile phase A (%)	Mobile phase B (%)

Note: Adjustment of the operating parameters may be required, depending on the instrument and column conditions as well as the resolution of the chromatographic peaks.

OR

GC Column:	NAME BRAND AND TYPE OF COLUMN, XX m x X.XX mm ID
Coating:	NAME TYPE OF COATING or equivalent
Column Temperature:	XXX °C (isothermal, or include a table with temperature program)
Injection Temperature:	XXX °C
Detector Temperature:	XXX °C
Carrier Gas:	NAME GAS TYPE at a flow rate of X.X mL/min
Injection Volume:	X.X µl
Injection Mode:	Split X:XX

Note: Adjustment of the operating parameters may be required, depending on the instrument and column conditions as well as the resolution of the chromatographic peaks.

16.2 Expected Retention Times

    16.2.1 For the conditions described here, the expected sequence of elution will be NAME CONTENT(S) IN ORDER OF EXPECTED RT

    16.2.2 Differences in e.g. FOR HPLC: flow rate, mobile phase concentration and age of the column may alter retention times. FOR GC: temperature, gas flow rate and age of the column may alter retention times.

    16.2.3 Elution order and retention times must be verified before analysis is begun.

    16.2.4 Using the above conditions, the expected total analysis time will be about XX minutes. (The analysis time may be extended to optimize performance)

    16.2.5 IF NEEDED SPECIFIC INSTRUMENT SETTINGS HAVING SPECIFIC INFLUENCE ON THE RETENTION TIMES CAN BE INCLUDED HERE

16.3 NAME CONTENT (GROUP) Determination

The sequence of determination will be in accordance with individual laboratory practices. This section illustrates an example of a sequence of determination for Name Content (Group).

16.3.1 Inject two (IF NEEDED) replicate aliquots of the standard solutions and sample extracts under identical conditions.

16.3.2 Condition the system just prior to use by injecting two X μl aliquots of a sample solution as a primer. (IF NEEDED)

16.3.3 Inject a check standard (blank with labelled internal standards) under the same conditions as the samples to verify the performance of the HPLC-MS/MS / GC system.

16.3.4 Inject a blank solution (extraction solution minus internal standard(s)) to check for any contamination in the system or reagents.

16.3.5 Inject an aliquot of each standard solution into the HPLC-MS/MS / GC.

16.3.6 Record the peak areas of NAME CONTENTS and the internal standard(s).

16.3.7 Calculate the relative response ratio (RF) of the NAME CONTENTS peak to the internal standard peak (RF = $A_{nicotine} / A_{IS}$) for each of the content standard solutions including the solvent blanks.

16.3.8 Plot the graph of the concentration of NAME CONTENTS (X axis) in accordance with the area ratios (Y axis).

16.3.9 The intercept should not be statistically significantly different from zero.

16.3.10 Linearity of the standard curve should extend over the entire standard range.

16.3.11 Calculate a linear regression equation (Y = a + bx) from this data, and use both the slope (b) and the intercept (a) of the linear regression. If the linear regression is less than 0.99, then the calibration should be repeated. If an individual calibration point differs by more than 10% from the expected value (estimated by linear regression), this point should be omitted.

16.3.12 Inject the QCs and samples and determine peak areas using the appropriate software.

16.3.13 The signal (peak area) obtained for all test portions must fall within the working range of the calibration curve, otherwise solutions should be adjusted as necessary.

Note: See Appendix X for representative chromatograms.

17. DATA ANALYSIS AND CALCULATIONS

17.1 Calculate the relative response ratio from the peak areas for each of the TSNAs for each of the calibration standards [**10.2.3**]:

$$RF = \frac{A_a}{A_{IS}}$$

where:

RF = relative response ratio

A_a = area of the target analyte

A_{IS} = area of the corresponding internal standard.

Plot a graph for each of the NAME CONTENTS (GROUP), of the relative response factor, A_a / A_{IS}, (Y axis) versus concentration (X axis). Calculate a linear regression equation ($RF = kX + m$) from these data, and use both the slope (k) and the intercept (m) of the linear regression

17.2 The linearity of the standard curve should cover the entire standard range.

17.3 The content M_{ts} of a given COMPONENT of the target analyte (ng/cigarette) is determined from the calculated relative response ratio for the test portion, the slope and intercept obtained from the appropriate calibration curve, and the equation:

(EXAMPLE)

$$M_{ts} = \frac{Y - m}{k} \times \frac{V}{N}$$

where:

M_{ts} = calculated content, in ng/cigarette

V = volume of extraction solution (40 mL)

N = number of cigarettes smoked onto each pad.

ALTERNATIVE CALCULATION PROCEDURES MAY ALSO BE USED IF APPLICABLE!

18. SPECIAL PRECAUTIONS

18.1 After installing a new column, condition it as specified by the manufacturer followed by injecting a tobacco sample extract under the specified instrument conditions. Injections should be repeated until the peak areas (or heights) of NAME CONTENTS and internal standards are reproducible.

18.2

18.3

19. DATA REPORTING

19.1 Report individual measurements for each of the samples evaluated.

19.2 Results should be reported as specified by method specifications.

19.3 For more information, see World Health Organization Tobacco Laboratory Network, *Standard operating procedure for validation of analytical methods of tobacco product contents and emissions (WHO TobLabNet SOP_02* [**2.6**].

20. QUALITY CONTROL

20.1 Control Parameters

Note: If the control measurements are outside the tolerance limits of the expected values, appropriate investigation and action must be taken.

Note: Additional laboratory quality assurance procedures should be carried out in compliance with the policies of the individual laboratory.

20.2 Laboratory Reagent Blank

To detect potential contamination during sample preparation and analysis, include a laboratory reagent blank as described in **16.3.4**. The blank consists of all reagents and materials used in analysing test samples and is analysed like a test sample. The assessment of the blank should be in accordance with the practices of the individual laboratory.

20.3 Quality Control Sample

To verify consistency of the entire analytical process, analyse a reference cigarette in accordance with the practices of the individual laboratory.

21. METHOD PERFORMANCE SPECIFICATIONS

21.1 Limit of reporting

The limit of reporting is set to the lowest concentration of the calibration standards used, recalculated to mg/g / mg/cig (X.X mg/g / mg/cig with XX mg/L as lowest calibration solution). Data outside the calibration range should be reported as below limit of quantification (BLOQ) or above limit of Quantification (ALOQ) as necessary.

21.2 Lab fortified matrix recovery.

Recovery of analyte spiked onto the matrix is used as a surrogate measure of accuracy. Recovery is determined by spiking known amounts of standards .... ADD PROCEDURE FOR PERFORMING RECOVERY DETERMINATION. The recovery is calculated from the following formula, as shown in Table X:

Recovery = 100*(analytical result-unspiked result/spiked amount)

Table 5. Mean and recovery of COMPONENT GROUP spiked onto the matrix

Spiked amount (ng)	COMPONENT		COMPONENT		COMPONENT	
	Mean (ng)	Recovery (%)	Mean (ng)	Recovery (%)	Mean (ng)	Recovery (%)
XX.X	XX.X	XX.X	XX.X	XX.X	XX.X	XX.X
XX.X	XX.X	XX.X	XX.X	XX.X	XX.X	XX.X
XX.X	XX.X	XX.X	XX.X	XX.X	XX.X	XX.X

21.3

21.4 Analytical Specificity

DESCRIBE ANALYTICAL SPECIFICITY OF THE METHOD / TECHNIQUE USED, e.g. The retention time of the analyte of interest is used to verify the analytical specificity. An established range of ratios of the response of the component to that of the internal standard component of QC mainstream cigarette smoke/cigarette tobacco filler is used to verify the specificity of the results from an unknown sample.

21.5 Linearity

The CONTENT(S) calibration curves established is/are linear over the standard concentration range of XX ng/mL to XX ng/mL.

21.6 Possible interferences

The presence of XXX can cause interference since the retention time of this component can be similar to the retention time of COMPONENT / INTERNAL STANDARD. This interference is most likely to occur with samples containing XXX.

OR (example)

There are no known components that can cause interference by having a similar retention time as COMPONENT(S), or the internal standard(s).

22. REPEATABILITY AND REPRODUCIBILITY

An international collaborative study conducted in 2012, performed according to World Health Organization Tobacco Laboratory Network, *Standard operating procedure for validation of analytical methods of tobacco product contents and emissions (WHO TobLabNet SOP_02)* [**2.6**] gave the following precision limits for this method:

The difference between two single results found on matched cigarette samples by the same operator using the same apparatus within the shortest feasible time interval will exceed the repeatability, r, on average not more than once in 20 cases in the normal and correct operation of the method.

Single results for matched cigarette samples reported by two laboratories will differ by more than the reproducibility, R, on average not more than once in 20 cases in the normal and correct operation of the method.

The test results were subjected to statistical analysis in accordance with ISO 5725-1 [23.2] and ISO 5725-2 [2.4] to give the precision data shown in Table X.

FOR METHOD WITH SMOKING

For the purpose of calculating r and R, one test result under ISO smoking conditions was defined as the average of seven individual replicates of the mean yield of four Cambridge Filter Pads (CFPs) (five cigarettes smoked per pad) from linear smoking machines and one CFP (20 cigarettes smoked per pad) for rotary smoking machines. Under intense smoking conditions, one test result was defined as the average of seven individual replicates of the mean yield of seven CFPs (three cigarettes smoked per pad) from linear smoking machines and two CFPs (10 cigarettes smoked per pad) for rotary smoking machines.

Table X: Precision limits for the determination of CONTENTS in mainstream cigarette smoke (mg/cig) from the reference test pieces.

Reference cigarette	ISO smoking regimen				ISO smoking regimen			
	N	m_{cig}	r_{limit}	R_{limit}	N	m_{cig}	r_{limit}	R_{limit}
1R5F	X	XX.XX	XX.XX	XX.XX	X	XX.XX	XX.XX	XX.XX
3R4F	X	XX.XX	XX.XX	XX.XX	X	XX.XX	XX.XX	XX.XX
CM6	X	XX.XX	XX.XX	XX.XX	X	XX.XX	XX.XX	XX.XX

Note: The levels of the commercial brand were inside the range of the reference pieces and are therefore not reported in this table.

OR FOR METHOD WITHOUT SMOKING

Table X: Precision limits for the determination of CONTENTS in Cigarette Tobacco Filler (mg/g) from the reference test pieces.

Reference product	Tobacco Filler			
	N	m_{cig}	r_{limit}	R_{limit}
1R5F	X	XX.XX	XX.XX	XX.XX
3R4F	X	XX.XX	XX.XX	XX.XX
CM6	X	XX.XX	XX.XX	XX.XX

Note: The levels of the commercial brand were inside the range of the reference pieces and are therefore not reported in this table.

23. BIBLIOGRAPHY

23.1 World Health Organization Tobacco Laboratory Network, *Method Validation Report of World Health Organization TobLabNet Official Method XXXX: Determination of CONTENT (GROUP) in Mainstream Cigarette Smoke / Cigarette Tobacco Filler.*

23.2 ISO 5725-1. *Accuracy (trueness and precision) of measurement methods and results – Part 1: General principles and definitions.*

23.3 Reference to basic method of which the method has been deducted (if applicable).

23.4 *ISO 8243: Cigarettes—Sampling*

23.5 United Nations Office on Drugs and Crime (UNODC); Guidelines of representative drug sampling (http://www.unodc.org/documents/scientific/Drug_Sampling.pdf).

23.6 ISO Standards – Products by TC (Website for ordering methods): ISO/TC 126.

23.7

Annex 2. Example / Template validation protocol contents

WHO TobLabNet

Validation Protocol for the analytical method for determining content(s) in tobacco

Method:	Protocol for the validation of analytical method for the determination of contents(s) in cigarette tobacco filler
Analytes:	Target analyte(s)
Matrix:	Tobacco
Last update:	Month yyyy

FOREWORD

This document was prepared by members of the World Health Organization (WHO) Tobacco Laboratory Network (TobLabNet) as a standard operating procedure (SOP) for validation of analytical methods for determining tobacco contents and cigarette tobacco filler contents.

INTRODUCTION

In order to establish comparable measurements for testing tobacco products globally, consensus methods are required for measuring specific contents and emissions of cigarettes. The Conference of the Parties (COP) to the WHO Framework Convention on Tobacco Control (WHO FCTC) at its third session in Durban, South Africa, in November 2008, recalling its decisions FCTC/COP1(15) and FCTC/COP2(14) on the elaboration of guidelines for implementation of Articles 9 *(Regulation of the contents of tobacco products)* and 10 *(Regulation of tobacco product disclosures)* of the WHO FCTC, noting the information contained in the report of the working group to the third session of the Conference of the Parties on the progress of its work ... requested the Convention Secretariat to invite WHO's Tobacco Free Initiative to ... validate, within five years, the analytical chemical methods for testing and measuring cigarette contents and emissions (FCTC/COP/3/REC/1).

Using the criteria for prioritization set at its third meeting in Ottawa, Canada, in October 2006, the working group on Articles 9 and 10 identified the following contents for which methods for testing and measurement (analytical chemistry) should be validated as a priority:

- nicotine
- ammonia
- humectants (propane-1,2-diol, glycerol (propane-1,2,3-triol) and triethylene glycol (2,2-ethylenedioxybis(ethanol)).

Measurement of these contents will require validation of three methods: one for nicotine, one for ammonia and one for humectants.

Using the criteria for prioritization set at the meeting in Ottawa mentioned above, the working group identified the following emissions in mainstream smoke for which methods for testing and measurement (analytical chemistry) should be validated as a priority:

- 4-(methylnitrosamino)-1-(3-pyridyl)-1-butanone (NNK)
- *N*-nitrosonornicotine (NNN)
- acetaldehyde
- acrylaldehyde (acrolein)

- benzene
- benzo[a]pyrene
- 1,3-butadiene
- carbon monoxide
- formaldehyde

Measurement of these emissions with the two smoking regimens described below will require validation of five methods: one for tobacco-specific nitrosamines (NNK and NNN), one for benzo[a]pyrene, one for aldehydes (acetaldehyde, acrolein and formaldehyde) and one for volatile organic compounds (benzene, 1,3-butadiene) and one for carbon monoxide.

1. SCOPE

To introduce a standardized method to determine contents(s) in tobacco and cigarette filler in non-tobacco affiliated industry laboratories and to determine the degree of agreement on the content among the participants.

2. SCHEDULES

Introduction of standardized method

A standardized method is described, based on the source of the method.

2.1 The standardized method will be sent to all TobLabNet members, together with this protocol, and an invitation to participate in this study, which will be closed on dd Month yyyy.

2.2 Test cigarettes

Test cigarettes or test articles will be procured and sent to all participants by a TobLabNet member. The total number of cigarettes needed for this study will be determined after the call for participants is closed. dd Month yyyy the samples will be distributed to all participants, together with the SOP, study protocol and a Microsoft Excel data sheet for the test results. Please DO NOT modify any data sheet for the convenience of data analysis.

2.3 Scope and time for measurements

The content measurements shall be executed between dd Month yyyy and dd Month yyyy. Participants should carry out each of the seven determinations per sample on a separate day (more than one sample per day can be measured).

2.4 Reporting of results

Test results of the study shall be reported to WHO by using the designated data sheet no later than dd Month yyyy to:

WHO TobLabNet Focal Person
Geneva, Switzerland

WHO will assign code numbers to each participating laboratory in confidence.

2.5 Data analysis and statistical evaluation

The reported data will be statistically evaluated by TobLabNet, as described in World Health Organization Tobacco Laboratory Network, *Standard operating procedure for validation of analytical method of tobacco product contents and emissions (WHO TobLabNet SOP_02)* in combination with *ISO 5725 part 2 – Accuracy (trueness and precision) of measurement methods and results - Part 2: Basic method for the determination of repeatability and reproducibility of a standard measurement method.*

A summary report of the results and the statistical evaluation will be sent to all participants by TobLabNet before dd Month yyyy. The statistical evaluation will be discussed at the first TobLabNet meeting after the report has been sent to the participants.

3. TEST ITEMS

3.1 Description

For the validation study, the number of samples has been set to five, divided into three so-called reference cigarettes and two commercially available brands. For efficiency reasons, the samples to be used for the validation study of all content determinations will be sent to all participants.

Table 1: Example test items for tobacco specific nitrosamines study

Sample code	Product type	Sample name
A	Reference cigarette	1R5F
B	Reference cigarette	2R4F
C	Reference cigarette	CM6
D	Cigarette	Commercial brand name
E	Cigarette	Commercial brand name

3.2 Shipment and receipt

a) TobLabNet will send ## cigarettes or tobacco samples of each sample to each participating laboratory.

b) TobLabNet will inform all participating laboratories that samples have been sent.

c) Upon receipt, each participating laboratory will notify TobLabNet by email of the receipt of the samples and any problems with or damage to the cigarettes.

d) If problems were encountered during shipment, TobLabNet will send replacement samples as soon as practicable after the notification.

e) After receipt of the samples, each laboratory will complete the analyses and report the data within the designated time schedule.

f) Handle the samples carefully to avoid damage. Discard cigarettes that are obviously damaged, or missing significant amounts of tobacco.

3.3 Storage

Samples can be stored at room temperature for up to three months. For longer periods, the samples are to be stored in the original packaging or airtight containers in a freezer at −20 °C or lower.

3.4 Sample preparation

Sample preparation shall be done according to the SOP (grinding and sieving as required).

3.5 Number of test items to be analysed

For each individual sample seven replicate measurements are to be executed. Each of the seven determinations per sample shall be executed on a separate day (more than one sample per day can be measured).

4. REPORTING TEST RESULTS

Test results shall be reported using the Microsoft Excel template file provided by email. Please DO NOT modify any data sheet for the convenience of data analysis.

If Microsoft Excel is not available, laboratories can report data electronically by using an ASCII format or if this is not possible, by hard copy.

Test results shall be given according to the details given in Table 2.

Table 2: Data and formats for reporting

Parameter	Unit	Report to the nearest ...
Amount of tobacco used	gram	#.###1 gram
Content	mg/gram - mg/gram	#,##1 mg/gram

Participants shall also report an overview of the major equipment used in the study (type, model, manufacturer). Please refer to the data sheet for further details.

If, for any reason, there has been a deviation from the SOP or the study protocol, please note the deviation and the reason for this in the Microsoft Excel data sheet or individual participating laboratory report.

5. PROJECTED TIMELINE

Content(s) method validation schedule	Timeline
Sending of SOP, Study protocol and call for participants	Dd Month yyyy
Returning comments and participation inquiry	Dd Month yyyy
Shipping samples to participants	Dd Month yyyy
Carry out analyses and report results	Dd Month yyyy
Statistical evaluation of results	Dd Month yyyy
Report evaluation summary to participants	Dd Month yyyy
Report summary results to COP Working Group Articles 9 & 10 via TFI	After discussion at TobLabNet meeting

Annex 3. Example / Template validation protocol emission

WHO TobLabNet

Validation protocol for analytical methodology for determination of contents(s) of mainstream cigarette smoke under ISO and intense smoking conditions

Method:	Protocol for the validation of analytical method for the determination of content(s) of mainstream cigarette smoke under ISO and intense smoking conditions
Analytes:	Not applicable
Matrix:	Mainstream cigarette smoke
Last update:	Month yyyy

FOREWORD

This document was prepared by members of the World Health Organization (WHO) Tobacco Laboratory Network (TobLabNet) as a standard operating procedure (SOP) for validation of analytical methods of tobacco cigarette mainstream smoke contents and cigarette tobacco filler contents.

INTRODUCTION

In order to establish comparable measurements for testing tobacco products globally, consensus methods are required for measuring specific contents and emissions of cigarettes. The Conference of the Parties (COP) to the WHO Framework Convention on Tobacco Control (WHO FCTC) at its third session in Durban, South Africa, in November 2008, recalling its decisions FCTC/COP1(15) and FCTC/COP2(14) on the elaboration of guidelines for implementation of Articles 9 (*Regulation of the contents of tobacco products*) and 10 (*Regulation of tobacco product disclosures*) of the WHO FCTC, noting the information contained in the report of the working group to the third session of the Conference of the Parties on the progress of its work ... requested the Convention Secretariat to invite WHO's Tobacco Free Initiative to ... validate, within five years, the analytical chemical methods for testing and measuring cigarette contents and emissions (FCTC/COP/3/REC/1).

Using the criteria for prioritization set at its third meeting in Ottawa, Canada, in October 2006, the working group on Articles 9 and 10 identified the following contents for which methods for testing and measurement (analytical chemistry) should be validated as a priority:

- nicotine
- ammonia
- humectants (propane-1,2-diol, glycerol (propane-1,2,3-triol) and triethylene glycol (2,2-ethylenedioxybis(ethanol)).

Measurement of these contents will require validation of three methods: one for nicotine, one for ammonia and one for humectants.

Using the criteria for prioritization set at the meeting in Ottawa mentioned above, the working group identified the following emissions in mainstream smoke for which methods for testing and measurement (analytical chemistry) should be validated as a priority:

- 4-(methylnitrosamino)-1-(3-pyridyl)-1-butanone (NNK)
- *N*-nitrosonornicotine (NNN)
- acetaldehyde
- acrylaldehyde (acrolein)

- benzene
- benzo[a]pyrene
- 1,3-butadiene
- carbon monoxide
- formaldehyde

Measurement of these emissions with the two smoking regimens described below will require validation of five methods: one for tobacco-specific nitrosamines (NNK and NNN), one for benzo[a]pyrene, one for aldehydes (acetaldehyde, acrolein and formaldehyde), one for volatile organic compounds (benzene, 1,3-butadiene) and one for carbon monoxide.

The table below sets out the two smoking regimens for validation of the test methods referred above.

Smoking regimen	Puff volume (mL)	Puff frequency	Filter ventilation holes
ISO regimen: ISO 3308; Routine analytical cigarette smoking machine – definitions and standard conditions	35	Once every 60 s	No modifications
Intense regimen: Same as ISO 3308, but modified as indicated	55	Once every 30 s	All ventilation holes must be blocked 100% as described in WHO TobLabNet SOP 01.

1. **SCOPE**

    To introduce a standardized method for the determination of content(s) of mainstream cigarette smoke in non-tobacco industry laboratories and on the content.

2. **SCHEDULES**

2.1 **Introduction of standardized method.**

    A standardized method is described, based on the source of the method.

    The standardized method will be sent to all TobLabNet members, together with this protocol, including a call for participants in this study, which will be closed on dd Month yyyy.

2.2 **Test cigarettes.**

    Test cigarettes will be purchased and sent to all participants by TobLabNet. The total number of cigarettes needed for this study will depend on the number of participants and will be determined after the call for participants is closed. dd Month yyyy the samples will be distributed to all participants, together with the SOP, study protocol and a Microsoft Excel data sheet for the test results. Please DO NOT modify any data sheet for the convenience of data analysis.

2.3 Scope and time for measurements.

The content group to be determined by the study participants shall be contents in mainstream tobacco smoke. A smoking machine, using both ISO and intense regimen, shall be used to generate mainstream smoke. More details related to the type of smoking machines, including smoking plans, are given in Tables 3 to 6.

The content measurements shall be executed between dd Month yyyy and dd Month yyyy. Participants are kindly asked to carry out each of the seven determinations per sample on a separate day (more than one sample per day can be measured).

2.4 Reporting of results.

Test results of the study shall be reported to WHO using the designated data sheet no later than dd Month yyyy to:

> WHO TobLabNet Focal Person
> Geneva, Switzerland

WHO will assign code numbers to each participating laboratory in confidence.

2.5 Data analysis and statistical evaluation.

The reported data will be statistically evaluated by TobLabNet, as described in World Health Organization Tobacco Laboratory Network, *Standard operating procedure for validation of analytical methods of tobacco product contents and emissions (WHO TobLabNet SOP_02)*, in combination with ISO 5725-2: *Accuracy (trueness and precision) of measurement methods and results – Part 2: Basic method for the determination of repeatability and reproducibility of a standard measurement method*.

A summary report of the results and the statistical evaluation will be sent to all participants by TobLabNet before dd Month yyyy. The statistical evaluation will be discussed at the first TobLabNet meeting after the report has been sent to the participants.

3. TEST ITEMS

3.1 Description

For the validation study the number of samples has been set to five, divided into three so-called reference cigarettes and two commercially available brands. For efficiency reasons the samples to be used for the validation study of all emissions determinations will be sent to all participants at once.

Table 1: Test items for tobacco specific nitrosamines study

Sample code	Product type	Sample name
A	Reference cigarette	1R5F
B	Reference cigarette	2R4F
C	Reference cigarette	CM6
D	Cigarette	Commercial brand name
E	Cigarette	Commercial brand name

3.2 Shipment and receipt

a) TobLabNet will send ## cigarettes of each sample to each participating laboratory.

b) TobLabNet will inform all participating laboratories that samples have been sent.

c) Upon receipt, each participating laboratory will notify TobLabNet by email of the receipt of the samples and any problems with or damage to the cigarettes.

d) If problems were encountered during shipment, TobLabNet will send replacement samples as soon as practicable after notification.

e) After receipt of the samples, each laboratory will complete the analyses and report the data within the designated time schedule.

f) Handle the samples carefully to avoid damage. Discard cigarettes that are obviously damaged, or missing significant amounts of tobacco.

3.3 Storage

Samples can be stored at room temperature for up to three months. For longer periods, the samples are to be stored in the original packaging or airtight containers in a freezer at −20 °C or lower.

3.4 Sample preparation

Sample preparation shall be done according to the SOP (selection and storage). Mark the samples for further identification. If the tobacco products are removed from the original packages, store them in airtight containers just large enough to contain the sample and keep them at **−20 °C or lower until needed.**

3.5 Number of test items to be analysed

For each individual sample, seven replicate measurements are to be executed. Each of the seven determinations per sample shall be executed on a separate day (more than one sample per day can be measured).

4. REPORTING TEST RESULTS

Test results shall be reported using the Microsoft Excel template file provided by email. Please DO NOT modify any data sheet for the convenience of data analysis.

If Microsoft Excel is not available, laboratories can report data electronically by using an ASCII format or if this is not possible, by hard copy.

Test results shall be given according to the details given in Table 2.

Table 2: Data and formats for reporting

Parameter	Unit	Report to the nearest ...
Cigarette weight	gram	#.###1 gram
Component	ng/cigarette	#,##1 ng/cigarette
Component	ng/cigarette	#,##1 ng/cigarette
Component	ng/cigarette	#,##1 ng/cigarette

Participants shall also report an overview of the major equipment used in the study (type, model, manufacturer). Please refer to the data sheet for further details.

If, for any reason, there has been a deviation from the SOP or the study protocol, please note the deviation and the reason for this in the Microsoft Excel data sheet.

5. PROJECTED TIME LINE

Content(s) method validation schedule	Timeline
Sending of SOP, study protocol and call for participants	Dd Month yyyy
Returning comments and participation inquiry	Dd Month yyyy
Shipping samples to participants	Dd Month yyyy
Carry out analyses and report results	Dd Month yyyy
Statistical evaluation of results	Dd Month yyyy
Report evaluation summary to participants	Dd Month yyyy
Report summary results to COP Working Group Articles 9 & 10 via TFI	After discussion at TobLabNet meeting

6. SMOKING PLANS

The described smoking plans are based on five different samples (cigarettes brands/types) and seven replicates per sample. If a different number of samples or replicates are used, the plans should be adjusted accordingly.

6.1 ISO Regime

Table 3: Smoking plan for rotary smoking machine (20 cigarettes per filter pad) (Example)

Day	Run	Sample	Day	Run	Sample
1	1	A	5	21	C
1	2	B	5	22	D
1	3	C	5	23	E
1	4	D	5	24	A
1	5	E	5	25	B
2	6	D	6	26	D
2	7	E	6	27	C
2	8	A	6	28	B
2	9	B	6	29	A
2	10	C	6	30	E
3	11	B	7	31	C
3	12	C	7	32	B
3	13	D	7	33	A
3	14	E	7	34	E
3	15	A	7	35	D
4	16	E			
4	17	A			
4	18	B			
4	19	C			
4	20	D			

Note: Each filter pad should be extracted individually with xx mL extraction solution, which will be further analysed. One result is the result of each individual filter pad / extraction solution (20 cigarettes).

Table 4: Smoking plan for 20-port linear smoking machine (5 cigarettes per filter pad) (Example)

Day	Port number																			
	1	2	3	4	5	6	7	8	9	10	11	12	13	14	15	16	17	18	19	20
1	A	A	A	A	B	B	B	B	C	C	C	C	D	D	D	D	E	E	E	E
2	D	D	D	D	E	E	E	E	A	A	A	A	B	B	B	B	C	C	C	C
3	B	B	B	B	C	C	C	C	D	D	D	D	E	E	E	E	A	A	A	A
4	E	E	E	E	A	A	A	A	B	B	B	B	C	C	C	C	D	D	D	D
5	C	C	C	C	D	D	D	D	E	E	E	E	A	A	A	A	B	B	B	B
6	D	D	D	D	C	C	C	C	B	B	B	B	A	A	A	A	E	E	E	E
7	C	C	C	C	B	B	B	B	A	A	A	A	E	E	E	E	D	D	D	D

Note: Four filter pads should be combined and extracted with xx mL extraction solution, which will be further analysed. One result is the result of one extraction solution combining a set of four filter pads in each extraction solution (20 cigarettes).

6.2 Intense Regime

Table 5: Smoking plan for rotary smoking machine (10 cigarettes per filter pad) (Example)

Day	Run	Sample	Day	Run	Sample
8	36	A	12	76	C
	37	B		77	D
	38	C		78	E
	39	D		79	A
	40	E		80	B
	41	A		81	C
	42	B		82	D
	43	C		83	E
	44	D		84	A
	45	E		85	B
9	46	D	13	86	D
	47	E		87	C
	48	A		88	B
	49	B		89	A
	50	C		90	E
	51	D		91	D
	52	E		92	C
	53	A		93	B
	54	B		94	A
	55	C		95	E
10	56	B	14	96	C
	57	C		97	B
	58	D		98	A
	59	E		99	E
	60	A		100	D
	61	B		101	C
	62	C		102	B
	63	D		103	A
	64	E		104	E
	65	A		105	D
11	66	E			
	67	A			
	68	B			
	69	C			
	70	D			
	71	E			
	72	A			
	73	B			
	74	C			
	75	D			

Note: Each filter pad should be extracted individually with xx mL extraction solution, which will be further analysed. One result is the average result of two individual filter pads / extraction solutions (20 cigarettes).

Table 6: Smoking plan for 20-port linear smoking machine (3 cigarettes per filter pad) (Example)

Day	Run	1	2	3	4	5	6	7	8	9	10	11	12	13	14	15	16	17	18	19	20
		\multicolumn{20}{c}{Port number}																			
8	8	A	A	A		B	B	B		C	C	C		D	D	D			E	E	E
	9			B	B	B	C	C	C	D	D	D	E	E	E	A	A	A			
9	10	D	D	D		E	E	E		A	A	A		B	B	B			C	C	C
	11			E	E	E	A	A	A	B	B	B	C	C	C	D	D	D			
10	12	B	B	B		C	C	C		D	D	D		E	E	E			A	A	A
	13			C	C	C	D	D	D	E	E	E	A	A	A	B	B	B			
11	14	E	E	E		A	A	A		B	B	B		C	C	C			D	D	D
	15			A	A	A	B	B	B	C	C	C	D	D	D	E	E	E			
12	16	C	C	C		D	D	D		E	E	E		A	A	A			B	B	B
	17			D	D	D	E	E	E	A	A	A	B	B	B	C	C	C			
13	18	D	D	D		C	C	C		B	B	B		A	A	A			E	E	E
	19			C	C	C	B	B	B	A	A	A	E	E	E	D	D	D			
14	20	C	C	C		B	B	B		A	A	A		E	E	E			D	D	D
	21			B	B	B	A	A	A	E	E	E	D	D	D	C	C	C			

Note: Three filter pads are combined and extracted with xx mL extraction solution, which will be further analysed. One result is the average of two sets of combined filter pads / extraction solutions (18 cigarettes = 2 × 9 cigarettes).

WHO TobLabNet
Official Method
SOP 03

Standard Operating Procedure for Determination of Tobacco-Specific Nitrosamines in Mainstream Cigarette Smoke under ISO and Intense Smoking Conditions

Method:	Determination of tobacco-specific nitrosamines in mainstream cigarette smoke under ISO and intense smoking conditions
Analytes:	3-(1-Nitrosopyrrolidin-2-yl)pyridine(CAS# 16543-55-8)
	4-(Methylnitrosamino)-1-(3-pyridyl)-1-butanone (CAS# 64091-91-4)
	N-Nitrosoanatabine (CAS# 71267-22-6)
	N-Nitrosoanabasine (CAS# 37620-20-5)
Matrix:	Tobacco cigarette mainstream smoke particulate matter
Last update:	June 2014

The mention of specific companies or of certain manufacturers' products does not imply that they are endorsed or recommended by WHO in preference to others of a similar nature that are not mentioned. Errors and omissions excepted, the names of proprietary products are distinguished by initial capital letters.

> No machine smoking regimen can represent all human smoking behaviour: machine smoking testing is useful for characterizing cigarette emissions for design and regulatory purposes, but communication of machine measurements to smokers can result in misunderstanding about differences between brands in exposure and risk. Data on smoke emissions from machine measurements may be used as inputs for product hazard assessment, but they are not intended to be nor are they valid as measures of human exposure or risks. Representing differences in machine measurements as differences in exposure or risk is a misuse of testing with WHO TobLabNet standards.

FOREWORD

This document was prepared by members of the World Health Organization (WHO) Tobacco Laboratory Network (TobLabNet) as an analytical method standard operating procedure (SOP) for measuring tobacco-specific nitrosamines (TSNAs) in mainstream cigarette smoke under International Organization for Standardization (ISO) and intense smoking conditions.

INTRODUCTION

In order to establish comparable measurements for testing tobacco products globally, consensus methods are required for measuring specific contents and emissions of cigarettes. The Conference of the Parties to the WHO Framework Convention on Tobacco Control (FCTC) at its third session in Durban, South Africa, in November 2008, "recalling its decisions FCTC/COP1(15) and FCTC/COP2(14) on the elaboration of guidelines for implementation of Articles 9 (*Regulation of the contents of tobacco products*) and 10 (*Regulation of tobacco product disclosures*) of the WHO FCTC, noting the information contained in the report of the working group to the third session of the Conference of the Parties on the progress of its work ... requested the Convention Secretariat to invite WHO's Tobacco Free Initiative to ... validate, within five years, the analytical chemical methods for testing and measuring cigarette contents and emissions" (FCTC/COP/3/REC/1).

Using the criteria for prioritization set at its third meeting in Ottawa, Canada, in October 2006, the working group on Articles 9 and 10 identified the following contents for which methods for testing and measurement (analytical chemistry) should be validated as a priority:

- nicotine
- ammonia
- propylene glycol (propane-1,2-diol)
- glycerol (propane-1,2,3-triol)
- triethylene glycol (2,2-ethylenedioxybis(ethanol)).

Measurement of these contents will require validation of three methods: one for nicotine, one for ammonia and one for humectants.

Using the criteria for prioritization set at the meeting in Ottawa mentioned above, the working group identified the following emissions in mainstream smoke for which methods for testing and measurement (analytical chemistry) should be validated as a priority:

- 4-(methylnitrosamino)-1-(3-pyridyl)-1-butanone (NNK)
- *N*-nitrosonornicotine (NNN)
- acetaldehyde
- acrylaldehyde (acrolein)
- benzene
- benzo[*a*]pyrene
- 1,3-butadiene

- carbon monoxide
- formaldehyde

Measurement of these emissions with the two smoking regimens described below will require validation of five methods: one for tobacco-specific nitrosamines (NNK and NNN), one for benzo[a]pyrene, one for aldehydes (acetaldehyde, acrolein and formaldehyde), one for volatile organic compounds (benzene, 1,3-butadiene) and one for carbon monoxide.

The table below sets out the two smoking regimens for validation of the test methods referred to above.

Smoking regimen	Puff volume (mL)	Puff frequency	Filter ventilation holes
ISO regimen: ISO 3308; *Routine analytical cigarette smoking machine—definitions and standard conditions*	35	Once every 60 s	No modifications
Intense regimen: Same as ISO 3308, but modified as indicated	55	Once every 30 s	All ventilation holes must be blocked 100% as described in WHO TobLabNet SOP 01.

This SOP was prepared to describe the procedure for the determination of TSNAs in mainstream cigarette smoke under ISO and intense smoking conditions.

1 SCOPE

1.1 This method is suitable for quantitative determination of the following four TSNAs in mainstream cigarette smoke: 3-(1-nitrosopyrrolidin-2-yl)pyridine(NNN), 4-(methylnitrosamino)-1-(3-pyridyl)-1-butanone (NNK), *N*-nitrosoanatabine (NAT) and *N*-nitrosoanabasine (NAB) by combined high-performance liquid chromatography-tandem mass spectrometry (HPLC–MS-MS).

Note: The Conference of the Parties recommended that only NNN and NNK be measured. Information on the analysis of NAT and NAB is included for laboratories that choose to conduct those measurements.

1.2 NNN and NNK are potent carcinogens; NAB is a weaker carcinogen, while NAT is not carcinogenic. NNN and NNK are not originally present in tobacco leaves but are formed from the nitrosation of nicotine during tobacco curing and storage; NAB is formed from nitrosation of anabasine and NAT from nitrosation of anatabine. After absorption into the body, NNN and NNK can be hydroxylated to compounds that form adducts with haemoglobin or DNA.

2 REFERENCES

2.1 *ISO 3402: Tobacco and tobacco products—Atmosphere for conditioning and testing.*

2.2 ISO 4387: *Cigarettes—Determination of total and nicotine-free dry particulate matter using a routine analytical smoking machine.*

2.3 World Health Organization. *Standard operating procedure for intense smoking of cigarettes.* Geneva, Tobacco Laboratory Network (WHO TobLabNet SOP 01).

2.4 ISO 3308: *Routine analytical cigarette-smoking machine—Definitions and standard conditions.*

2.5 ISO 8243: *Cigarettes—Sampling*

2.6 United Nations Office on Drugs and Crime. *Guidelines on representative drug sampling.* Vienna, Laboratory and Scientific Section, 2009 (http://www.unodc.org/documents/scientific/Drug_Sampling.pdf).

2.7 World Health Organization. *Standard operating procedure for validation of analytical methods of tobacco product contents and emissions.* Geneva, Tobacco Laboratory Network (WHO TobLabNet SOP 02).

2.8 World Health Organization. *Method validation report of World Health Organization TobLabNet official method: Determination of tobacco-specific nitrosamines in mainstream cigarette smoke under ISO and intense smoking conditions.* Geneva, Tobacco Laboratory Network, forthcoming.

2.9 ISO 5725-1. *Accuracy (trueness and precision) of measurement methods and results—Part 1: General principles and definitions.*

2.10 ISO 5725-2. *Accuracy (trueness and precision) of measurement methods and results—Part 2: Basic method for the determination of repeatability and reproducibility of a standard measurement method.*

3 TERMS AND DEFINITIONS

3.1 *TPM*: total particulate matter

3.2 *TSNAs*: tobacco-specific nitrosamines

3.3 *NNN*: 3-(1-nitrosopyrrolidin-2-yl)pyridine

3.4 *NNK*: 4-(methylnitrosamino)-1-(3-pyridyl)-1-butanone

3.5 *NAT*: *N*-nitrosoanatabine

3.6 *NAB*: *N*-nitrosoanabasine

3.7 *Tobacco products*: Products entirely or partly made of leaf tobacco as the raw material that are manufactured to be used for smoking, sucking, chewing or snuffing (Article 1(f) of the WHO FCTC)

3.8 *Intense regimen*: Parameters for smoking tobacco products that include a 55-mL puff volume, a 30-s puff interval, a 2-s puff duration and 100% blocking of filter ventilation holes

3.9 *ISO regimen*: Parameters for smoking tobacco products that include a 35-mL puff volume, a 60-s puff interval, a 2-s puff duration and no blocking of filter ventilation holes

3.10 *Laboratory sample*: Sample intended for testing in a laboratory, consisting of a single type of product delivered to the laboratory at one time or within a specified period

3.11 *Test sample*: Product to be tested, taken at random from the laboratory sample.

3.12 *Test portion*: Random portion from the test sample to be used for a single determination.

4 METHOD SUMMARY

4.1 Mainstream smoke from the cigarette test sample is trapped onto a glass-fibre filter pad (commonly referred to as a Cambridge filter pad).

4.2 The number of cigarettes may have to be adjusted to prevent breakthrough of the filter pad. If the TPM exceeds 600 mg for a 92-mm filter pad or 150 mg for a 44-mm filter pad, the number of cigarettes smoked onto each pad must be decreased.

4.3 A solution containing a mixture of two (or four) isotope-labelled internal standards is spiked onto the pad, which is extracted with ammonium acetate.

4.4 The extract is filtered and analysed by high-performance liquid chromatography–tandem mass spectrometry (HPLC–MS-MS) with electrospray ionization. Analyte ions are detected in the MS-MS mode.

4.5 The reconstructed ion chromatogram peak area ratio of native analyte to labelled internal standard is compared on a calibration curve created by analysis of standards with known concentrations of the TSNAs to determine the TSNA content of each test portion.

5 SAFETY AND ENVIRONMENTAL PRECAUTIONS

CAUTION: Nitrosamines are potent carcinogens. Precautions shall be taken to avoid human exposure.

Nitrosamines and their solutions should be handled in an adequately ventilated fume hood, glove box or equivalent.

The laboratory shall establish procedures for disposal of solutions containing nitrosamines.

5.1 Take routine safety and environmental precautions, as in any chemical laboratory activity.

5.2 The testing and evaluation of certain products with this test method may require the use of materials or equipment that could be hazardous or harmful to the environment; this document does not address all the safety aspects associated with its use. All persons using this method have the responsibility to consult the appropriate authorities and to establish health and safety practices as well as environmental precautions in conjunction with any existing applicable regulatory requirements prior to its use.

5.3 Special care should be taken to avoid inhalation or dermal exposure to harmful chemicals. Use a chemical fume hood, and wear an appropriate laboratory coat, gloves and safety goggles when preparing or handling undiluted materials, standard solutions, extraction solutions or collected samples.

6 APPARATUS AND EQUIPMENT

Usual laboratory apparatus, in particular:

6.1 Equipment required to condition cigarettes as specified in ISO 3402 [**2.1**]

6.2 Equipment required to mark butt length as specified in ISO 4387 [**2.2**]

6.3 Equipment required to cover filter ventilation holes for the intense regimen as specified in WHO TobLabNet SOP 01 [**2.3**]

6.4 Equipment required to perform smoking of tobacco products as specified in ISO 3308 [**2.4**]

6.5 Analytical balance capable of measurements to at least four decimal places

6.6 Vortexer, wrist-action shaker or equivalent

6.7 Pipettes and tips capable of accurately dispensing volumes of 100–1000 µL

6.8 Volumetric pipette(s) or equivalent, 2 mL

6.9 Glass graduated cylinders, 25 mL and 2 L

6.10 Teflon-lined septa or equivalent

6.11 Volumetric flasks, 10 mL, 50 mL, 100 mL, 200 mL, 1 L and 2 L

6.12 Amber glass bottles, 60 mL, 1 L, or suitable flasks

6.13 25 mL × 0.45 µm GD/X nylon syringe filters or equivalent

6.14 5-mL syringe or equivalent

6.15 47 mm × 0.45 µm nylon membrane filters or equivalent

6.16 Glass transfer pipettes or equivalent

6.17 Autosampler vials, 2 mL or equivalent

6.18 HPLC system interfaced to MS-MS

6.19 MS-MS triple quadrupole mass spectrometer

6.20 HPLC column capable of distinct separation of TSNA and isotope-labelled TSNA peaks from those of other cigarette emission components, e.g. Agilent Zorbax Eclipse XDB-C18 (2.1 × 150 mm, 3.5-µm particle size)

6.21 Opti-Guard RP C_{18} guard column or appropriate equivalent (optional)

7 REAGENTS AND SUPPLIES

All reagents shall be of at least analytical reagent grade unless otherwise noted. When possible, reagents are identified by their Chemical Abstract Service (CAS) registry numbers.

7.1	3-(1-Nitrosopyrrolidin-2-yl)pyridine (NNN) [16543-55-8]	
7.2	Deuterium-labelled 3-(1-nitrosopyrrolidin-2-yl)pyridine (NNN-d_4)	
7.3	4-(Methylnitrosamino)-1-(3-pyridyl)-1-butanone (NNK) [64091-91-4]	
7.4	Deuterium-labelled 4-(methylnitrosamino)-1-(3-pyridyl)-1-butanone (NNK-d_4)	
7.5	N-Nitrosoanatabine (NAT) [71267-22-6]	
7.6	Deuterium-labelled N-nitrosoanatabine (NAT-d_4)	
7.7	N-Nitrosoanabasine (NAB) [37620-20-5]	
7.8	Deuterium-labelled N-nitrosoanabasine (NAB-d_4)	
7.9	Ammonium acetate, HPLC-grade [631-61-8]	
7.10	Glacial acetic acid [64-19-7]	
7.11	Water Type 1 [7732-18-5], deionized water or equivalent	
7.12	Acetonitrile, HPLC-grade [75-05-8]	
7.13	Methanol, HPLC-grade [67-56-1]	

8　PREPARATION OF GLASSWARE

Clean and dry glassware in a manner to avoid contamination.

9　PREPARATION OF SOLUTIONS

The method for preparing the solutions described below is for reference purpose and can be adjusted if necessary.

9.1　Ammonium acetate (100 mmol/L): extraction solution

9.1.1　Weigh approximately 7.7 g of ammonium acetate.

9.1.2　Dissolve the measured ammonium acetate in water in a 1-L volumetric flask or other suitable flask (e.g. beaker).

9.1.3　Cap tightly, and shake the contents to ensure complete mixing.

9.1.4　Transfer to a 1-Lamber glass bottle, and store at 4–10 °C.

9.2　Mobile phase A (0.1% acetic acid in water (v/v))

9.2.1　Transfer approximately 1.4 L of water into a 2-L volumetric flask or other suitable flask (e.g. beaker).

9.2.2　Pipette 2 mL of glacial acetic acid into the flask.

9.2.3　Mix and add water to volume.

9.2.4　Cap and mix the solution, and store at 4–10 °C.

9.3　Mobile phase B (0.1% acetic acid in methanol (v/v))

9.3.1　Transfer approximately 1.4 L of methanol into a 2-L volumetric flask or other suitable flask (e.g. beaker).

9.3.2　Pipette 2 mL of glacial acetic acid into the flask.

9.3.3　Mix, and add methanol to volume.

9.3.4　Cap and mix the solution, and store at 4–10 °C.

10 PREPARATION OF STANDARDS

The method for preparing standard solutions described below is for reference purposes and can be adjusted if necessary.

10.1 Isotope-labelled internal standards

10.1.1 TSNA internal standard stocks (0.4 mg/mL)

10.1.1.1 Weigh approximately 4 mg (m_{IS}) of each isotope-labelled TSNA into individual 10-mL volumetric flasks on a four-decimal-place balance, and record the weight to the nearest 0.1mg.

10.1.1.2 Dissolve each of the measured TSNA internal standards in 10 mL of acetonitrile, and mix well.

10.1.1.3 Label, and store at −20 ± 5 °C in amber vials. The solution is stable for 2 years.

10.1.2 TSNA mixed internal standard solution (5 μg/mL)

10.1.2.1 Pipette 1.25 mL (V_{IS}) of each internal standard stock from **10.1.1** into a 100-mL volumetric flask.

10.1.2.2 Dilute to the mark with 100 mmol/L ammonium acetate, and mix well.

10.1.2.3 Label and store at −20 ± 5 °C in amber vials. The solution is stable for 2 weeks.

10.2 Native standards

Individual standard stock solutions are prepared in acetonitrile and stored at −20 ± 5 °C. Intermediate mixed standard solutions are made from the individual standard stocks.

10.2.1 TSNA standard stock solutions (0.4 mg/mL)

10.2.1.1 Weigh approximately 4 mg (m_s) of each of the four TSNAs on a four-decimal-place analytical balance, and record the weight.

10.2.1.2 Dissolve each of the measured TSNA standards in approximately 8 mL of acetonitrile in a 10-mL volumetric flask.

10.2.1.3 Fill to the mark with acetonitrile, and mix well.

10.2.1.4 Label, and store at −20 ± 5 °C in amber vials. The solution is stable for 2 years.

10.2.2 TSNA intermediate mixed standard solution (1 μg/mL)

10.2.2.1 Pipette 25 μL (V_{ss}) of each of the four standard stocks from **10.2.1** into a 10-mL volumetric flask.

10.2.2.2 Dilute to the mark with 100 mmol/L ammonium acetate, and mix well.

10.2.2.3 Label, and store at −20 ± 5 °C in an amber vial. The solution is stable for 2 weeks.

10.2.3 TSNA final mixed standard solutions

10.2.3.1 Prepare final standard solutions according to Table 1 below.

10.2.3.2 Spike variable volumes of the intermediate mixed standard solution (**10.2.2**) into a 10-mL volumetric flask.

10.2.3.3 Add 100 μL of the internal standard solution (**10.1.2**), fill to the mark with 100 mmol/L ammonium acetate, and mix well.

10.2.3.4 The final concentrations of the internal standards can be determined from:

Final concentration (IS) (ng/mL)

$$= \frac{m_{IS}}{10} \times \frac{V_{IS}}{100} \times \frac{0.1}{10} \times 10^6$$

where
m_{IS} is the original weight (in mg) of the internal standard determined in 10.1.1.1
V_{IS} is the volume of the mixed internal standard solution (in mL) pipetted in **10.1.2.1**.

Note: 10^6 is the factor to converting mg to ng

10.2.3.5 The final concentrations of standards are determined from:

Final concentration (ng/mL)

$$= \frac{m_S}{10} \times \frac{V_{SS}}{10} \times \frac{V_{MIX}}{10} \times 10^6$$

where
m_S is the original weight (in mg) of the standard determined in 10.2.1.1,
V_{SS} is the volume of the mixed standard solution (in mL) pipetted in 10.2.2.1,
V_{MIX} is the volume of the mixed standard solution (in mL) indicated in Table 1

10.2.3.6 Additional, higher concentrations of standards may be required to extend the calibration range to cover products with high levels of TSNAs.

Table 1. TSNA calibration standards

Standard	Volume of TSNA intermediate mixed standard solution (1μg/mL) (mL) (V_{MIX})	Volume of mixed TSNA internal standard solution (mL) (V_{IS})	Total volume (mL)	Approximate concentration of each TSNA in final mixed standard solution (ng/mL)
1	0.005	0.100	10	0.5
2	0.010	0.100	10	1.0
3	0.020	0.100	10	2.0
4	0.050	0.100	10	5.0
5	0.200	0.100	10	20.0
6	0.500	0.100	10	50.0
7	1.000	0.100	10	100.0
8	2.000	0.100	10	200.0

The range of the standard solutions may be adjusted, depending on the equipment used and the samples to be tested, keeping in mind the possible effect on the sensitivity of the method.

All solvents and solutions must be adjusted to room temperature before use.

11 SAMPLING

11.1 Sample cigarettes according to ISO 8243 [**2.5**] or alternative approaches may be used to obtain a representative laboratory sample in accordance with individual laboratory practice or when required by specific regulation or availability of samples.

11.2 Constitution of test sample

11.2.1 Divide the laboratory sample into separate units (e.g. packet, container), if applicable.

11.2.2 Take an equal amount of product for each test sample from at least √n [**2.6**] of the individual units (e.g. packet, container).

12 CIGARETTE PREPARATION

12.1 Condition all cigarettes to be smoked in accordance with ISO 3402 [**2.1**].

12.2 Mark cigarettes at a butt length in accordance with ISO 4387 [**2.2**] and WHO TobLabNet SOP01 [**2.3**].

12.3 Prepare test samples to be smoked in accoudance to either ISO or intense smoking conditions as specified in WHO TobLabNet SOP 01 [**2.3**].

13 PREPARATION OF THE SMOKING MACHINE

13.1 Ambient conditions

The ambient conditions for smoking are specified in ISO 3308 [**2.4**].

13.2 Smoking machine specifications

Follow ISO 3308 [**2.4**] smoking machine specifications, except for intense smoking, for which the smoking machine should be prepared as described in WHO TobLabNet SOP01 [**2.3**].

14 SAMPLE GENERATION

Smoke a sufficient number of cigarettes on the specified smoking machine such that breakthrough does not occur and the concentrations of the NNN, NNK, NAT and NAB fall within the calibration range prepared for the analysis.

14.1 Smoke the cigarette test samples and collect the TPM as specified in ISO 4387 [**2.2**] or in WHO TobLabNet SOP 01 [**2.3**].

14.2 Include at least one reference test sample to be used for quality control, if applicable.

14.3 When testing sample types for the first time, breakthrough should be evaluated. The number of cigarettes might have to be adjusted to prevent breakthrough of the filter pad. If the TPM exceeds 600 mg for a 92-mm filter pad or 150 mg for a 44-mm filter pad, the number of cigarettes smoked onto each pad must be decreased.

14.4 The number of cigarettes to be smoked per result for linear and rotary smoking machines at ISO and intense smoking regimens are shown in Table 2.

Table 2. Number of cigarettes to be smoked for one measurement in linear and rotary smoking machines

	ISO smoking regimen		Intense smoking regimen	
	Linear	Rotary	Linear	Rotary
No. of cigarettes per pad (n)	5	20	3	10
No. of pads per result	4	1	7	2

14.5 Record the number of cigarettes and total puffs for each filter pad.

14.6 After smoking the required number of test samples, perform five clearing puffs, and remove the pad holder from the smoking machine.

15 SAMPLE PREPARATION
15.1 Extraction of filter pads

15.1.1 Remove the pads from the holders. Fold each pad loosely in half and then in half again, with the TPM on the inside. Use the side opposite to that on which TPM is collected to wipe the inner surface of the pad holder, thus including any particulate matter that may have been left in the holder. Place each filter pad in a 60-mL amber bottle or equivalent.

15.1.2 Spike each pad with the TSNA mixed internal standard solution (**10.1.2**) containing isotope-labelled TSNA (200 μL for a 44-mm pad, 500 μl for a 92-mm pad).

15.1.3 Pipette 100 mmol/L aqueous ammonium acetate extraction solution (20 mL for 44-mm pads, 50 mL for 92-mm pads) into the amber bottle.

15.1.4 Using a wrist-action or circular shaker or equivalent device, extract the TSNAs from the pad by shaking at 250 rpm for 30 min.

15.1.5 Remove aliquots (e.g. 1–2 mL) of the extract, filter the aliquots with at most 0.45 µm membrane filter, and place in an autosampler vial.

16   SAMPLE ANALYSIS

This method for quantifying TSNAs in mainstream cigarette smoke involves HPLC coupled with a triple quadrupole mass spectrometer. The analytes are resolved from other potential interfering substances on an HPLC column. Further selectivity is obtained by use of a triple quadrupole mass spectrometer with electrospray ionization, in MS-MS mode. Comparison of the area ratio (native analyte area to isotope-labelled analyte area) of the unknowns with the area ratio (native analyte area to isotope-labelled analyte area) of the known standard concentrations yields the concentrations of individual analytes.

16.1   HPLC and mass spectrometer operating conditions

HPLC operating conditions: example

HPLC column:	Agilent Zorbax Eclipse XDB-C18 (2.1 × 150 mm, 3.5-µm particle size) or equivalent
Injection volume:	5 µL
Column oven temperature:	40 ± 1 °C (Alter as appropriate for the column used.)
Degasser:	On
Binary pump flow:	0.2 mL/min
Mobile phase A:	0.1 % acetic acid in water
Mobile phase B:	0.1 % acetic acid in methanol
Gradient:	Shown in Table 3
Total analysis time:	12 min

Table 3.   HPLC gradients for TSNA separation

Time (min)	Flow (µL/min)	Mobile phase A (%)	Mobile phase B (%)
0	200	50	50
3.0	200	10	90
4.0	200	0	100
5.0	200	0	100
5.5	200	50	50
12.0	200	50	50

Mass spectrometer operating conditions

Dwell time:	40 ms
Ionization / mode:	Electrospray ionization source / positive ion mode
Ionspray voltage:	1500 V
Turbolonspray temperature:	450 °C
Curtain gas:	Nitrogen
Collision-induced dissociation gas:	Nitrogen
Nebulizing gas:	Nitrogen

Note: The operating parameters may have to be adjusted, depending on the instrument and column conditions and the resolution of the chromatographic peaks.

Quantification and confirmation ion transitions for NNN, NNK, NAT and NAB are shown in Table 4.

Table 4. Quantification and confirmation ion pairs for each TSNA

Analyte	Ion pair Q1/Q3 (m/z)
NNN:Quantification	178/148
NNN: Confirmation	178/120
NNN-d_4	182/152
NNK:Quantification	208/122
NNK: Confirmation	208/106
NNK-d_4	212/126
NAT:Quantification	190/160
NAT: Confirmation	190/106
NAT-d_4	194/164
NAB:Quantification	192/162
NAB: Confirmation	192/133
NAB-d_4	196/166

Note: The confirmation ion ratio can be a useful guide for determining individual analyte integrity.

16.2 Expected retention times

16.2.1 For the conditions described here, the expected sequence of elution will be NNN, NNN-d_4, NNK, NNK-d_4, NAT, NAT-d_4, NAB, NAB-d_4.

16.2.2 Differences in e.g. flow rate, mobile phase concentration and age of the column can be expected to alter retention times.

16.3 Determination of TSNAs

The sequence of determination will be in accordance with individual laboratory practice. This section gives an example of a sequence of determination for TSNAs.

16.3.1 Inject a blank solution (extraction solution minus labelled internal standards(s)) to check for any contamination in the system or reagents.

16.3.2 Inject the blank solution (blank with labelled internal standards(s)) to verify the performance of the HPLC-MS-MS system.

16.3.3 Inject the calibration standards, quality control and samples.

16.3.4 Record the peak areas of each TSNA and the internal standard(s).

16.3.5 Calculate the relative response ratio (RF) of the peak to the internal standard peak (RF = A_{TSNA} / A_{IS}) for each of the component standard solutions for each TSNA, including the standard blanks.

16.3.6 Plot the graph of the concentration of each TSNA (X axis) in accordance with the area ratios (Y axis).

16.3.7 The intercept should not be statistically significantly different from zero.

16.3.8 The standard curve should be linear over the entire standard range.

16.3.9 Calculate the linear regression ($Y = a + bx$) from these data, and use both the slope (b) and the intercept (a) of the linear regression. If the linear regression R^2 is less than 0.99, the calibration should be repeated. If an individual calibration point differs by more than 10% from the expected value (estimated by linear regression), the point should be omitted.

16.3.10 Inject the quality controls and samples, and determine peak areas with the appropriate software.

16.3.11 The peak ratio obtained for all test portions must fall within the working range of the calibration curve; otherwise, the standard solutions or test portions concentrations should be adjusted.

See Annex 1 for representative chromatograms.

17 DATA ANALYSIS AND CALCULATIONS

The slope and intercept are determined from the relative responses of the analyte and the labelled standard versus the relative concentrations of the analyte and the labelled standard.

17.1 Calculate the relative response ratio from the peak areas for each of the TSNAs for each of the calibration standards (**10.2.3**):

$$RF = \frac{A_a}{A_{IS}}$$

where
RF is the relative response ratio,
A_a is the area of the target analyte
A_{IS} is the area of the corresponding internal standard.

Plot a graph of the relative response factor, A_a/A_{IS}, (Y axis) versus concentration (X axis) for each TSNA. Calculate the linear regression (Y = a + bX) from these data, and use both the slope (b) and the intercept (a) of the linear regression.

17.2 The standard curve should be linear over the entire standard range.

17.3 The content M_{ts} of a given TSNA of the target analyte (ng/cigarette) is determined from the calculated relative response ratio for the test portion, the slope and intercept obtained from the appropriate calibration curve, and the equation:

$$M_{ts} = \frac{Y - a}{b} \times \frac{V}{n}$$

where
M_{ts} is the calculated content in nanogram per cigarette,
Y is the relative response ratio (A_a/A_{IS})
a is the intercept of the linear regression obtained from the standard calibration curve
b is the slope of the linear regression obtained from the standard calibration curve
V is the volume of extraction solution 20mL for 44mm Cambridge Filter Pad used or 50mL for 92 mm Cambridge Filter Pad used.
n is the number of cigarettes smoked on to each pad (refer to Table 2)

Alternative calculation procedures may be used if applicable.

18 SPECIAL PRECAUTIONS

18.1 After installing a new column, condition it as specified by the manufacturer by injecting a tobacco sample extract under the specified instrument conditions. Injections should be repeated until the peak areas (or heights) for each TSNA and the internal standards are reproducible.

19 DATA REPORTING

19.1 Report individual measurements for each sample evaluated.

19.2 Report results as ng/cigarette or as required.

20 QUALITY CONTROL

20.1 Control parameters

Note: If the control measurements are outside the tolerance limits of the expected values, appropriate investigation and action must be taken.

Note: Additional laboratory quality assurance procedures should be carried out if necessary in order to comply with the policies of individual laboratories.

20.2 Laboratory reagent blank

To detect potential contamination during sample preparation and analysis, include a laboratory reagent blank, as described in **16.3.2**. The blank consists of all reagents and materials used in analysing test samples and is analysed like a test sample. The blank should be assessed in accordance with the practices of individual laboratories.

20.3 Quality control sample

To verify the consistency of the entire analytical process, analyse a reference cigarette (or an appropriate quality control sample) in accordance with the practices of individual laboratories.

20.4 A new calibration curve is made after every 20 runs (20 sample extracts).

21 METHOD PERFORMANCE SPECIFICATIONS

21.1 Limit of reporting

The limit of reporting is set to the lowest concentration of the calibration standards used, recalculated to ng/cigarette (e.g. 20 ng/cigarette corresponds to the lowest calibration standard concentration of 0.5 ng/mL).

21.2 Laboratory-fortified matrix recovery

Recovery of analyte spiked onto the matrix is used as a surrogate measure of accuracy. Recovery is determined by spiking known amounts of standards (after smoking) onto a filter pad with cigarette smoke and extracting the pad by the same method as used for samples. Unspiked pads are also analysed. The recovery is calculated from the following equation and is shown in Table 5:

Recovery = 100 × (analytical result − unspiked result/spiked amount)

Table 5. Mean and recovery of TSNAs spiked onto the matrix

Spiked amount (ng)	NNN		NNK		NAT		NAB	
	Mean (ng)	Recovery (%)	Mean (ng)	Recovery (%)	Mean (ng)	Recovery (%)	Mean (ng)	Recovery (%)
150	146.3	97.5	145.3	96.8	161.3	107.5	149.1	99.4
500	452.8	90.6	473.8	94.8	516.5	103.3	509.4	101.9
1000	962.5	96.3	973.8	97.4	1015.8	101.6	1001.1	100.1

21.3 Analytical specificity

LC−MS-MS provides excellent analytical specificity. The retention time and confirmation ion ratio, an established range of ratios of the response of the quantification ion to that of the confirmation ion and quality control samples, are used to verify the specificity of the results for an unknown sample.

21.4 Linearity

The TSNA calibration curves established are linear over the standard concentration range of 0.5−200 ng/mL.

21.5 Possible interference

No known components have both similar retention time and the same confirmation ion pair ratio as TSNAs or internal standards.

22 REPEATABILITY AND REPRODUCIBILITY

An international collaborative study [**2.8**] conducted between 2009 and 2012, involving nine laboratories and three reference cigarettes (1R5F; 3R4F and CM6) and two commercial brands, gave the precision limits for this method indicated in Table 6,7,8, and 9 below.

The difference between two single results found for matched cigarette samples by the same operator using the same apparatus within the shortest feasible time will exceed the repeatability, r, on average not more than once in 20 cases in the normal, correct operation of the method.

Single results for matched cigarette samples reported by two laboratories will differ by more than the reproducibility, R, on average not more than once in 20 cases with normal, correct application of the method.

The test results were analysed statistically in accordance with ISO 5725-1 [**2.9**] and ISO 5725-2 [**2.10**] to give the precision data shown in Tables 6–9.

For the purpose of calculating r and R, one test result under ISO smoking conditions was defined as the average of seven individual replicates of the mean yield of four Cambridge filter pads (five cigarettes smoked per pad) from linear smoking machines and one Cambridge filter pad (20 cigarettes smoked per pad) from rotary smoking machines. Under intense smoking conditions, one test result was defined as the average of seven replicates of the mean yield of seven Cambridge filter pads (three cigarettes smoked per pad) from linear smoking machines and two Cambridge filter pads (10 cigarettes smoked per pad) from rotary smoking machines. For more information, see reference [**2.8**].

Note: The levels of TSNAs in the commercial brands were within the range of those in the reference test cigarettes and therefore not reported in this table.

Table 6. Precision limits for determination of NNN (ng/cigarette) in mainstream cigarette smoke from reference cigarettes

Reference cigarette	ISO smoking regimen				Intense smoking regimen			
	N	m_{cig}	r_{limit}	R_{limit}	N	m_{cig}	r_{limit}	R_{limit}
1R5F	9	45.55	6.15	11.22	9	256.72	23.44	53.37
3R4F	9	113.76	12.24	20.02	9	302.24	26.94	60.77
CM6	8	21.02	3.97	8.05	8	40.26	6.77	18.18

Table 7. Precision limits for determination of NNK (ng/cigarette) in mainstream cigarette smoke from reference cigarettes

Reference cigarette	ISO smoking regimen				Intense smoking regimen			
	N	m_{cig}	r_{limit}	R_{limit}	N	m_{cig}	r_{limit}	R_{limit}
1R5F	8	23.31	6.17	11.90	9	128.28	21.23	49.05
3R4F	8	95.81	12.65	33.78	9	259.25	34.27	84.70
CM6	9	28.47	6.63	13.46	9	55.91	14.03	29.35

Table 8. Precision limits for determination of NAT (ng/cigarette) in mainstream cigarette smoke from reference cigarettes

Reference cigarette	ISO smoking regimen				Intense smoking regimen		
	N	m_{cig}	r_{limit}	R_{limit}	N	m_{cig}	r_{limit}	R_{limit}
1R5F	7	102.43	11.52	48.66	7	264.63	24.41	129.21
3R4F	6	39.7	6.95	12.97	7	225.58	19.36	116.45
CM6	7	32.03	4.45	15.41	7	62.48	8.84	30.59

Table 9. Precision limits for determination of NAB (ng/cigarette) in mainstream cigarette smoke from reference cigarettes

Reference cigarette	ISO smoking regimen				Intense smoking regimen		
	N	m_{cig}	r_{limit}	R_{limit}	N	m_{cig}	r_{limit}	R_{limit}
1R5F	6	12.85	1.87	3.46	6	31.95	5.65	10.38
3R4F	6	6.36	1.87	2.12	6	29.48	3.49	7.62
CM6	7	4.67	1.55	6.21	7	8.98	2.88	10.93

where
1R5F, 3R4F and CM6 are three reference cigarettes analysed in this study
ng is nanogram
N is the number of laboratories participated
m_{cig} is the mean value of respective nitorsamine
r_{limit} is the repeatability limit of the respective introsamine
R_{limit} is the reproducibility limit of the respective introsamine

23 Test report

The following information shall be included in the test report:

    (a) A reference to this method i.e. WHO TobLabNet SOP 03

    (b) Date of receipt of the sample

    (c) The result and its units

Annex 1. Typical chromatograms obtained in the determination of tobacco-specific nitrosamines in mainstream cigarette smoke

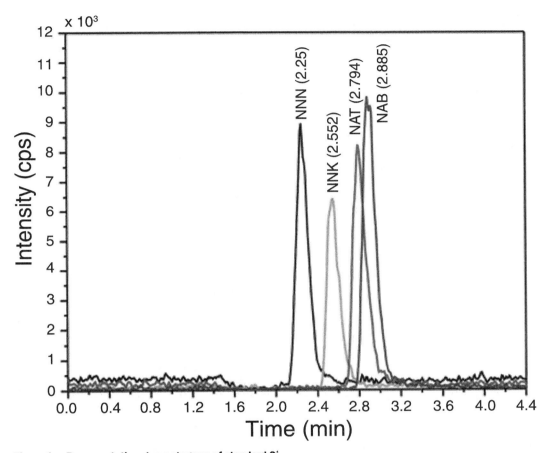

Figure 1. Representative chromatogram of standard 2[1]

---

[1] Derived from an international collaborative study conducted in 2009 and provided by the Centers for Disease Control and Prevention for information only.

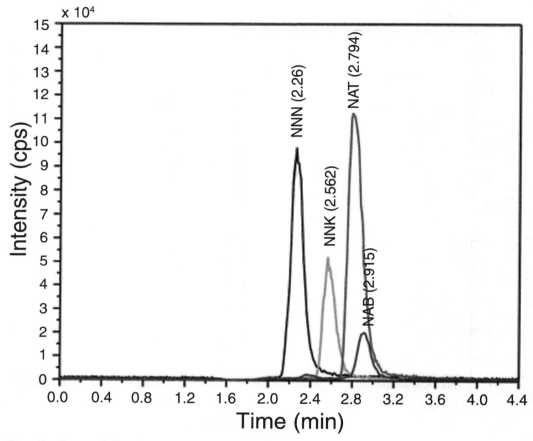

Figure 2. Representative chromatogram of TSNAs in cigarette mainstream smoke[1]

[1] Derived from an international collaborative study conducted in 2009 and provided by the Centers for Disease Control and Prevention for information only.

WHO TobLabNet
Official Method
SOP 04

Standard Operating Procedure for Determination of Nicotine in Cigarette Tobacco Filler

Method:	Determination of nicotine in cigarette tobacco filler
Analytes:	Nicotine(3-[(2S)-1-methylpyrrolidin-2-yl]pyridine) (CAS # 54-11-5)
Matrix:	Cigarette tobacco filler
Last update:	June 2014

The mention of specific companies or of certain manufacturers' products does not imply that they are endorsed or recommended by WHO in preference to others of a similar nature that are not mentioned. Errors and omissions excepted, the names of proprietary products are distinguished by initial capital letters.

> No machine smoking regimen can represent all human smoking behaviour: machine smoking testing is useful for characterizing cigarette emissions for design and regulatory purposes, but communication of machine measurements to smokers can result in misunderstanding about differences between brands in exposure and risk. Data on smoke emissions from machine measurements may be used as inputs for product hazard assessment, but they are not intended to be nor are they valid as measures of human exposure or risks. Representing differences in machine measurements as differences in exposure or risk is a misuse of testing with WHO TobLabNet standards.

FOREWORD

This document was prepared by members of the World Health Organization (WHO) Tobacco Laboratory Network (TobLabNet) as an analytical method standard operating procedure (SOP) for measuring nicotine in cigarette tobacco filler.

INTRODUCTION

In order to establish comparable measurements for testing tobacco products globally, consensus methods are required for measuring specific contents and emissions of cigarettes. The Conference of the Parties to the WHO Framework Convention on Tobacco Control (FCTC) at its third session in Durban, South Africa, in November 2008, "recalling its decisions FCTC/COP1(15) and FCTC/COP2(14) on the elaboration of guidelines for implementation of Articles 9 (*Regulation of the contents of tobacco products*) and 10 (*Regulation of tobacco product disclosures*) of the WHO FCTC, noting the information contained in the report of the working group to the third session of the Conference of the Parties on the progress of its work ... requested the Convention Secretariat to invite WHO's Tobacco Free Initiative to ... validate, within five years, the analytical chemical methods for testing and measuring cigarette contents and emissions" (FCTC/COP/3/REC/1).

Using the criteria for prioritization set at its third meeting in Ottawa, Canada, in October 2006, the working group on Articles 9 and 10 identified the following contents for which methods for testing and measurement (analytical chemistry) should be validated as a priority:

- nicotine
- ammonia
- propylene glycol (propane-1,2-diol)
- glycerol (propane-1,2,3-triol)
- triethylene glycol (2,2-ethylenedioxybis(ethanol)).

Measurement of these contents will require validation of three methods: one for nicotine, one for ammonia and one for humectants.

Using the criteria for prioritization set at the meeting in Ottawa mentioned above, the working group identified the following emissions in mainstream smoke for which methods for testing and measurement (analytical chemistry) should be validated as a priority:

- 4-(methylnitrosamino)-1-(3-pyridyl)-1-butanone (NNK)
- *N*-nitrosonornicotine (NNN)
- acetaldehyde
- acrylaldehyde (acrolein)
- benzene
- benzo[*a*]pyrene
- 1,3-butadiene
- carbon monoxide
- formaldehyde

Measurement of these emissions with the two smoking regimens described below will require validation of five methods: one for tobacco-specific nitrosamines (NNK and NNN), one for benzo[a]pyrene, one for aldehydes (acetaldehyde, acrolein and formaldehyde), one for volatile organic compounds (benzene, 1,3-butadiene) and one for carbon monoxide.

The table below sets out the two smoking regimens for validation of the test methods referred above.

Smoking regimen	Puff volume (mL)	Puff frequency	Filter ventilation holes
ISO regimen: ISO 3308; *Routine analytical cigarette smoking machine—definitions and standard conditions*	35	Once every 60 s	No modifications
Intense regimen: Same as ISO 3308 but modified as indicated	55	Once every 30 s	All ventilation holes must be blocked 100% as described in WHO TobLabNet SOP 01.

This SOP was prepared to describe the procedure for the determination of nicotine in cigarette tobacco filler.

1 SCOPE

This method is suitable for the quantitative determination of nicotine in cigarette tobacco filler by gas chromatography (GC).

2 REFERENCES

2.1 ISO 3402: *Tobacco and tobacco products—Atmosphere for conditioning and testing.*

2.2 ISO 13276: *Tobacco and tobacco products—Determination of nicotine purity—Gravimetric method using tungstosilicic acid.*

2.3 ISO 8243: *Cigarettes—Sampling*

2.4 United Nations Office on Drugs and Crime. *Guidelines on representative drug sampling.* Vienna, Laboratory and Scientific Section, 2009 (http://www.unodc.org/documents/scientific/Drug_Sampling.pdf).

2.5 World Health Organization. *Standard operating procedure for validation of analytical methods of tobacco product contents and emissions.* Geneva, Tobacco Laboratory Network, (WHO TobLabNet SOP 02 in preparation).

2.6 World Health Organization. *Method validation report of World Health Organization TobLabNet official method: Determination of nicotine in cigarette tobacco filler.* Geneva, Tobacco Laboratory Network, forthcoming.

2.7 ISO 5725-1: *Accuracy (trueness and precision) of measurement methods and results—Part 1: General principles and definitions.*

2.8 ISO 5725-2: *Accuracy (trueness and precision) or measurement methods and results—Part 2: Basic method for the determination of repeatability and reproducibility of a standard measurement method.*

3 TERMS AND DEFINITIONS

3.1 *Nicotine content*: Total amount of nicotine in cigarette tobacco filler, expressed as milligrams per gram cigarette tobacco filler

3.2 *Cigarette tobacco filler*: Tobacco-containing part of a cigarette, including reconstituted tobacco, stems, expanded tobacco and additives

3.3 *Tobacco products*: Products entirely or partly made of leaf tobacco as the raw material that are manufactured to be used by smoking, sucking, chewing or snuffing (Article 1(f) of the WHO FCTC)

3.4 *Laboratory sample*: Sample intended for testing in a laboratory, consisting of a single type of product delivered to the laboratory at one time or within a specified period

3.5 *Test sample*: Product to be tested, taken at random from the laboratory sample

3.6 *Test portion*: Random portion from the test sample to be used for a single determination

4 METHOD SUMMARY

4.1 After conditioning, the cigarette tobacco filler is ground and mixed.

4.2 Nicotine is extracted from the cigarette tobacco filler with a mixture of *n*-hexane, sodium hydroxide solution and water.

4.3 The organic layer is analysed by GC with a flame ionization detector.

4.4 The ratio of nicotine peak area to internal standard is compared on a calibration curve created by analysis of standards with known concentrations of nicotine to determine the nicotine content of each test portion.

5 SAFETY AND ENVIRONMENTAL PRECAUTIONS

5.1 Follow routine safety and environmental precautions, as in any chemical laboratory activity.

5.2 The testing and evaluation of certain products with this test method may require the use of materials or equipment that could be hazardous or harmful to the environment; this document does not address all the safety aspects associated with its use. All persons using this method have the responsibility to consult the appropriate authorities and to establish health and safety practices as well as environmental precautions in conjunction with any existing applicable regulatory requirements prior to its use.

5.3 Special care should be taken to avoid inhalation or dermal exposure to harmful chemicals. Use a chemical fume hood, and wear an appropriate laboratory coat, gloves and safety goggles when preparing or handling undiluted materials, standard solutions, extraction solutions or collected samples.

6 APPARATUS AND EQUIPMENT

Usual laboratory apparatus, in particular:

6.1 Equipment required to condition cigarette tobacco filler as specified in ISO 3402 [**2.1**].

6.2 Extraction flasks: Erlenmeyer flasks (250-mL) with stoppers, 100-mL Pyrex bottles with crimp seals and septa, 100-mL culture tubes with Teflon-lined caps or other suitable flasks

6.3 Linear shaker configured to hold the extraction flasks in position

6.4 Capillary GC equipped with a flame ionization detector

6.5 Capillary GC column capable of distinct separation of peaks for the solvent, the internal standard, nicotine and other tobacco components (e.g. Varian WCOT Fused Silica, 25 m × 0.25 mm ID; coating: CP-WAX 51)

6.6 Ultrasonic bath

7 REAGENTS AND SUPPLIES

All reagents shall be of at least analytical reagent grade unless otherwise noted. When possible, reagents are identified by their Chemical Abstract Service (CAS) registry numbers.

7.1 Carrier gas: Helium [7440-59-7] of high purity (> 99.999%)

7.2 Auxiliary gases: Air and hydrogen [1333-74-0] of high purity (> 99.999%) for the flame ionization detector

7.3 n-Hexane [110-54-3], GC grade, with a maximum water content of 1.0 g/L

7.4 –(–)Nicotine [54-11-5] of known purity not less than 98%. Nicotine salicylate [29790-52-1] of known purity not less than 98% may be used. The laboratory may verify the nicotine purity if necessary [**2.2**]

7.5 Sodium hydroxide [1310-73-2]

7.6 Internal standard: n-Heptadecane (purity not less than 98% of mass fraction) [629-78-7]. Quinaldine [91-63-4], isoquinoline [119-65-3], quinoline [91-22-5] or other suitable alternatives may be used.

8 PREPARATION OF GLASSWARE

8.1 Clean and dry glassware in a manner to avoid contamination.

9 PREPARATION OF SOLUTIONS

9.1 Sodium hydroxide solution (2 mol/L)

    9.1.1 Weigh approximately 80 g of sodium hydroxide.

    9.1.2 Dissolve measured sodium hydroxide in water and dilute with water to 1 L.

9.2 Extraction solution (0.5 mg/mL)

    9.2.1 Weigh approximately 0.5 g (to 0.001-g accuracy) of n-heptadecane or alternative internal standard.

9.2.2 Dissolve measured *n*-heptadecane or alternative internal standard in *n*-hexane, and dilute to 1 L with *n*-hexane.

10 PREPARATION OF STANDARDS

The method for preparing standard solutions described below is for reference purposes and can be adjusted if necessary.

10.1 Nicotine standard stock solution (2 g/L)

10.1.1 Weigh approximately 200 mg nicotine or 370 mg nicotine salicylate to 0.0001-g accuracy into a 250-mL flask.

10.1.2 Dissolve the measured nicotine in 50 mL of water.

10.1.3 Pipette 100 mL of extraction solution (**9.2.2**), and add 25 mL of 2 mol/L sodium hydroxide solution.

10.1.4 Shake the two-phase mixture obtained vigorously for 60 ± 2 min in a shaker. Take care to mix the phases well.

10.1.5 Separate the supernatant organic phase, and store this solution, protected from light, at 4–8 °C.

10.2 Nicotine standard solutions

10.2.1 Pipette 0.5 mL, 2.5 mL, 5.0 mL, 7.5 mL and 10.0 mL of the standard stock solution prepared in **10.1.5** into 20-mL volumetric flasks.

10.2.2 Fill the volumetric flasks to the mark with extraction solution (**9.2.2**).

10.2.3 Store the standard solutions, protected from light, at 4–8 °C.

10.2.4 Determine the final nicotine concentrations in the standard solutions from:

Final concentration (mg/L)

$$= x \times y \times \frac{1000}{100 \times 20} \times \text{purity of standard}$$

where
x is the original weight (in mg) of nicotine as weighed in **10.1.1**
y is the volume of the standard stock solution as pipetted in **10.2.1**

The final nicotine concentrations in the standard solutions are shown in Table 1.

Table 1. Concentrations of nicotine in standard solutions

Standard	Volume of nicotine standard solution (2g/L) (mL) (y)	Volume of internal standard solution	Total volume (mL)	Approximate nicotine concentration in final standard solution (mg/L)	Approximate level equivalent to unknown levels in cigarette tobacco filler (mg/g)
1	0.5		20	50	1.3
2	2.5	Not applicable, included in extraction solution	20	250	6.7
3	5.0		20	500	13.3
4	7.5		20	750	20.0
5	10.0		20	1000	26.7

The range of the standard solutions may be adjusted, depending on the equipment used and the samples to be tested, keeping in mind the possible effect on the sensitivity of the method.

All solvents and solutions must be adjusted to room temperature before use.

11 SAMPLING

11.1 Sample cigarettes according to ISO 8243 [**2.3**] or alternative approaches may be used to obtain a representative laboratory sample in accordance with individual laboratory practice, or when required by specific regulation or availability of samples.

11.2 Constitution of test sample

11.2.1 Divide the laboratory sample into separate units (e.g. packet, container), if applicable.

11.2.2 Take an equal amount of product for each test sample from at least √n [**2.4**] of the individual units (e.g. packet, container).

12 CIGARETTE PREPARATION

12.1 Remove the tobacco filler from the cigarettes and quality control samples (when applicable) in one packet (e.g. containing 20 cigarettes), or use at least 15 g of processed cigarette tobacco filler.

12.2 Combine and mix sufficient cigarette tobacco filler to constitute at least 10 g for each test portion. At least three replicates of test portions are prepared.

12.3 Mix and grind the cigarette tobacco filler until it is fine enough to pass through a 4-mm screen.

12.4 Condition the ground cigarette tobacco filler as required for the tobacco product according to ISO 3402 [**2.1**].

13 PREPARATION OF THE SMOKING MACHINE

Not applicable.

14 SAMPLE GENERATION

Not applicable.

15 SAMPLE PREPARATION

15.1 For each test portion, take 1.5 g of the well-mixed, ground, conditioned test portion, and weigh it to 0.001-g accuracy into an extraction vessel.

15.2 Mix the test portion with 20 mL of water, 40 mL of extraction solution (V_e) (**9.2.2**) and 10 mL of 2 mol/L sodium hydroxide solution.

15.3 Shake the flask for 60 ± 2 min on a shaker.

15.4 Leave the sample flask to stand for another 20 min to allow visible, clear separation of the phases. After separation of the phases, analyse the organic (upper) phase as rapidly as possible by GC. To facilitate separation of the phases, the Erlenmeyer flask can be placed in an ultrasonic bath.

15.5 If the extracted sample is to be stored, keep it protected from light at 4–8 °C.

16 SAMPLE ANALYSIS

GC coupled with a flame ionization detector is used to quantify nicotine in cigarette tobacco filler. The analytes are resolved from other potential interferences on the GC column. Comparison of the area ratio of the unknowns with the area ratio of the known standard concentrations yields individual analyte concentrations.

16.1 GC operating conditions: example

GC column:	Varian WCOT fusedsilica, 25 m × 0.25 mm ID
Coating:	CP-WAX 51 or equivalent
Column temperature:	170 °C (isothermal)
Injection temperature:	270 °C
Detector temperature:	270 °C
Carrier gas:	Helium at a flow rate of 1.5 mL/min
Injection volume:	1.0 µl
Injection mode:	Split 1:10

Note: The operating parameters might have to be adjusted, depending on the instrument and column conditions and the resolution of chromatographic peaks.

16.2 Expected retention times

16.2.1 For the conditions described here, the expected sequence of elution will be *n*-heptadecane, nicotine.

16.2.2 Differences in e.g. temperature, gas flow rate and age of the column may alter retention times.

16.2.3 Under the above conditions, the expected total analysis time will be about 10 min. The analysis time may be extended to optimize performance.

16.3 Determination of nicotine

The sequence of determination will be in accordance with individual laboratory practices. This section illustrates an example of a sequence of determination for nicotine in cigarette tobacco filler.

16.3.1 Inject a blank solution (extraction solution minus internal standard) to check for any contamination in the system or reagents.

16.3.2 Inject the standard blank (blank with internal standard) to verify the performance of the Gas-chromatography system.

16.3.3 Inject the calibration standards, quality control, and samples.

16.3.4 Assess the retention times and responses (area counts) of the standards. If the retention times are similar (± 0.2 min) to the retention times in previous injections, and the responses are within 20% of typical responses in previous injections, the system is ready to perform the analysis. If the responses are outside specifications, seek corrective action according to your laboratory policy.

16.3.5 Record the peak areas of nicotine and the internal standard.

16.3.6 Calculate the relative response ratio (RF) of the nicotine peak to the internal standard peak ($RF = A_{nicotine} / A_{IS}$) for each of the nicotine standard solutions, including the standard blanks.

16.3.7 Plot a graph of the concentration of nicotine (X axis) against the area ratios (Y axis).

16.3.8 The intercept should not be statistically significantly different from zero.

16.3.9 The standard curve should be linear over the entire standard range.

16.3.10 Calculate a linear regression equation ($Y = a + bx$) from this data, and use both the slope (b) and the intercept (a) of the linear regression. If the linear regression coefficient R^2 is less than 0.99, the calibration should be repeated. If an individual calibration point differs by more than 10% from the expected value (estimated by linear regression), this point should be omitted.

16.3.11 Inject 1 µl of each of the quality control samples and test portions, and determine the peak areas with the appropriate software.

16.3.12 The signal (peak ratios) obtained for all test portions must fall within the working range of the calibration curve; otherwise, standard solutions or test portions concentrations should be adjusted.

See Appendix 1 for representative chromatograms.

17 DATA ANALYSIS AND CALCULATIONS

17.1 For each test portion, calculate the ratio of the nicotine response to the internal standard response (Y_t) from the peak area.

17.2 Calculate the nicotine concentration in mg/L (milligram per litre) for each test portion aliquot using the coefficients of the linear regression ($m_t = (Y_t - a) / b$).

where

m_t is the concentration of nicotine in the test solution in milligram per litre

Y_t is the ratio of nicotine response to the internal standard response from the peak area

a is the intercept of the linear regression obtained from the standard calibration curve

b is the slope of the linear regression obtained from the standard calibration curve

17.3 Calculate the nicotine content, m_n, of the tobacco sample expressed in mg/g (milligram per gram tobacco) using the following equation:

$$m_n = \frac{m_t \times V_e}{m_o \times 1000}$$

where

m_t is the concentration of nicotine in the test portion, in mg/L (milligram per litre);

V_e is the volume of the extraction solution used, in mL (millilitre);

m_o is the mass of the test portion (15.1), in g (gram)

18 SPECIAL PRECAUTIONS

18.1 After installing a new column, condition it by injecting a tobacco sample extract under the GC conditions described. Injections should be repeated until the peak areas (or heights) of both the nicotine and the internal standard are reproducible.

18.2 It is recommended to purge high-boiling-point components from the GC column after each sample set (series) by raising the column temperature to 220 °C for 30 min.

18.3 When the peak areas (or heights) for the internal standard are significantly higher than expected, it is recommended that the tobacco sample be extracted without internal standard in the extraction solution. This makes it possible to determine whether any component co-elutes with the internal standard, which would cause artificially lower values for nicotine.

19 DATA REPORTING

19.1 Report individual measurements for each sample evaluated.

19.2 Report results as mg/g tobacco or as required.

20 QUALITY CONTROL

20.1 Control parameters

Note: If the control measurements are outside the tolerance limits of the expected values, appropriate investigation and action must be taken.

Note: Additional laboratory quality assurance procedures should be carried out if necessary in order to comply with the policies of individual laboratories.

20.2 Laboratory reagent blank

To detect potential contamination in solvents or chemicals, include a laboratory reagent blank as described in **16.3.1**. The blank consists of all reagents and materials used in preparing test portions and is analysed like a test sample. The result should be less than the limit of detection.

20.3 Quality control sample

To verify the consistency of the entire analytical process, analyse a reference cigarette (or an appropriate quality control sample) in accordance with the practices of individual laboratories.

20.4

Quality control samples should be run after every 20 runs. A new calibration curve is made after every 24 hours.

21 METHOD PERFORMANCE SPECIFICATIONS

21.1 Limit of reporting

The limit of reporting is set to the lowest concentration of the calibration standards used, recalculated to mg/g (e.g. 1.3 mg/g corresponds to the lowest calibration standard concentration of 50 mg/L).

21.2 Laboratory-fortified matrix recovery

Recovery of analyte spiked onto the matrix is used as a surrogate measure of accuracy. Recovery is determined by spiking known amounts of standards (before extraction) into an Erlenmeyer flask with tobacco and extracting the nicotine by the same method as for samples. Unspiked tobacco is also analysed. The recovery is calculated from the following equation and as shown in Table 2:

Recovery (%) = 100 × (analytical result − unspiked result) / spiked amount

Table 2. Mean and recovery of nicotine spiked onto the matrix

Spiked amount (mg/g)	Nicotine (mg/g)	
	Mean (mg/g)	Recovery (%)
4.92	4.81	97.8
5.34	5.48	102.7
8.01	7.75	96.8
9.84	9.72	98.8
13.37	12.72	95.1
19.68	19.60	99.6

21.3 Analytical specificity

The retention time of the analyte of interest is used to verify the analytical specificity. An established range of ratios of the response of the component to that of the internal standard component of quality control cigarette tobacco filler is used to verify the specificity of the results for an unknown sample.

21.4 Linearity

The nicotine calibration curves established are linear over the standard concentration range of 50–1000 mg/L (1.3–26.7 mg/g).

21.5 Possible interference

The presence of eugenol can cause interference, as its retention time is similar to that of nicotine. This interference is most likely to occur with samples containing clove. The laboratory may need to resolve the interference by adjusting the analytical instrument parameters.

22 REPEATABILITY AND REPRODUCIBILITY

An international collaborative study [**2.6**] conducted in 2010, involving eighteen laboratories and three reference cigarettes (1R5F, 3R4F, and CM6) and two commercial brands, gave the precision limits for this method indicated in Table 3.

The difference between two single results found on matched cigarette tobacco filler samples by the same operator using the same apparatus within the shortest feasible time will exceed the repeatability, r, on average not more than once in 20 cases with normal, correct application of the method.

Single results for matched cigarette tobacco filler samples reported by two laboratories will differ by more than the reproducibility, R, on average no more than once in 20 cases with normal, correct application of the method.

The test results were analysed statistically in accordance with ISO 5725-1 [**2.7**] and ISO 5725-2 [**2.8**] to give the precision data shown in Table 3.

Table 3. Precision limits for determination of nicotine (mg/g) in cigarette tobacco filler

Reference cigarette	N	m_{cig}	r_{limit}	R_{limit}
1R5F	15	15.92	0.977	2.243
3R4F	17	17.16	1.152	2.414
CM6	17	18.77	1.378	2.635

Note: The levels of nicotine in the filler of the commercial brands were within the range of the reference cigarettes and are therefore not reported in this table.

where
1R5F, 3R4F, and CM6 are three reference cigarettes analysed in this study
mg/g is milligram per gram tobacco
N is the number of laboratories participated
m_{cig} is the mean value of nicotine content in cigarette tobacco filler
r_{limit} is the repeatability limit of nicotine content in cigarette tobacco filler
R_{limit} is the reproducibility limit of nicotine content in cigarette tobacco filler

23 Test report

The following information shall be included in the test report:
(a) A reference to this method i.e. WHO TobLabNet SOP 04
(b) Date of receipt of the sample
(c) The results and its units

Appendix 1. Typical chromatograms obtained in the analysis of cigarette tobacco filler for nicotine content

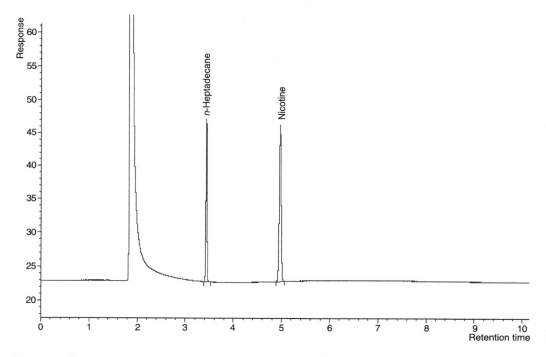

Figure 1. Example of a chromatogram of a standard solution with a nicotine concentration of 50 mg/L

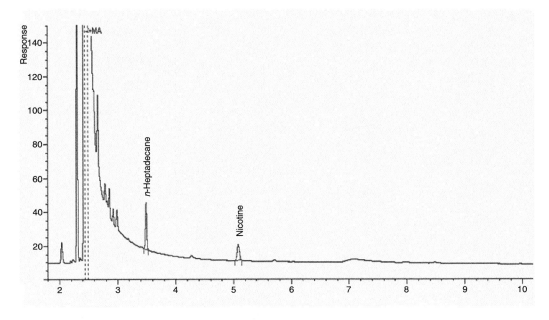

Figure 2. Example of a chromatogram of a test solution

WHO TobLabNet
Official Method
SOP 05

Standard Operating Procedure for Determination of Benzo[a]pyrene in Mainstream Cigarette Smoke under ISO and Intense Smoking Conditions

Method:	Determination of benzo[a]pyrene in mainstream cigarette smoke under ISO and intense smoking conditions
Analytes:	Benzo[a]pyrene (CAS # 50-32-8)
Matrix:	Tobacco cigarette mainstream smoke particulate matter
Last update:	February 2015

The mention of specific companies or of certain manufacturers' products does not imply that they are endorsed or recommended by WHO in preference to others of a similar nature that are not mentioned. Errors and omissions excepted, the names of proprietary products are distinguished by capital letters.

> No machine smoking regimen can represent all human smoking behaviour: machine smoking testing is useful for characterizing cigarette emissions for design and regulatory purposes, but communication of machine measurements to smokers can result in misunderstanding about differences between brands in exposure and risk. Data on smoke emissions from machine measurements may be used as inputs for product hazard assessment, but they are not intended to be nor are they valid as measures of human exposure or risk. Representing differences in machine measurements as differences in exposure or risk is a misuse of testing with WHO TobLabNet-recommended methods.

FOREWORD

This document was prepared by members of the World Health Organization (WHO) Tobacco Laboratory Network (TobLabNet) as an analytical method standard operating procedure (SOP) for measuring benzo[a]pyrene (B[a]P) in mainstream cigarette smoke under International Organization for Standardization (ISO) and intense smoking conditions.

INTRODUCTION

In order to establish comparable measurements for testing tobacco products globally, consensus methods are required for measuring specific contents and emissions of cigarettes. The Conference of the parties to the WHO Framework Convention on Tobacco Control (WHO FCTC) at its third session in Durban, South Africa, in November 2008, "recalling its decisions FCTC/COP1(15) and FCTC/COP2(14) on the elaboration of guidelines for implementation of Articles 9 (Regulation of the contents of tobacco products) and 10 (Regulation of tobacco product disclosures) of the WHO FCTC, noting the information contained in the report of the working group to the third session of the Conference of the Parties on the progress of its work ... requested the Convention Secretariat to invite WHO's Tobacco Free Initiative to ... validate, within five years, the analytical chemical methods for testing and measuring cigarette contents and emissions" (FCTC/COP/3/REC/1).

Using the criteria for prioritization set at its third meeting in Ottawa, Canada, in October 2006, the working group on Articles 9 and 10 identified the following contents for which methods for testing and measurement (analytical chemistry) should be validated as a priority:

- nicotine
- ammonia
- propylene glycol propane-1,2-diol)
- glycerol (propane-1,2,3-triol)
- triethylene glycol (2,2-ethylenedioxybis(ethanol)).

Measurement of these contents will require validation of three methods: one for nicotine, one for ammonia and one for humectants.

Using the criteria for prioritization set at the meeting in Ottawa mentioned above, the working group identified the following emissions in mainstream smoke for which methods for testing and measurement (analytical chemistry) should be validated as a priority:

- 4-(methylnitrosamino)-1-(3-pyridyl)-1-butanone (NNK)
- N-nitrosonornicotine (NNN)
- acetaldehyde
- acrylaldehyde (acrolein)
- benzene
- benzo[a]pyrene
- 1,3-butadiene
- carbon monoxide
- formaldehyde

Measurement of these emissions with the two smoking regimens described below will require validation of five methods: one for tobacco-specific nitrosamines (NNK and NNN), one for B[a]P, one for aldehydes (acetaldehyde, acrolein and formaldehyde), one for volatile organic compounds (benzene, 1,3-butadiene) and one for carbon monoxide).

The table below sets out the two smoking regimens for validation of the test methods referred to above.

Smoking regimen	Puff volume (ml)	Puff frequency	Filter ventilation holes
ISO regimen: *ISO 3308: Routine analytical cigarette smoking machine—definitions and standard conditions*	35	Once every 60 s	No modification
Intense regimen: Same as ISO 3308, but modified as indicated	55	Once every 30 s	All ventilation holes must be blocked 100% as described in WHO TobLabNet SOP 01.

This SOP was prepared to describe the procedure for the determination of B[a]P in mainstream cigarette smoke under ISO and intense smoking conditions.

1 **SCOPE**

This method is suitable for the quantitative determination of B[a]P in mainstream cigarette smoke by gas chromatography coupled with mass spectrometry (GC–MS).

B[a]P is a polycyclic aromatic hydrocarbon. It was classified as carcinogenic to humans (Group 1) by the International Agency for Research on Cancer (IARC) in 2012. It is formed during incomplete combustion of organic matter. Inhalation, oral ingestion and dermal absorption are important routes of entry.

2 REFERENCES

2.1 ISO 3402: *Tobacco and tobacco products—Atmosphere for conditioning and testing.*

2.2 ISO 4387: *Cigarettes—Determination of total and nicotine-free dry particulate matter using a routine analytical smoking machine.*

2.3 World Health Organization. *Standard operating procedure for intense smoking of cigarettes.* Geneva: Tobacco Laboratory Network (WHO TobLabNet SOP 01).

2.4 ISO 3308: *Routine analytical cigarette-smoking machine—Definitions and standard conditions.*

2.5 United Nations Office on Drugs and Crime. *Guidelines on representative drug sampling.* Vienna: Laboratory and Scientific Section; 2009 (http://www.unodc.org/documents/scientific/Drug_Sampling.pdf).

2.6 ISO 8243: *Cigarettes—Sampling.*

2.7 ISO 5725-1: *Accuracy (trueness and precision) of measurement methods and results—Part 1: General principles and definitions.*

2.8 ISO 5725-2: *Accuracy (trueness and precision) of measurement methods and results—Part 2: Basic method for the determination of repeatability and reproducibility of a standard measurement method.*

2.9 World Health Organization. *Method validation report on World Health Organization TobLabNet official method: Determination of benzo[a]pyrene in mainstream cigarette smoke under ISO and Intense smoking conditions.* Geneva: Tobacco Laboratory Network, forthcoming

3 TERMS AND DEFINITIONS

3.1 *TPM*: total particulate matter

3.2 *B[a]P*: benzo[a]pyrene

3.3 *B[a]P-d12*: deuterium-labelled B[a]P

3.4 *Tobacco products*: Products made entirely or partly of leaf tobacco as the raw material that are manufactured for smoking, sucking, chewing or snuffing (Article 1(f) of the WHO FCTC)

3.5 *Intense regimen*: Parameters for smoking tobacco products that include a 55-mL puff volume, a 30-s puff interval, a 2-s puff duration and 100% blocking of the filter ventilation holes

3.6 *ISO regimen*: Parameters for smoking tobacco products that include a 35-mL puff volume, a 60-s puff interval, a 2-s puff duration and no blocking of the filter ventilation holes

3.7 *Laboratory sample*: Sample intended for testing in a laboratory, consisting of a single type of product delivered to the laboratory at one time or within a specified period

3.8 *Test sample*: Product to be tested, taken at random from the laboratory sample

3.9 *Test portion*: Random portion from the test sample to be used for a single determination

4 METHOD SUMMARY

4.1 Mainstream smoke total particulate matter from the cigarette test sample is trapped onto a glass-fibre filter pad (commonly referred to as a Cambridge filter pad).

4.2 The number of cigarettes may have to be adjusted to prevent breakthrough of the filter pad. If the TPM exceeds 600 mg for a 92-mm filter pad or 150 mg for a 44-mm filter pad, the number of cigarettes smoked onto each pad must be decreased.

4.3 A solution containing an isotope-labelled internal standard is spiked onto the pad, which is extracted with cyclohexane.

4.4 The cyclohexane extract is eluted through a silica solid-phase extraction cartridge. The eluent collected is analysed by gas chromatography–mass spectrometry (GC–MS) with electron ionization detection.

4.5 A calibration curve is created by plotting the peak area ratio of B[a]P to B[a]P-d12 from the reconstructed ion chromatogram against known concentrations of B[a]P in standards. The B[a]P concentration in the sample is obtained from the calibration curve.

5 SAFETY AND ENVIRONMENTAL PRECAUTIONS

CAUTION: Benzo[a]pyrene is a human carcinogen. Precautions shall be taken to avoid human exposure.

B[a]P and its solutions should be handled in an adequately ventilated fume hood, glove box or equivalent.

The laboratory shall establish procedures for disposal of solutions containing B[a]P.

5.1 Take routine safety and environmental precautions, as in any chemical laboratory activity.

5.2 The testing and evaluation of certain products with this test method may require the use of materials and or equipment that could be hazardous or harmful to the environment; this document does not address all the safety aspects associated with use of the method. All persons using this method have the responsibility to consult the appropriate authorities and to establish health and safety practices as well as environmental precautions in conjunction with any existing applicable regulatory requirements prior to its use.

5.3 Special care should be taken to avoid inhalation or oral or dermal exposure to harmful chemicals. Use a chemical fume hood, and wear an appropriate laboratory coat, gloves and safety goggles when preparing or handling undiluted materials, standard solutions, extraction solutions or collected samples.

6 APPARATUS AND EQUIPMENT

Usual laboratory apparatus, in particular:

6.1 Equipment required to condition cigarettes as specified in ISO 3402 [**2.1**]

6.2 Equipment required to mark butt length as specified in ISO 4387 [**2.2**]

6.3 Equipment required to cover filter ventilation holes for the intense regimen as specified in WHO TobLabNet SOP 01 [**2.3**]

6.4 Equipment required to perform smoking of tobacco products as specified in ISO 3308 [**2.4**]

6.5 Analytical balance capable of measurement to at least four decimal places

6.6 Vortexer or wrist-action shaker or equivalent

6.7 Pipettes and tips capable of accurately dispensing volumes of 10–1000 μL

6.8 Volumetric pipette(s) or equivalent, 10 mL, 40 mL and 100 mL

6.9 Volumetric flasks, 10 mL, 25 mL and 100 mL

6.10 Erlenmeyer flasks, 100 mL, 200 mL or suitable flask

6.11 Silica solid-phase extraction manifold and cartridges (500 mg): pure silica unbound phase

Note: e.g. Sep-pak Vac silica cartridge (Waters), Bond Elut Jr SI cartridge (Agilent), Supelclean LC-Si cartridge (Supelco), Strata SI-1 Silica (Phenomenex) or equivalent

6.12 Glass transfer pipettes or equivalent

6.13 Rotary evaporator with suitable flasks, or Turbovap with test tubes or equivalent

6.14 GC–MS system equipped with a computerized control, data acquisition and processing system. The system must be capable of operating the mass spectrometer in order to obtain chromatographic data in single-ion monitoring detection mode. The gas chromatograph must be configured to perform splitless injections on a capillary column. It is recommended that the gas chromatograph be equipped with an autosampler for sample injection.

6.15 Column: Fused silica capillary column with a methylphenyl (5%) polysiloxane stationary phase, a 30-m column with 0.25-mm internal diameter and 0.25-μm film thickness is suitable for this analysis.

Note: e.g. DB-5ms (Agilent) or equivalent

6.16 Autosampler vials, 2 mL or equivalent

7 REAGENTS AND SUPPLIES

7.1 All reagents shall be of at least analytical reagent grade unless otherwise noted. Reagents are identified by their Chemical Abstract Service (CAS) registry numbers when available.

7.2 Cyclohexane (110-82-7)

7.3 B[a]P (50-32-8)

7.4 B[a]P-d12 (63466-71-7)

Note: B[a]P and B[a]P-d12 are carcinogenic to humans. Appropriate safety precautions shall be taken when manipulating these compounds or any solution containing these compounds.

8 PREPARATION OF GLASSWARE

Clean and dry glassware in a manner to avoid contamination.

9 PREPARATION OF SOLUTIONS

Not applicable

10 PREPARATION OF STANDARDS

The method for preparing standard solutions described below is for reference purposes and can be adjusted if necessary.

10.1 Isotope-labelled internal standard

Prepare in accordance with either option A or B.

Option A

10.1.1 Primary B[a]P-d12 stock solution (200 μg/mL)

10.1.1.1 Weigh approximately 0.005 g of B[a]P-d12 into a 25-mL volumetric flask on a four-decimal place balance, and record the weight to the nearest 0.0001 g.

10.1.1.2 Dissolve B[a]P-d12 internal standard in 25 mL of cyclohexane, and mix well.

10.1.1.3 Label, and store at −20 ± 5 °C in amber vials.

10.1.2 B[a]P-d12 spiking solution (2 μg/mL)

10.1.2.1 Pipette 1 mL of the primary B[a]P-d12 stock solution (**10.1.1**) into a 100-mL volumetric flask.

10.1.2.2 Dilute to the mark with cyclohexane, and shake well.

10.1.2.3 Label and store at −20 ± 5 °C.

Option B

10.1.3 Obtain primary B[a]P-d12 stock solution of 1000 μg/mL in methylene chloride from a commercial supplier, label, and store at −20 ± 5 °C.

10.1.4 Prepare B[a]P-d12 spiking solution (2 μg/mL).

10.1.4.1 Pipette 200 μL of 1000 μg/mL B[a]P-d12 stock solution (**10.1.3**) into a 100-mL volumetric flask.

10.1.4.2 Dilute to the mark with cyclohexane, and shake well.

10.1.4.3 Label and store at −20 ± 5 °C.

10.2 Native standards

10.2.1 Primary B[a]P stock solution (200 μg/mL)

10.2.1.1 Weigh approximately 0.005 g B[a]P into a 25-mL volumetric flask on a four-decimal place balance, and record the weight to the nearest 0.0001 g.

10.2.1.2 Dissolve the B[a]P standard in 25 mL of cyclohexane, and mix well.

10.2.1.3 Label and store at −20 ± 5 °C.

10.2.2 Secondary B[a]P stock solution (1 μg/mL)

10.2.2.1 Pipette 0.5 mL of the primary B[a]P stock solution (**10.2.1**) into a 100-mL volumetric flask.

10.2.2.2 Dilute to the mark with cyclohexane, and mix well.

10.2.2.3 Label and store at −20 ± 5 °C.

The standard solutions are stable for up to 6 months if stored at −20 ± 5 °C.

Note: Sonication may be used in the preparation of primary stock solutions to help dissolve B[a]P and B[a]P-d12 standards completely. In this case, first dissolve the standards in about 15 mL of solvent. After sonication, let the solution stand until it cools to working temperature (e.g. room temperature), then dilute to the mark with cyclohexane.

10.3 B[a]P working standard solutions

Prepare working standard solutions as stated below and summarized in Table 1.

Table 1. Working standard solutions of B[a]P

Standard	Volume of secondary B[a]P stock standard solution (1 µg/mL) (µL)	Volume of B[a]P-d12 spiking solution (2 µg/mL from 10.1.2 or 10.1.4) (µL)	Total volume (mL)	Approximate concentration of B[a]P working standard solution (ng/mL)
1	20	100	10	2
2	40	100	10	4
3	80	100	10	8
4	200	100	10	20
5	400	100	10	40
6	600	100	10	60

Note: All solvents and solutions must be equilibrated to room temperature before use.

10.3.1 Spike variable volumes of the secondary B[a]P stock solution from 10.2.2 into a 10-mL volumetric flask.

10.3.2 Add 100 µL of the B[a]P-d12 spiking solution prepared in **10.1.2** or **10.1.4**.

10.3.3 If option A is chosen, the concentrations of B[a]P-d12 spiking solution (**10.1.2**) can be determined from the equation:

Final concentration (µg/mL) = $x \times 400$,

where x is the original weight (in g) of standard as determined in **10.1.1.1**.

10.3.4 When option B is chosen, the concentration of B[a]P-d12 spiking solution (**10.1.4**) is computed according to the value provided by the supplier.

10.3.5 Fill to the mark with cyclohexane, and mix well.

10.3.6 The final concentrations of standard are determined from the equation:

Final concentration (ng/mL) = $x \times y \times 20$.

Note: The equation in 10.3.6 is derived as follows:

The concentrations of B[a]P working standard solutions (in ng/mL) can be calculated from:

$$\frac{x \text{ (g)}}{25 \text{ mL}} \times \frac{0.5 \text{ mL}}{100 \text{ mL}} \times \frac{y \text{ (μL)}}{10 \text{ mL}} \times \frac{1 \text{ mL}}{1000 \text{ μL}} \times \frac{1\,000\,000\,000 \text{ ng}}{1 \text{ g}}$$

$$= \frac{x \times 0.5 \times 1000 \times y \text{ (ng)}}{25 \text{ mL}}$$

$$= x \times y \times 20$$

where x is the original weight (in g) of standard as determined in **10.2.1.1**, and y is the volume (in μL) of secondary B[a]P stock solution as described in **10.2.2** and Table 1.

10.3.7 The range of the standard solutions may be adjusted in accordance with the equipment used and the samples to be tested, keeping in mind a possible effect on the sensitivity limits of the method.

11 SAMPLING

11.1 Sample cigarettes according to ISO 8243 [**2.6**]. Alternative approaches may be used to obtain a representative laboratory sample in accordance with individual laboratory practice or when required by specific regulation or the availability of samples.

11.2 Constitution of test sample

11.2.1 Divide the laboratory sample into separate units (e.g. packet, container), if applicable.

11.2.2 Take an equal amount of product for each test sample from at least $\sqrt{n}$ [**2.5**] of the individual units (e.g. packet, container).

12 CIGARETTE PREPARATION

12.1 Condition all cigarettes to be smoked in accordance with ISO 3402 [**2.1**].

12.2 Mark cigarettes at a butt length in accordance with ISO 4387 [**2.2**] and WHO TobLabNet SOP 01 [**2.3**].

12.3 Prepare test samples to be smoked in accordance with either ISO or intense smoking conditions as specified in WHO TobLabNet SOP 01 [**2.3**].

13 PREPARATION OF THE SMOKING MACHINE

13.1 Ambient conditions

The ambient conditions for smoking are specified in ISO 3308 [**2.4**].

13.2 Smoking machine specifications

Follow ISO 3308 [**2.4**] smoking machine specifications, except for intense smoking, for which the smoking machine should be prepared as described in WHO TobLabNet SOP 01 [**2.3**].

14 SAMPLE GENERATION

Smoke a sufficient number of cigarettes on the specified smoking machine such that breakthrough does not occur and the concentrations of the B[a]P fall within the calibration range prepared for the analysis.

14.1 Smoke the cigarette test samples, and collect the TPM as specified in ISO 4387 [**2.2**] or WHO TobLabNet SOP 01 [**2.3**].

14.2 Include at least one reference test sample for quality control, if applicable.

14.3 When testing sample types for the first time, evaluate breakthrough of the filter pad. The number of cigarettes might have to be adjusted. If the TPM exceeds 600 mg for a 92-mm filter pad or 150 mg for a 44-mm filter pad, decrease the number of cigarettes smoked onto each pad.

14.4 The numbers of cigarettes to be smoked per measurement for linear and rotary smoking machines in ISO and intense smoking regimes are shown in Table 2.

Table 2. Suggested numbers of cigarettes to be smoked for one measurement on linear and rotary smoking machines

	ISO smoking regimen		Intense smoking regimen	
	Linear	Rotary	Linear	Rotary
No. of cigarettes per pad	5	20	3	10
No. of pads per result	4	1	3	1

Note: The number of cigarettes to be smoked should be adjusted such that breakthrough does not occur and the concentrations of B[a]P fall within the calibration range prepared for the analysis.

14.5 Record the number of cigarettes and total puffs for each filter pad.

14.6 After smoking the required number of test samples, perform five clearing puffs, and remove the pad holder from the smoking machine.

15 SAMPLE PREPARATION

15.1 Extraction of filter pads

15.1.1 Remove the pads from the holders. Fold each pad loosely in half and then in half again, with the TPM on the inside. Use the side opposite to that on which TPM is collected to wipe the inner surface of the pad holder, thus including any particulate matter that may have been left in the holder. Transfer the filter pad(s) into a 100-mL Erlenmeyer or suitable flask for a 44-mm pad or a 200-mL Erlenmeyer or suitable flask for a 92-mm pad.

Note: In the ISO regimen, four 44-mm pads for a linear smoking machine or one 92-mm pad for a rotary smoking machine are extracted into one flask. In the intense regimen, three 44-mm pads for a linear smoking machine or one 92-mm pad for a rotary smoking machine are extracted into one flask.

15.1.2 Add cyclohexane (40 mL for 44-mm pads, 100 mL for 92-mm pads) to the flask.

15.1.3 Spike each flask with B[a]P-d12 spiking solution (**10.1.2** or **10.1.4**) (40 μL for a 44-mm pad, 100 μL for a 92-mm pad)

15.1.4 Using a vortexer, wrist-action or equivalent device, extract the B[a]P from the pad by shaking at 200 rpm for 60–80 min.

15.2 Sample clean-up

15.2.1 Condition a solid-phase extraction cartridge with 10 mL of cyclohexane, and discard the eluate. Do not let the cartridge run dry.

15.2.2 Pipette 10 mL of sample extract onto the cartridge, and allow it to pass through the cartridge. Collect the eluate.

15.2.3 Elute the cartridge with two further 15-mL aliquots of cyclohexane, allowing the cartridge to run dry after the last aliquot is passed through.

15.2.4 Evaporate the cyclohexane solution almost to dryness.

Note: The conditions of the rotary evaporator are to be set at 55 °C with a vacuum pressure of 30 kPa (300 mbar) (approximately 10 min). The vacuum pressure can be adjusted for the equipment. The conditions of the Turbovap are to be set at 30 °C in a steady stream of nitrogen (approximately 15 min).

15.2.5 Pipette 1 mL of cyclohexane into the flask to dissolve the extract. Cap the flask, and rotate it carefully to rinse the inner surface. Take an aliquot for GC–MS testing.

Important note: The final concentration of B[a]P-d12 internal standard in the sample is approximately 20 ng/mL.

16 SAMPLE ANALYSIS

This method for quantifying B[a]P in mainstream cigarette smoke involves GC–MS. The analytes are resolved from other potentially interfering substances on a GC column. Comparison of the area ratio (native analyte to isotope-labelled analyte) of the unknowns with the area ratio (native analyte to isotope-labelled analyte) of the known standard concentrations yields the concentrations of the analyte.

16.1 GC–MS operating conditions: example

GC column	Fused silica capillary column with a methylphenyl (5%) polysiloxane stationary phase, a 30-m column with a 0.25-mm internal diameter and 0.25-μm film thickness is suitable (DB-5ms Agilent or equivalent).
Injector temperature	280 °C
Mode	Constant flow
Flow rate	1.2 mL/min
Injection	1 μL or 2 μL splitless
Column temperature	150 °C for 0 min; 6 °C/min to 260 °C, hold at 260 °C for 7 min; 50 °C/min to 290 °C, hold at 290 °C for 20 min
Transfer line temperature	280 °C
MS source	230 °C
Ion traces	B[a]P: m/z 252
	B[a]P-d12: m/z 264
Dwell time	50 ms
Ionization mode	Electron ionization

Note: The operating parameters may have to be adjusted to the instrument and column conditions and the resolution of the chromatographic peaks.

16.2 General analytical information

16.2.1 For the conditions described here, the expected sequence of elution will be B[a]P-d12, B[a]P.

16.2.2 Differences in e.g. temperature, gas flow rate and the age of the column can be expected to alter retention times.

16.2.3 The sequence of determination of B[a]P will be in accordance with individual laboratory practice. This section gives an example.

16.2.4 Inject a blank solution (extraction solution minus labelled internal standard) to check for any contamination in the system or reagents.

16.2.5 Inject a standard blank (blank with labelled internal standard) to verify the performance of the GC–MS system.

16.2.6 Inject the calibration standards, the quality control and the samples.

16.2.7 Record the peak areas of B[a]P and B[a]P-d12.

16.2.8 Calculate the relative response ratio of the B[a]P peak to the B[a]P-d12 peak ($A_{B[a]P} / A_{B[a]P\text{-}d12}$) for each standard solution, including standard blanks.

16.2.9 Plot a graph of the concentration of B[a]P (X axis) against the area ratios (Y axis).

16.2.10 The intercept should not be statistically significantly different from zero.

16.2.11 The standard curve should be linear over the entire standard range.

16.2.12 Calculate the linear regression ($Y = a + bx$) from these data, and use both the slope (b) and the intercept (a) of the linear regression.

If the linear regression, R^2, is less than 0.99, the calibration should be repeated. If an individual calibration point differs by more than 10% from the expected value (estimated by linear regression), the point should be omitted.

16.2.13 Inject the quality controls and samples, and determine the peak areas with the appropriate software.

16.2.14 The peak ratio obtained for all test portions must fall within the working range of the calibration curve; otherwise, the concentrations of the standard and standard solutions or test portion solutions should be adjusted.

See Annex 1 for representative chromatograms.

17 DATA ANALYSIS AND CALCULATIONS

17.1 The slope and intercept are determined from the relative responses of B[a]P and B[a]P-d12 versus the relative concentrations of B[a]P.

17.2 Calculate the relative response ratio from the peak areas for each of the calibration standards (**10.3**):

$$RF = A_{B[a]P} / A_{B[a]P\text{-}d12},$$

where RF is the relative response ratio, $A_{B[a]P}$ is the peak area of B[a]P (reconstructed ion chromatogram m/z 252) and $A_{B[a]P\text{-}d12}$ is the peak area of B[a]P-d12 (reconstructed ion chromatogram m/z 264).

17.3 Plot a graph of the relative response factor, $A_{B[a]P} / A_{B[a]P\text{-}d12}$ (Y axis), versus concentration (X axis) for each standard solution. Calculate the linear regression ($Y = a + bx$) from these data, and use both the slope (b) and the intercept (a) of the linear regression.

17.4 The standard curve should be linear over the entire standard range.

17.5 The content of B[a]P (ng/cigarette) is determined from the calculated relative response ratio for the test portion, the slope and the intercept obtained from the appropriate calibration curve and the equation:

$$M = \frac{Y - a}{b} \times \frac{V \times V_e}{V_c \times n}$$

where M is the calculated content of B[a]P in nanograms per cigarette, Y is the relative response ratio $A_{B[a]P} / A_{B[a]P\text{-}d12}$, a is the intercept of the linear regression obtained from the standard calibration curve, b is the slope of the linear regression obtained from the standard calibration curve, V is the volume of the sample solution (1 mL), V_e is the volume of the extraction solution (40 mL for a 44-mm Cambridge filter pad, 100 mL for a 92-mm Cambridge filter pad), V_c is the volume of the aliquot of the extraction solution used in clean-up (10 mL), and n' is the number of cigarettes smoked (see Table 2).

Alternative calculation procedures may be used if applicable.

18 SPECIAL PRECAUTIONS

After installing a new column, condition it as specified by the manufacturer by injecting a tobacco sample extract under the specified instrument conditions. Injections should be repeated until the peak areas (or heights) of both B[a]P and B[a]P-d12 are reproducible.

19 DATA REPORTING

19.1 Report individual measurements for each sample evaluated.

19.2 Report results as nanograms per cigarette or as required.

20 QUALITY CONTROL

20.1 Control parameters

Note: If the control measurements are outside the tolerance limits of the expected values, appropriate investigation and action must be undertaken.

Note: Additional quality assurance procedures should be used if necessary in order to comply with the policies of individual laboratories.

20.2 Laboratory reagent blank

To detect any contamination during sample preparation and analysis, include a laboratory reagent blank, as described in **16.2.5** The blank consists of all the reagents and materials used in analysing test samples and is analysed like a test sample. The blank should be assessed in accordance with the practices of individual laboratories.

20.3 Quality control sample

To verify the consistency of the entire analytical process, analyse a reference cigarette (or an appropriate quality control sample) in accordance with the practices of the individual laboratory.

20.4 Make a new calibration curve after every 20 injections of test sample (20 sample extracts).

21 METHOD PERFORMANCE SPECIFICATIONS

21.1 Limit of reporting

The limit of reporting is set to the lowest concentration of the calibration standards used, recalculated to nanograms per cigarette

21.2 Laboratory-fortified matrix recovery

Recovery of analyte spiked onto the matrix is used as a surrogate measure of accuracy. Recovery is determined by spiking a known amount of standard (after smoking) onto a filter pad with cigarette smoke and extracting the pad by the same method used for extracting samples. Unspiked pads are also analysed. The recovery is calculated from the following equation and is shown in Table 3:

$$\text{Recovery (\%)} = 100 \times (\text{analytical result} - \text{unspiked result} / \text{spiked amount})$$

Table 3. Mean and recovery of B[a]P spiked onto the matrix

Spiked amount (ng/cig)	Mean (ng/cig)	Recovery (%)
2.91	2.93	100.58
6.79	6.82	100.39
9.70	10.08	103.92

21.3 Analytical specificity

GC–MS provides analytical specificity. The retention time and molecular mass to charge ratio (m/z) are used to verify the specificity of the results for an unknown sample.

21.4 Linearity

The B[a]P calibration curves established are linear over the standard concentration range of 2–60 ng/mL.

21.5 Possible interference

No known components have both similar retention times and m/z as B[a]P or B[a]P-d12.

22 REPEATABILITY AND REPRODUCIBILITY LIMITS

An international collaborative study [2.9] conducted in 2012, involving testing of three reference cigarettes (1R5F, 3R4F and CM6) and two commercial brands by eight laboratories, gave the precision limits for the method indicated in Table 4.

The difference between two single results found for matched cigarette samples by the same operator using the same apparatus within the shortest feasible time will exceed the repeatability, r, on average not more than once in 20 cases in the normal, correct application of the method.

Single results for matched cigarette samples reported by two laboratories will differ by more than the reproducibility, R, on average no more than once in 20 cases with normal, correct application of the method.

The test results were analysed statistically in accordance with ISO 5725-1 [2.7] and ISO 5725-2 [2.8] to give the precision data shown in Table 4.

Table 4. Precision limits for the determination of B[a]P (ng/cigarette) in mainstream cigarette smoke from reference cigarettes

Reference cigarette	ISO smoking regimen				Intense smoking regimen			
	n	m_{cig}	r_{limit}	R_{limit}	n	m_{cig}	r_{limit}	R_{limit}
1R5F	8	1.46	0.47	0.99	8	6.47	1.52	2.91
3R4F	8	5.99	1.03	2.09	8	14.51	2.52	3.15
CM6	7	13.70	1.43	5.83	7	25.86	2.81	8.45

Where 1R5F, 3R4F and CM6 are three reference cigarettes analysed in this study, n is the number of laboratories that participated, m_{cig} is the mean value of B[a]P per cigarette, r_{limit} is the repeatability limit of B[a]P, and R_{limit} is the reproducibility limit of B[a]P.

For the purpose of calculating r and R, one test result under ISO smoking conditions was defined as the average of seven individual replicates of the mean yield of four Cambridge filter pads (five cigarettes smoked per pad) from linear smoking machines and one Cambridge filter pad (20 cigarettes smoked per pad) from rotary smoking machines. Under intense smoking conditions, one test result was defined as the average of seven individual replicates of the mean yield of two sets of three Cambridge filter pads (three cigarettes smoked per pad) from linear smoking machines and two sets of one Cambridge filter pad (10 cigarettes smoked per pad) from rotary smoking machines. For more information, see reference [2.9].

23 TEST REPORT

The following information shall be included in the test report:

- A reference to this method, i.e. WHO TobLabNet SOP 05
- Date of receipt of the sample
- The results and their units

ANNEX 1. Typical chromatograms obtained in the determination of benzo[a]pyrene in mainstream cigarette smoke

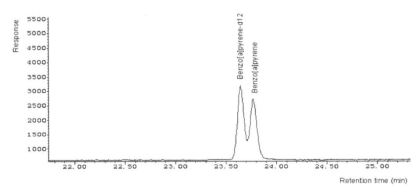

Figure 1. Representative chromatogram of a standard solution

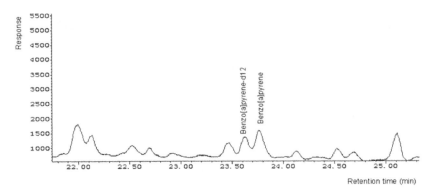

Figure 2. Representative chromatogram of a sample solution

WHO TobLabNet
Official Method
SOP 06

Standard Operating Procedure for Determination of Humectants in Cigarette Tobacco Filler

Method:	Determination of humectants in cigarette tobacco filler
Analytes:	Propylene glycol (propane-1,2-diol) (CAS # 57-55-6) Glycerol (propane-1,2,3-triol) (CAS # 56-81-5) Triethylene glycol (2,2´-ethylenedioxybis(ethanol)) (CAS # 112-27-6)
Matrix:	Cigarette tobacco filler
Last update:	June 2016

The mention of specific companies or of certain manufacturers' products does not imply that they are endorsed or recommended by WHO in preference to others of a similar nature that are not mentioned. Errors and omissions excepted, the names of proprietary products are distinguished by initial capital letters.

> No machine smoking regimen can represent all human smoking behaviour: machine smoking testing is useful for characterizing cigarette emissions for design and regulatory purposes, but communication of machine measurements to smokers can result in misunderstanding about differences between brands in exposure and risk. Data on smoke emissions from machine measurements may be used as inputs for product hazard assessment, but they are not intended to be nor are they valid as measures of human exposure or risks. Representing differences in machine measurements as differences in exposure or risk is a misuse of the results of testing with WHO TobLabNet standards.

FOREWORD

This document was prepared by members of the WHO Tobacco Laboratory Network (TobLabNet) as an analytical method standard operating procedure (SOP) for measuring humectants in cigarette tobacco filler.

INTRODUCTION

In order to establish comparable measurements for testing tobacco products globally, consensus methods are required for measuring specific contents and emissions of cigarettes. The Conference of the Parties to the WHO Framework Convention on Tobacco Control (WHO FCTC) at its third session in Durban, South Africa, in November 2008, "recalling its decisions FCTC/COP1(15) and FCTC/COP2(14) on the elaboration of guidelines for implementation of Articles 9 (Regulation of the contents of tobacco products) and 10 (Regulation of tobacco product disclosures) of the WHO FCTC, noting the information contained in the report of the working group to the third session of the Conference of the Parties on the progress of its work ... requested the Convention Secretariat to invite WHO's Tobacco Free Initiative to ... validate, within five years, the analytical chemical methods for testing and measuring cigarette contents and emissions" (FCTC/COP/3/REC/1).

Using the criteria for prioritization set at its third meeting in Ottawa, Canada, in October 2006, the working group on Articles 9 and 10 identified the following contents for which methods for testing and measurement (analytical chemistry) should be validated as a priority:

- nicotine
- ammonia
- propylene glycol (propane-1,2-diol)
- glycerol (propane-1,2,3-triol)
- triethylene glycol (2,2-ethylenedioxybis(ethanol)).

Measurement of these contents will require validation of three methods: one for nicotine, one for ammonia and one for humectants.

Using the criteria for prioritization set at the meeting in Ottawa mentioned above, the working group identified the following emissions in mainstream smoke for which methods for testing and measurement (analytical chemistry) should be validated as a priority:

- 4-(methylnitrosamino)-1-(3-pyridyl)-1-butanone (NNK)
- N'-nitrosonornicotine (NNN)

- acetaldehyde
- acrylaldehyde (acrolein)
- benzene
- benzo[a]pyrene (B[a]P)
- 1,3-butadiene
- carbon monoxide
- formaldehyde

Measurement of these emissions with the two smoking regimens described below will require validation of five methods: one for tobacco-specific nitrosamines (NNK and NNN), one for B[a]P, one for aldehydes (acetaldehyde, acrolein and formaldehyde), one for volatile organic compounds (benzene and 1,3-butadiene) and one for carbon monoxide.

The table below sets out the two smoking regimens for validation of the test methods referred to above.

Smoking regimen	Puff volume (mL)	Puff frequency	Filter ventilation holes
ISO regimen: *ISO 3308: Routine analytical cigarette smoking machine—definitions and standard conditions*	35	Once every 60 s	No modification
Intense regimen: Same as ISO 3308, but modified as indicated	55	Once every 30 s	All ventilation holes must be blocked 100% as described in WHO TobLabNet SOP 01.

This SOP was prepared to guide participating laboratories in analysing humectants in cigarette tobacco filler.

1 **SCOPE**

This standard operating procedure is suitable for quantitative determination of the humectants propylene glycol, glycerol (propane-1,2,3-triol) and triethylene glycol in cigarette tobacco filler by gas chromatography.

2 REFERENCES

2.1 *ISO 8243: Cigarettes – Sampling.*

2.2 *Guidelines on representative drug sampling.* New York: United Nations Office on Drugs and Crime; 2009 (http://www.unodc.org/documents/scientific/Drug_Sampling.pdf).

2.3 *Method validation report of WHO TobLabNet official method: Determination of humectants in cigarette tobacco filler.* Geneva: World Health Organization, Tobacco Laboratory Network (in preparation)

2.4 *Standard operating procedures for validation of analytical methods of tobacco product contents and emissions.* Geneva: World Health Organization, Tobacco Laboratory Network (WHO TobLabNet SOP 02).

2.5 *ISO 5725-1: Accuracy (trueness and precision) of measurement methods and results – Part 1: General principles and definitions.*

2.6 *ISO 5725-2: Accuracy (trueness and precision) of measurement methods and results – Part 2: Basic method for the determination of repeatability and reproducibility of a standard measurement method.*

3 TERMS AND DEFINITIONS

3.1 *Humectant content:* Individual amounts of glycerol, propylene glycol and triethylene glycol in cigarette tobacco filler, expressed as milligrams per gram of cigarette tobacco filler.

3.2 *Cigarette tobacco filler:* Tobacco-containing part of a cigarette, including reconstituted tobacco, stems, expanded tobacco and additives.

3.3 *Tobacco products:* Products made entirely or partly of leaf tobacco as the raw material and that are manufactured for smoking, sucking, chewing or snuffing (Article 1(f) of the WHO FCTC)

3.4 *Laboratory sample:* Sample intended for testing in a laboratory, consisting of a single type of product delivered to the laboratory at one time or within a specified period.

3.5 *Test sample:* Product to be tested, taken at random from the laboratory sample. The number of products taken shall be representative of the laboratory sample.

3.6 *Test portion:* Random portion from the test sample to be used for a single determination. The number of portions taken shall be representative of the test sample.

4 METHOD SUMMARY

4.1 Humectants are extracted from the cigarette tobacco filler with a mixture of methanol and 1,3-butanediol solution.

4.2 The extract is analysed with a flame ionization or mass spectrometer detector.

4.3 The ratio of the peak area of each humectant to that of the internal standard is compared on a calibration curve created by analysing standards containing known concentrations of each humectant to determine the humectant content in each test sample.

5 SAFETY AND ENVIRONMENTAL PRECAUTIONS

5.1 Follow routine safety and environmental precautions, as in any chemical laboratory activity.

5.2 The testing and evaluation of certain products with this test method may require the use of materials or equipment that could be hazardous or harmful to the environment. This document does not address all the safety aspects associated with use of the method. All persons using the method are responsible for consulting the appropriate authorities and establishing health and safety practices as well as environmental precautions in conjunction with any existing applicable regulatory requirements prior to its use.

5.3 Special care should be taken to avoid inhalation or dermal exposure to harmful chemicals. Use a chemical fume hood, and wear an appropriate laboratory coat, gloves and safety goggles when preparing or handling undiluted materials, standard solutions, extraction solutions or collected samples.

6 APPARATUS AND EQUIPMENT

Usual laboratory apparatus, in particular:

6.1 Analytical balance capable of measurement to at least four decimal places

6.2 Extraction vessels: flasks with stoppers or equivalent glassware

6.3 Mechanical wrist-action shaker or equivalent

6.4 Gas chromatograph equipped with a flame ionization detector

6.5 Capillary gas chromatography column capable of distinct separation of solvent peaks, the peaks for the internal standard, propylene glycol, glycerol, triethylene glycol and other tobacco components (e.g. DB-Wax fused silica column 30 m × 0.32 mm × 1 μm, or equivalent)

6.7 Centrifuge

7 REAGENTS AND SUPPLIES

All reagents shall be of at least analytical reagent grade, unless otherwise noted. Reagents are identified by their Chemical Abstract Service [CAS] registry numbers, when possible.

7.1	Methanol, chromatographic purity [67-56-1]	
7.2	Glycerol [56-81-5]	
7.3	Triethylene glycol [112-27-6]	
7.4	Propylene glycol [57-55-6]	
7.5	1,3-Butanediol [107-88-0] (used as an internal standard)	
7.6	Carrier gas: Helium [7440-59-7] of adequate purity	
7.7	Auxiliary gases: Hydrogen [1333-74-0] and air of adequate purity for flame ionization	

8   PREPARATION OF GLASSWARE

Clean and dry glassware in a manner to avoid contamination from residues.

9   PREPARATION OF SOLUTIONS

1,3-Butanediol primary stock (approximately 200 mg/mL):

9.1   Weigh approximately 20 g (± 0. 05 g) of 1,3-butanediol.

9.2   Place measured 1,3-butanediol into a 100-mL volumetric flask, and make up to volume with methanol.

Extraction solution (approximately 2 mg/mL):

9.3   Pipette 20 mL of primary stock into a 2-L volumetric flask, and make up to volume with methanol. The volume of the solution can be scaled up or down as necessary. Mix well. Store at 4–8 °C.

10   PREPARATION OF STANDARDS

10.1   Glycerol primary standard (approximately 100 mg/mL):
Accurately weigh 10 g (± 0.05 g) of glycerol into a 100-mL volumetric flask, and make up to volume with extraction solution [**9.3**].

10.2   Propylene glycol primary standard (approximately 50 mg/mL):
Accurately weigh 5 g (± 0.05 g) of propylene glycol into a 100-mL volumetric flask, and make up to volume with extraction solution [**9.3**].

10.3   Triethylene glycol primary standard (approximately 25 mg/mL):
Accurately weigh 2.5 g (± 0.05 g) of triethylene glycol into a 100-mL volumetric flask, and make up to volume with extraction solution [**9.3**].

10.4   Mixed secondary standard (4 mg/mL glycerol; 3 mg/mL propylene glycol; 1.5 mg/mL triethylene glycol):

Pipette a 4-mL aliquot of glycerol [**10.1**] and 6-mL aliquots each of the prepared primary stocks of propylene glycol [**10.2**] and triethylene glycol [**10.3**] into a 100-mL volumetric flask. Make up to volume with extraction solution [**9.3**].

Store all standard solutions at 4–8 °C.

10.5 Working standards
All standards are made up in volumetric flasks at the dilutions listed.

Table 1. Working standard solutions

No.	Mixed secondary standard solution (mL)	Final volume (mL)	Approximate glycerol concentration in working standard solution (mg/mL)	Approximate propylene glycol concentration in working standard solution (mg/mL)	Approximate triethylene glycol concentration in working standard solution (mg/mL)
1	0.2	10	0.08	0.06	0.03
2	0.5	10	0.20	0.15	0.08
3	1.0	10	0.40	0.30	0.15
4	2.0	10	0.80	0.60	0.30
5	4.0	10	1.60	1.20	0.60
6	8.0	10	3.20	2.40	1.20

Table 2. Working standard solutions for low yields of propylene glycol

No.	Mixed secondary standard solution (mL)	Final volume (mL)	Approximate propylene glycol concentration in working standard solution (mg/mL)
1	0.017	10	0.005
2	0.050	10	0.015
3	0.100	10	0.030
4	0.250	10	0.075
5	0.500	10	0.150
6	1.000	10	0.300

Make up to volume in volumetric flasks with extraction solution [**9.3**] containing the internal standard.

Note: The range of the standard solutions may be adjusted according to the equipment used and the samples to be tested, keeping in mind a possible effect on the sensitivity of the method.

A lower or higher concentration of standards can be prepared with the required volume of mixed secondary stock to make up to 10 mL in a volumetric flask with extraction solvent to a lower or higher limit of quantification, if necessary. See Table 2 for the example of propylene glycol; it is important to adjust the range of calibration if necessary. All solvents and solutions shall be at room temperature before use.

11 SAMPLING

11.1 Sample cigarettes according to ISO 8243 [**2.1**] or as required for specific application of the method. Alternative methods may be used to obtain a representative sample when required by specific regulation or the availability of samples.

11.2 Constitution of test sample

    11.2.1 Divide the laboratory sample into separate sales units, if applicable.

    11.2.2 Take an equal amount of product for each test sample from at least $\sqrt{n}$ [**2.2**] of the individual sales unit.

    11.2.3 If no individual sales units are available, combine the entire laboratory sample into one unit.

12 CIGARETTE PREPARATION

12.1 Remove the tobacco filler from the cigarettes or quality control samples (when applicable) in two packs, or use at least 10 g of processed cigarette tobacco filler.

12.2 Combine and mix sufficient cigarette tobacco filler to constitute at least 10 g for each test portion.

12.3 Homogenize the cigarette tobacco filler by grinding, and keep in an airtight amber bottle or container. Divide the sample into portions.

13 PREPARATION OF THE SMOKING MACHINE

Not applicable

14 SAMPLE GENERATION

Not applicable

15 SAMPLE PREPARATION

15.1 Take 4 g of well-mixed, ground test sample, and weigh it to 0.001 g accuracy into a suitable extraction flask.

15.2 Mix the test sample with 50 mL of extraction solution [**9.3**].

Note: Sample weight and dilution factor may be adjusted according to the concentrations of humectants or availability of sample.

15.3 Stopper the flasks, and place them on a wrist-action shaker or equivalent at a minimum rate of 210 rpm for at least 60 min.

15.4 Remove the samples from the shaker, and swirl the flasks to dissolve all the tobacco.

15.5 Let the samples stand undisturbed until the supernatant is clear. Alternatively, use a centrifuge.

15.6 Transfer the supernatant to an autosampler vial, and analyse on the gas chromatograph.

16 SAMPLE ANALYSIS

This method involves gas chromatography coupled with a flame ionization or mass spectrometry detector to quantify humectants in cigarette tobacco filler. The analytes are resolved from potential interference on the column. Comparison of the area of the unknowns with the area of the known standard concentrations yields the concentrations of individual analytes.

16.1 Gas chromatography operating conditions: example
Gas chromatograph column: DB-Wax fused silica, 30 m × 0.32 mm interior diameter × 25.0 µm
Coating: DB-Wax or equivalent
Carrier gas: Helium at a flow rate of 1.8 mL/min
Injection volume: 1 µL
Injection mode: split 25:1

Temperature programme:
Injection temperature: 220 °C
Detector temperature: 260 °C
Start temperature: 120 °C, hold for 3 min
Rate: 10 °C/min to 180 °C, hold for 11 min
Total run time: 20 min

Note: The operating parameters might have to be adjusted according to the instrument and column conditions and resolution of the chromatographic peaks.

16.2 Expected retention times

16.2.1 For the conditions described here, the expected sequence of elution will be propylene glycol, 1,3-butanediol (internal standard), glycerol and triethylene glycol.

16.2.2 Differences in e.g. temperature, gas flow rate or age of the column may alter retention times.

16.2.3 The elution order and retention times must be verified before analysis is begun.

16.3 Determination of humectants

The sequence of determination steps will be in accordance with individual laboratory practices. The following are provided as a guide:

16.3.1 Condition the system just before use by injecting two 1-μL aliquots of a sample solution as a primer.

16.3.2 Inject a standard (extraction solution [9.3]) under the same conditions as the samples to verify the performance of the gas chromatography system.

16.3.3 Inject a blank solution (solvent [7.1]) to check for contamination in the system or reagents.

16.3.4 Inject an aliquot of each standard solution [10.5] into the gas chromatograph.

16.3.5 Assess the retention times and responses (area counts) of the standards. If the retention times are similar (± 0.2 min) to the retention times in previous injections and the responses are within 20% of the typical responses in previous injections, the system is ready to perform the analysis. If the responses are outside the specifications, use follow-up actions according to individual laboratory policy.

16.3.6 Record the peak area of each humectant and the internal standard.

16.3.7 Calculate the relative response factor (RF) as the ratio between the peak areas of the humectants and that of the internal standard. RF = $A_{humectants} / A_{IS}$ for each humectant standard solution, including the solvent blanks.

16.3.8 Plot the concentration of added humectants (X axis) in accordance with the area ratios (RF, Y axis).

16.3.9 The intercept should not be statistically significantly different from zero.

16.3.10 The standard curve should be linear over the entire range.

16.3.11 Calculate a linear regression equation ($Y = a + bX$) from these data, and use both the slope (b) and the intercept (a). If the linear regression coefficient R^2 is < 0.99, the calibration should be repeated. If an individual calibration point differs by > 10% from the expected value (estimated by linear regression), that point should be omitted.
Inject the quality controls and samples, and determine the peak areas with appropriate instrument software.

16.3.12 The signals (peak areas) obtained for all test portions must fall within the working range of the calibration curve; otherwise, solutions should be adjusted.

See Annex 1 for a representative chromatogram.

17 DATA ANALYSIS AND CALCULATIONS

17.1 Inject two replicate aliquots of the sample extracts under identical conditions.

17.2 For each test portion, calculate the ratio of the response of each humectant (glycerol, propylene glycol, triethylene glycol) to that of the internal standard response (Y_t) from the peak area.

17.3 For each humectant (glycerol, propylene glycol, triethylene glycol), calculate the concentration in mg/mL for each test aliquot from the coefficients of the linear regression ($m_t = (Y_t - a) / b$).

17.4 For each humectant (glycerol, propylene glycol, triethylene glycol), calculate the content m_h, of the tobacco sample expressed in milligrams per gram from the following equation:

$$m_h = \frac{m_t * V_e}{m_o}$$

where:

m_h is the content of the humectant (glycerol, propylene glycol, triethylene glycol) in mg/g;

m_t is the concentration of each humectant (glycerol, propylene glycol, triethylene glycol) in the test solution, in mg/mL;

V_e is the volume of the extraction solution, in mL; and

m_o is the mass of the test portion, in g.

18 SPECIAL PRECAUTIONS

18.1 After installing a column, condition it as specified by the manufacturer, and then inject a tobacco sample extract under the instrument conditions described above. Injections should be repeated until the peak areas (or heights) of both the humectant (propylene glycol, glycerol, triethylene glycol) and the internal standard are reproducible. This will require approximately four injections.

18.2 It is recommended that high-boiling-point components be purged from the gas chromatography column after each sample set (series) by raising the column temperature to 220 °C for 30 min.

18.3 When the peak area (or height) of the internal standard is significantly higher than expected, it is recommended that the tobacco sample be extracted without internal standard in the extraction solution. This makes it possible to determine whether any component co-elutes with the internal standard, which would cause artificially lower values for humectants.

19   DATA REPORTING

19.1   Report individual measurements for each sample evaluated.

19.2   Report results as milligrams per gram of tobacco or as required.

20   QUALITY CONTROL

20.1   Control parameters
Note 1: If the control measurements are outside the tolerance limits of the expected values, make appropriate investigations and take action.

Note 2: Additional quality assurance procedures should be used in accordance with the practices of individual laboratories.

20.2   Laboratory reagent blank:
To detect potential contamination during sample preparation and analysis, include a laboratory reagent blank, as described in **16.3.2**. The blank consists of all reagents and materials used in analysing test samples and is analysed as a test sample. The result should be less than the limit of detection.

20.3   Quality control sample:
To verify the consistency of the entire analysis, analyse a reference cigarette in accordance with the practices of the individual laboratory.

20.4   Additional quality control samples may be added as required by individual laboratory policy.

21   METHOD PERFORMANCE

21.1   Limit of reporting:
The limit of reporting is set to the lowest concentration of the calibration standards, recalculated to mg/g (with 0.08 mg/mL for glycerol, 0.06 mg/mL for propylene glycol and 0.03 mg/mL for triethylene glycol as the lowest concentrations). The limits of reporting are therefore 1.0 mg/g for glycerol, 0.75 mg/g for propylene glycol and 0.375 mg/g for triethylene glycol.

For the propylene glycol working standard solutions (Table 2), the limit of reporting is set to the lowest concentration of the calibration standards, recalculated to 0.063 mg/g. (At 0.005 mg/mL, the limit of reporting is 0.063 mg/g.)

21.2   Laboratory-fortified matrix recovery:
Recovery of analyte spiked onto the matrix is used as a surrogate measure of accuracy. It is determined by spiking known amounts of standards into tobacco and extracting the tobacco in the same way as for samples. Unspiked tobacco is also analysed. Recovery is calculated from the following formula:

$$\text{Recovery} = 100 \times (\text{analytical result} - \text{unspiked result} / \text{spiked amount})$$

Table 3. Mean recovery of laboratory-fortified matrix

Propylene glycol			Glycerol			Triethylene glycol		
Spiked amount (mg/g)	Mean (mg/g)	Recovery (%)	Spiked amount (mg/g)	Mean (mg/g)	Recovery (%)	Spiked amount (mg/g)	Mean (mg/g)	Recovery (%)
5.00	4.94	99	5.00	5.47	109	5.00	5.70	114
10.00	9.87	99	10.01	10.04	100	10.00	11.05	110
20.00	19.64	98	20.02	19.74	99	20.01	20.76	104

21.3 Analytical specificity:
The retention time of the analyte of interest is used to verify the analytical specificity. An established range of ratios of the response of the component to that of the internal standard component of quality control cigarette tobacco filler is used to verify the specificity of the results for an unknown sample.

21.4 Linearity:
The calibration curves of the humectants are linear over the standard concentration range: 0.08–3.2 mg/mL for glycerol, 0.005–0.30 mg/mL for propylene glycol and 0.03–1.2 mg/mL for triethylene glycol.

21.5 Possible interference:
At the time of validation, there were no known components that would cause interference because of similar retention times as propylene glycol, triethylene glycol, glycerol or the internal standard (1,3-butanediol).

22 REPEATABILITY AND REPRODUCIBILITY LIMITS

An international collaborative study with flame ionization detection [**2.3**], conducted in 2012–2014 by 13 laboratories and with seven samples (five reference cigarettes and two commercial cigarette brands), performed according to WHO TobLabNet SOP 02 [**2.4**], gave the following values for this method:

> The difference between two single results found for matched cigarette tobacco filler samples by the same operator using the same apparatus within the shortest feasible time will exceed the repeatability, r, on average no more than once in 20 cases with normal, correct application of the method.

> Single results for matched cigarette tobacco filler samples reported by two laboratories will differ by no more than the reproducibility, R, on average no more than once in 20 cases with normal, correct application of the method.

A collaborative study with mass spectrometry detection, conducted in 2012–2014 by seven laboratories and with seven samples (five reference cigarettes and two commercial cigarette brands) gave equivalent values (Annex 2).

The test results were analysed statistically in accordance with ISO 5725-1 [**2.5**] and ISO 5725-2 [**2.6**] to determine the repeatability and reproducibility limits shown in Tables 4–6 for gas chromatography–flame ionization detection analysis. The repeatability and reproducibility limits for gas chromatography in mean standard deviation analyses are shown in Tables A2.2–A2.4 in Annex 2.

Table 4. Repeatability and reproducibility limits for the determination of propylene glycol in tobacco from reference test pieces and commercial cigarettes by gas chromatography-flame ionization detection

Reference cigarette	n*	$\hat{m}$ (mg/g)	Repeatability limit (r) (mg/g)	Reproducibility limit (R) (mg/g)
1R5F	12	0.231	0.001	0.005
3R4F	12	0.180	0.000	0.001
CM6	12	0.140	0.000	0.004
Commercial cigarette No. 1	12	5.496	0.244	1.972
Commercial cigarette No. 2	12	0.078	0.000	0.001

* Results in one data set were removed as they were outliers.

Table 5. Repeatability and reproducibility limits for the determination of glycerol in tobacco from reference test pieces and commercial cigarettes by gas chromatography-flame ionization detection

Reference cigarette	n*	$\hat{m}$ (mg/g)	Repeatability limit (r) (mg/g)	Reproducibility limit (R) (mg/g)
1R5F	11	21.603	1.490	6.082
3R4F	11	21.203	1.299	7.638
CM6	11	1.160	0.006	0.063
Commercial cigarette No. 1	11	14.376	0.648	3.621
Commercial cigarette No. 2	11	1.222	0.015	0.049

* Results in one data set were removed as they were outliers.

Table 6. Repeatability and reproducibility limits for the determination of triethylene glycol in tobacco from reference test pieces by gas chromatography-flame ionization detection

Reference cigarette	n*	$\hat{m}$ (mg/g)	Repeatability limit (r) (mg/g)	Reproducibility limit (R) (mg/g)
Sample TG No. 1	12	0.891	0.207	0.409
Sample TG No. 2	11	8.078	1.565	2.621

* Results in one data set were removed for sample TG No. 1, and results in two data sets were removed for sample TG No. 2, as they were outliers.

23   TEST REPORT

The following information shall be included in the test report:
    (a) A reference to this method, i.e. WHO TobLabNet SOP 06
    (b) Date of receipt of the sample
    (c) The results and its units.

ANNEX 1. Example of a gas chromatography–flame ionization detection chromatogram of a standard solution

Elution order of humectants:
1. Propylene glycol
2. 1,3 Butanediol (internal standard)
3. Glycerol
4. Triethylene glycol

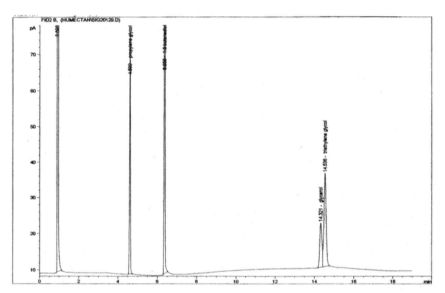

Instrument 1 11/23/2011 3:48:12 AM B. SOURABIE

ANNEX 2. Main points to consider when using a mass spectrometer detector instead of a flame ionization detector

A2.1.1 METHOD SUMMARY

The method is a gas chromatographic method with a silica column and a mass spectrometry detector. Extract 4 g of ground cigarette tobacco filler with 50 mL of a methanol-based extraction solvent on a mechanical shaker for 60 min. Pipette 2 mL of the sample extract onto a dispersive solid phase extraction cartridge containing 150 mg anhydrous $MgSO_4$ and 25 mg N-propylethylenediamine. Oscillate the cartridge in a vortex oscillator at 2000 rpm for 2 min, and centrifuge it at 10 000 rpm for 10 min if necessary. Transfer the supernatant to an autosampler vial, and analyse on the gas chromatograph.

A2.1.2 APPARATUS AND EQUIPMENT

Analytical balance capable of measurement to at least four decimal places

125-mL Erlenmeyer flasks with stoppers or equivalent

10-mL, 100-mL and 2000-mL volumetric flasks

20-mL pipette for preparation of extraction solvent

Volumetric pipettes for preparation of standard solutions (various sizes)

Disposable transfer pipettes Brinkmann Dispenser, 10–50 mL or equivalent

2-mL autosampler vials and caps with Teflon-faced septa

0.22-μm membrane filter

Vortex oscillator

Centrifuge

High-speed disintegrator

0.45-mm screen

Gas chromatograph equipped with mass spectrometer

Capillary gas chromatography column capable of distinct separation of peaks for the solvent, the internal standard, propylene glycol, glycerol, triethylene glycol and other tobacco components (e.g. DB-Wax fused silica column 30 m × 0.32 mm × 1 μm, or equivalent)

A2.1.3 REAGENTS AND SUPPLIES

Add the following if necessary.

A2.1.3.1 Anhydrous $MgSO_4$ [7487-88-9]

A2.1.3.2 N-Propylethylenediamine [111-39-7]

> Note: **A1.3.1** and **A1.3.2** can be replaced by a commercially dispersive solid-phase extraction cartridge containing 150 mg anhydrous $MgSO_4$ [7487-88-9] and 25 mg N-propylethylenediamine [111-39-7].

A2.1.4 PREPARATION OF STANDARDS

As for the gas chromatography-flame ionization detection method [**10.1–10.5**]

A2.1.5 SAMPLE PREPARATION

A2.1.5.1 Pipette 1–2 mL of the sample extract onto a dispersive solid-phase extraction cartridge containing 150 mg anhydrous $MgSO_4$ and 25 mg N-propylethylenediamine if needed.

A2.1.5.2 Oscillate the cartridge in a vortex oscillator at 2000 rpm for 2 min, and centrifuge it at 10 000 rpm for 10 min. Filter the supernatant through a 0.22-μm membrane of a syringe filter until it is clear.

A2.1.5.3 Transfer the supernatant to an autosampler vial, and analyse on the gas chromatograph.

A2.1.6 SAMPLE ANALYSIS

A2.1.6.1 Gas chromatograph operating conditions

> Injection mode: Split, ratio 100:1
>
> Column: DB-WAX, 30 m × 0.25 mm × 1.0 μm
>
> Detector: Mass spectrometer
>
> Carrier gas: Helium at a flow rate of 1.0 mL/min
>
> Temperature programme:
>
>> Injector: 250 °C
>>
>> Start temperature: 90 °C, hold for 0 min
>>
>> Rate: 15 °C/min to 180 °C, hold for 8 min, 50 °C/min to 230 °C, hold at 230 °C for 5 min, post run at 250 °C for 10 min
>
> Total run time: 17 min
>
> Autosampler conditions: Injection volume: 1.0 μL

A2.1.6.2 Mass spectrometer operating conditions

Transfer line temperature: 280 °C

Ionization mode: Electrospray ionization

Ionization voltage: 70 eV

Ion source temperature: 250 °C

Quadrupole temperature: 150 °C

Solvent delay: 3 min

Note: These gas chromatography–mass spectrometry operating conditions should be adapted to obtain correct resolution of the glycerol, triethylene glycol and propylene glycol peaks. A typical chromatogram (total ion chromatogram) is shown in Annex 3.

Table A2.1. Qualitative and quantitative ions of humectants and internal standard

Humectant	Quantitative ions	Secondary quantitative ions	Qualitative ions (abundance ratio)
Glycerol	61	43	61:43 (100:79)
Triethylene glycol	45	89	45:89 (100:19)
Propylene glycol	45	43	45:43 (100:20)
1,3-Butanediol	72	43	43:72 (100:28)

Table A2.2. Repeatability and reproducibility limits for determination of propylene glycol in tobacco in reference test pieces by gas chromatography–mass spectrometry

Reference cigarette	n*	$\hat{m}$ (mg/g)	Repeatability limit (r) (mg/g)	Reproducibility limit (R) (mg/g)
1R5F	6	0.215	0.001	0.003
3R4F	6	0.179	0.000	0.001
CM6	6	0.113	0.000	0.002
Commercial cigarette No.1	6	6.310	0.128	0.668
Commercial cigarette No.2	6	0.037	0.000	0.001

* Results in one data set were removed as they were outliers.

Table A2.3. Repeatability and reproducibility limits for determination of glycerol in tobacco in reference test pieces by gas chromatography–mass spectrometry

Reference cigarette	n*	$\hat{m}$ (mg/g)	Repeatability limit (r) (mg/g)	Reproducibility limit (R) (mg/g)
1R5F	5	21.147	0.659	1.941
3R4F	5	21.510	0.715	1.331
CM6	5	1.466	0.006	0.106
Commercial cigarette No.1	5	14.762	0.512	1.164
Commercial cigarette No.2	5	1.457	0.005	0.110

* Results in two data sets were removed as they were outliers.

Table A2.4. Repeatability and reproducibility limits for determination of triethylene glycol in tobacco in reference test pieces by gas chromatography–mass spectrometry

Reference cigarette	n*	$\hat{m}$ (mg/g)	Repeatability limit (r) (mg/g)	Reproducibility limit (R) (mg/g)
Sample TG No. 1	6	0.971	0.006	0.010
Sample TG No. 2	6	8.450	0.189	0.617

* Results in one data set were removed as they were outliers.

Annex 3. Example of chromatographs obtained with mass spectrometry

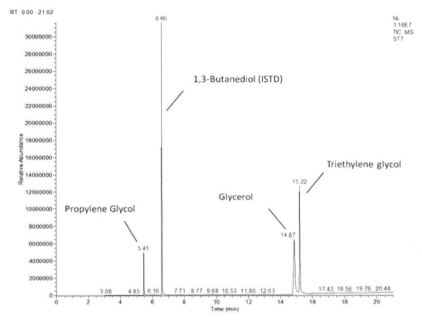

Fig. A3.1. Chromatogram of a standard solution

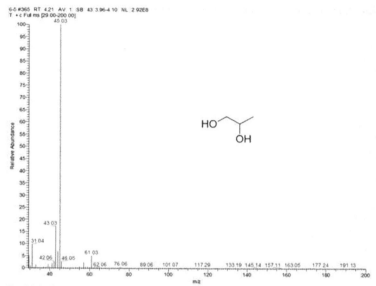

Fig. A3.2. Mass spectra of propylene glycol in scan mode

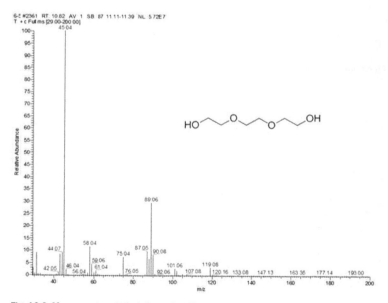

Fig. A3.3. Mass spectra of triethylene glycol in scan mode

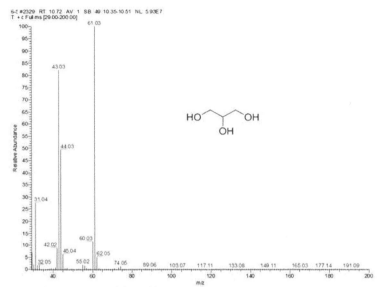

Fig. A3.4. Mass spectra of glycerol in scan mode

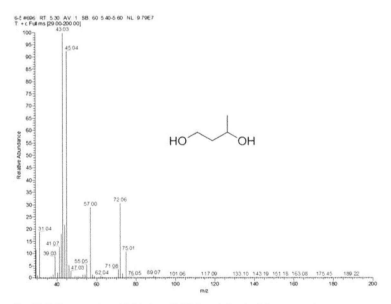

Fig. A3.5. Mass spectra of 1,3-butanediol (internal standard) in scan mode

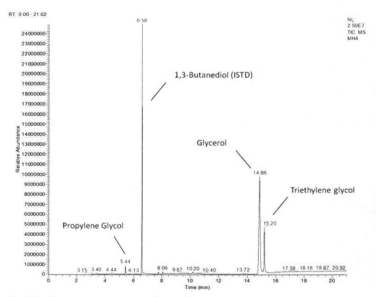

Fig. A3.6. Chromatogram of a sample solution

**WHO TobLabNet
Official Method
SOP 07**

Standard Operating Procedure for Determination of Ammonia in Cigarette Tobacco Filler

Method:	Determination of ammonia in cigarette tobacco filler
Analytes:	Ammonia (CAS # 7664-41-7)
Matrix:	Cigarette tobacco filler
Last update:	July 2016

The mention of specific companies or of certain manufacturers' products does not imply that they are endorsed or recommended by WHO in preference to others of a similar nature that are not mentioned. Errors and omissions excepted, the names of proprietary products are distinguished by initial capital letters.

> No machine smoking regimen can represent all human smoking behaviour: machine smoking testing is useful for characterizing cigarette emissions for design and regulatory purposes, but communication of machine measurements to smokers can result in misunderstanding about differences between brands in exposure and risk. Data on smoke emissions from machine measurements may be used as inputs for product hazard assessment, but they are not intended to be nor are they valid as measures of human exposure or risk. Representing differences in machine measurements as differences in exposure or risk is a misuse of the results of testing with WHO TobLabNet-recommended methods.

FOREWORD

This document was prepared by members of the World Health Organization (WHO) Tobacco Laboratory Network (TobLabNet) as an analytical method standard operating procedure (SOP) for measuring ammonia in cigarette tobacco filler.

INTRODUCTION

In order to establish comparable measurements for testing tobacco products globally, consensus methods are required for measuring specific contents and emissions of cigarettes. The Conference of the Parties to the WHO Framework Convention on Tobacco Control (WHO FCTC) at its third session, in Durban, South Africa, in November 2008, "recalling its decisions FCTC/COP1(15) and FCTC/COP2(14) on the elaboration of guidelines for implementation of Articles 9 (Regulation of the contents of tobacco products) and 10 (Regulation of tobacco product disclosures) of the WHO FCTC, noting the information contained in the report of the working group to the third session of the Conference of the Parties on the progress of its work ... requested the Convention Secretariat to invite WHO's Tobacco Free Initiative to ... validate, within five years, the analytical chemical methods for testing and measuring cigarette contents and emissions" (FCTC/COP/3/REC/1).

Using the criteria for prioritization set at its third meeting in Ottawa, Canada, in October 2006, the working group on Articles 9 and 10 identified the following contents for which methods for testing and measuring (analytical chemistry) should be validated as a priority:

- nicotine
- ammonia
- propylene glycol (propane-1,2-diol)
- glycerol (propane-1,2,3-triol)
- triethylene glycol (2,2-ethylenedioxybis(ethanol)).

Measurement of these contents will require validation of three methods: one for nicotine, one for ammonia and one for humectants.

Using the criteria for prioritization set at the meeting in Ottawa mentioned above, the working group identified the following emissions in mainstream smoke for which methods for testing and measurement (analytical chemistry) should be validated as a priority:

- 4-(methylnitrosamino)-1-(3-pyridyl)-1-butanone (NNK)
- *N*-nitrosonornicotine (NNN)

- acetaldehyde
- acrylaldehyde (acrolein)
- benzene
- benzo[a]pyrene
- 1,3-butadiene
- carbon monoxide
- formaldehyde.

Measurement of these emissions with the two smoking regimens described below will require validation of five methods: one for tobacco-specific nitrosamines (NNK and NNN), one for benzo[a]pyrene, one for aldehydes (acetaldehyde, acrolein and formaldehyde), one for volatile organic compounds (benzene and 1,3-butadiene) and one for carbon monoxide.

The table below sets out the two smoking regimens for validation of the test methods referred to above.

Smoking regimen	Puff volume (mL)	Puff frequency	Filter ventilation holes
ISO regimen: *ISO 3308: Routine analytical cigarette smoking machine – definitions and standard conditions*	35	Once every 60 s	No modification
Intense regimen: Same as ISO 3308, but modified as indicated	55	Once every 30 s	All ventilation holes must be blocked 100% as described in WHO TobLabNet SOP 01.

This SOP was prepared to describe the procedure for the determination of ammonia in cigarette tobacco filler.

1 SCOPE

This standard operating procedure is suitable for the quantitative determination of ammonia in cigarette tobacco filler by ion chromatography.

2 REFERENCES

2.1 *ISO8243: Cigarettes – Sampling*.

2.2 United Nations Office on Drugs and Crime. *Guidelines on representative drug sampling*. Vienna: Laboratory and Scientific Section; 2009 (http://www.unodc.org/documents/scientific/Drug_Sampling.pdf).

2.3 *ISO5725-1: Accuracy (trueness and precision) of measurement methods and results – Part 1: General principles and definitions.*

2.4 *ISO5725-2: Accuracy (trueness and precision) or measurement methods and results – Part 2: Basic method for the determination of repeatability and reproducibility of a standard measurement method.*

2.5 *ISO guide 34: General requirements for the competence of reference materials producers.*

2.6 *AOAC 966.02: Loss on drying (moisture) in tobacco.* In: *Official methods of analysis*, 20th edition. Rockville, Maryland: AOAC International; 2016.

3   TERMS AND DEFINITIONS

3.1 *Ammonia content*: amount of ammonia in cigarette tobacco filler, expressed as milligrams per gram of cigarette tobacco filler

3.2 *Cigarette tobacco filler*: Tobacco-containing part of a cigarette, including reconstituted tobacco, stems, expanded tobacco and additives

3.3 *Tobacco products*: Products made entirely or partly of leaf tobacco as the raw material that are manufactured for smoking, sucking, chewing or snuffing (Article 1(f) of the WHO FCTC)

3.4 *Laboratory sample*: Sample intended for testing in a laboratory, consisting of a single type of product delivered to the laboratory at one time or within a specified period

3.5 *Test sample*: Product to be tested, taken at random from the laboratory sample

3.6 *Test portion*: Random portion from the test sample to be used for a single determination

4   METHOD SUMMARY

4.1 Ammonia is extracted from the cigarette tobacco filler with a dilute sulfuric acid extraction solution, followed by centrifugation (and filtration if necessary).

4.2 The extract is analysed by ion chromatography coupled with conductivity detection.

4.3 Quantification is achieved by calibration against an external standard, by comparing the conductivity response of the analytes in the samples with those of the standards.

5 SAFETY AND ENVIRONMENTAL PRECAUTIONS

5.1 Follow routine safety and environmental precautions, as in any chemical laboratory activity.

5.2 The testing and evaluation of certain products with this test method may require the use of materials or equipment that could be hazardous or harmful to the environment. This document does not address all the safety aspects associated with use of the method. All persons using this method are responsible for consulting the appropriate authorities and establishing health and safety practices as well as environmental precautions in conjunction with any existing applicable regulatory requirements prior to its use.

5.3 Special care should be taken to avoid inhalation or dermal exposure to harmful chemicals. Use a chemical fume hood, and wear an appropriate laboratory coat, gloves and safety goggles when preparing or handling undiluted materials, standard solutions, extraction solutions or collected samples.

6 APPARATUS AND EQUIPMENT

Usual laboratory apparatus, in particular:

6.1 Analytical balance capable of measurement to at least four decimal places

6.2 Erlenmeyer flasks, 250 mL, with stoppers, or equivalent

6.3 Beakers (various sizes)

6.4 Volumetric flasks (various sizes)

6.5 Volumetric pipettes for preparation of standard solutions (various sizes)

6.6 Autosampler

6.7 Autosampler vials, 2 mL, or equivalent

6.8 Mechanical wrist-action shaker or equivalent

6.9 Ion chromatograph equipped with a conductivity detector

6.10 Cation exchange column (250 × 4 mm)

6.11 Cation exchange guard column (50 × 4 mm)

6.12 Renewable cation suppressor (optional)

Note 1: The above is the typical instrument configuration for this analysis. Alternative configurations (for example, without a suppressor) are expected to yield similar results.

Note 2: Samples should be either filtered or centrifuged before analysis.

6.13 Filtration equipment

6.13.1 Syringe, 10 mL, or equivalent

6.13.2 Screen, 0.45 mm

6.13.3 Water-phase membrane filters, 15 mm × 0.45 μm, or equivalent

6.13.4 Ashless quantitative filter paper, for example Whatman No. 40, 8 μm, or equivalent

6.14 Centrifugation equipment

6.14.1 Micro-17 centrifuge, or equivalent

6.14.2 Microcentrifuge vials, or equivalent

7 REAGENTS AND SUPPLIES

All reagents shall be of at least analytical reagent grade unless otherwise noted. Reagents are identified by their Chemical Abstract Service [CAS] registry numbers when available.

7.1 Methanesulfonic acid [75-75-2], chromatographic purity

7.2 Concentrated sulfuric acid [7664-93-9], mass percentage: 95% to ~98%

7.3 Ammonium sulfate [7783-20-2] or, alternatively, a standard solution endorsed by ISO Guide 34: [**2.5**] of 1000 mg/L ammonia

7.4 Water, Type 1 [7732-18-5]

8 PREPARATION OF GLASSWARE

8.1 Clean and dry glassware in a manner to avoid contamination.

Note: It is recommended that detergents not be used for cleaning, in order to minimize interference.

9 PREPARATION OF SOLUTIONS

The method for preparing extraction and mobile phase solutions described below is for reference purposes and can be adjusted if necessary.

9.1 Extraction solution (dilute sulfuric acid, approximately 0.0125 mol/L)

9.1.1 To obtain 0.0125 mol/L sulfuric acid solution, use the equation $ρ = m/V$, where $ρ$ is the density of sulfuric acid (1.84 g/mL) and m is the mass of sulfuric acid.

Note: 1.29 g sulfuric acid added to 1 L of Type 1 water [**7.4**] gives a 0.0125 mol/L dilution of sulfuric acid.

9.1.2 Label as "dilute sulfuric acid extraction solution", and store in a refrigerator.

9.2 Mobile phase (methanesulfonic acid in water, 0.02 mol/L)

Accurately transfer 1.32 mL of methanesulfonic acid [**7.1**] into a 1-L volumetric flask [**6.4**], and make up to volume with Type 1 water [**7.4**].

10 PREPARATION OF STANDARDS

The method for preparing standard solutions described below is for reference purposes and can be adjusted if necessary.

10.1 Ammonia standard solution (approximately 100 µg/mL)

10.1.1 Accurately weigh 0.092 g (± 0.002 g) of ammonium sulfate [**7.3**] into a 100-mL beaker, and dissolve it in extraction solution [**9.1**]. All solvents and solutions must be at room temperature.

10.1.2 Transfer the ammonium sulfate solution [**10.1.1**] into a 250-mL volumetric flask, and make up to volume with extraction solution [**9.1**].

10.2 Working standards

Table 1. Ammonia working standard solutions

Standard	Volume of standard solution [**10.1**] (mL)	Final volume (mL)	Approximate ammonia concentration in working standard solution (µg/mL)
1	0.01	10	0.1
2	0.05	10	0.5
3	0.2	10	2
4	0.5	10	5
5	1	10	10

Note 1: All standards are made in volumetric flasks at the dilutions described. Volumetric flasks are made up to volume with extraction solution [**9.1**].

Note 2: The range of the standard solutions may be adjusted, depending on the equipment used (with or without suppression equipment) and samples to be tested, keeping in mind a possible effect on the sensitivity of the method.

Note 3: The concentration of standard solutions can change significantly during storage, which will change the intercept of the calibration curve. They should not be stored for longer than 2 weeks.

Note 4: The purity of ammonia in the standard used must be considered in calculating the concentration.

11 SAMPLING

11.1 Sample cigarettes according to ISO 8243 [**2.1**]. Alternative approaches may be used to obtain a representative laboratory sample in accordance with individual laboratory practice or when required by specific regulation or the availability of samples.

11.2 Constitution of test sample

11.2.1 Divide the laboratory sample into separate units (e.g. packet, container), if applicable.

11.2.2 Take an equal amount of product for each test sample from at least $\sqrt{n}$ [**2.2**] of the individual units (e.g. packet, container).

12 CIGARETTE PREPARATION

12.1 Remove all tobacco filler from all test cigarettes or quality control samples (when applicable) in a pack (e.g. containing 20 cigarettes). Alternatively, use at least 7 g of processed cigarette tobacco filler.

12.2 Combine and mix sufficient cigarette tobacco filler to constitute test portions of at least 7 g each. Prepare at least three replicates of test portions.

Note: Sample size may be adjusted if necessary.

13 PREPARATION OF THE SMOKING MACHINE

Not applicable.

14 SAMPLE GENERATION

Not applicable.

15 SAMPLE PREPARATION

15.1 For each test portion, take 0.7 g of well-mixed test portion, and weigh it to 0.0001 g into a 250-mL Erlenmeyer or other suitable flask.

15.2 Add 50 mL of the extraction solution [**9.1**] to the sample.

Note: Sample weight and dilution factor may be adjusted accordingly depending on the concentrations of ammonia or availability of sample.

15.3 Stopper the flasks, and place them on a wrist-action shaker or equivalent at a rate of at least 160 rpm for 30 min or at a rate suitable for the type of mechanical shaker used.

15.4 Remove the samples from the shaker, swirling each flask to disperse all the tobacco in the solvent.

15.5 Leave the samples for 30 min until the supernatant is clear.

15.6 Filter the extract through 8-μm ashless quantitative filter paper.

15.7 Syringe the extract [**15.6**] onto a 0.45-μm membrane filter.

15.8 Transfer the filtrate to an autosampler vial, and analyse it on the ion chromatograph.

Note: If the sample signal (peak area) does not fall within the working range of the calibration curve, the solutions should be adjusted accordingly, i.e. further diluted. All the final dilutions should be recorded in the sample report.

Alternatively, the following steps can be followed:

15.5 Centrifuge 1.5 mL of the supernatant at 13 000 rpm for 5 min.

15.6 Transfer the centrifuged extract into an autosampler vial, and analyse it on the ion chromatograph.

16 SAMPLE ANALYSIS

This method involves ion chromatography coupled with conductivity detection to quantify ammonia in cigarette tobacco filler. The analytes are resolved from potential interference on the column. Comparison of the area of the unknowns with the area of the known standard concentrations yields the concentrations of individual analytes.

16.1 Ion chromatography operating conditions

Mobile phase: Methanesulfonic acid in water, 0.02 mol/L [**9.2**]

Rate: 1.0 mL/min

Injection volume: 25 μL

Column temperature: 30 °C

These instrumental settings and other specifications (e.g. current setting of cation suppressor) shall be adapted for each equipment and column profile in order to obtain good resolution and chromatographic performance.

Note 1: For example, a more dilute solution of methanesulfonic acid in water is recommended for the mobile phase (up to 10-fold dilution) when working without suppression.

Note 2: The operating parameters may have to be adjusted to the instrument and column conditions and the resolution of chromatographic peaks.

16.2 Expected retention times

Note: Differences in, for example, the type and age of the column will alter retention times.

Under the above conditions, the expected total analysis time will be about 15 min.

16.3 Determination of ammonia

16.3.1 Condition the system before use, for example by injecting two 25-µL aliquots of sample solution as a primer. Other equilibrating procedures, as suggested by the manufacturer, can also be used.

Note: Depending on the stationary phase, a blank might have to be injected after the calibration standards. A blank should also be injected at the end of a batch, to reduce cross-contamination.

16.3.2 After conditioning, inject 25 µL of blank solution to check for any contamination in the system or the reagents.

16.3.3 Inject an aliquot of each ammonia standard solution [**10.2**] into the ion chromatograph.

16.3.4 Assess the retention times and responses (area counts) for the standards. The system is considered ready to perform the analysis when the retention times of the standards are similar (defined as ±0.2 min) to the between-run retention times in the same system, the calculated resolution is ≥ 1.5, and responses are within 20% of the typical responses in previous runs.

16.3.5 Record the peak areas of ammonia.

16.3.6 Plot a graph of the areas in accordance with the concentration of ammonia.

16.3.7 The following alternatives should be used for calculating ammonia concentrations.

For ion chromatography with a suppressor:

Calculate a quadratic regression equation ($Y = ax^2 + bx + c$) from the data.

If the quadratic regression coefficient R^2 is < 0.99, the calibration should be repeated. An individual calibration point that differs by > 10% from the expected value (estimated by quadratic regression) should be omitted.

For ion chromatography without a suppressor:

Calculate a linear regression equation ($Y = a + bx$) from both the slope (b) and the intercept (a) of the linear regression. If the linear regression coefficient R^2 is < 0.99, the calibration should be repeated. An individual calibration point that differs by > 10% from the expected value (estimated by linear regression) should be omitted.

- **16.3.8** In both cases, the intercept should not be statistically significantly different from zero.

- **16.3.9** Inject the quality control samples and the test samples, and determine the peak areas with the instrument software.

- **16.3.10** The signal (peak area) obtained for all test portions must fall within the working range of the calibration curve.

    Note: See Annex 1 for typical calibration lines and chromatograms.

17 DATA ANALYSIS AND CALCULATIONS

17.1 For each test portion, calculate the peak areas of ammonia.

17.2 Calculate the concentration of ammonium ion (NH_4^+) in mg/mL in each test portion from the coefficients of the quadratic or linear regression.

17.3 The amount of ammonia (in mg/g of tobacco) is determined from the following formula:

$$M = \frac{C \times V \times 17.03}{m \times 18.04}$$

where:

M is the concentration of ammonia (NH_3) in cigarette filler, expressed in mg/g.

C is the concentration of NH_4^+ in the sample solution obtained from the calibration line, expressed in μg/mL.

V is the volume of the sample solution, expressed in mL.

m is the mass of tobacco filler, expressed in mg.

Note 1: The relative molecular mass of NH_3 is 17.03.

Note 2: The relative molecular mass of NH_4^+ is 18.04.

18 SPECIAL PRECAUTIONS

After installing a new column, condition it by injecting a tobacco sample extract under the specified instrument conditions. Injections should be repeated until the peak areas (or heights) of ammonia are reproducible according to the acceptance criteria of individual laboratories.

19 DATA REPORTING

19.1 Report individual measurements for each sample evaluated.

19.2 Report results as milligrams per gram of tobacco or as required.

19.3 Results may be reported as is or on a dry weight basis

Note: Moisture can be measured with AOAC 966.02 [**2.6**] or a method of equivalent standard.

20 QUALITY CONTROL

20.1 Typical control parameters

If the control measurements are outside the tolerance limits of the expected values, appropriate investigation and action must be undertaken.

Additional quality assurance procedures should be used if necessary in order to comply with the policies of individual laboratories.

20.2 Laboratory reagent blank

To detect any contamination during sample preparation and analysis, include a laboratory reagent blank. The blank consists of all the reagents and materials used in analysing test samples and is analysed like a test sample. The result should be below the level of detection.

20.3 Quality control sample

To verify the consistency of the entire analytical process, analyse a reference cigarette, such as a University of Kentucky research cigarette (Lexington, Kentucky, USA) or CORESTA Monitor (Paris, France), with each analytical run.

21 METHOD PERFORMANCE

21.1 Limit of reporting

The limit of reporting is set to the lowest concentration of the calibration standards used, recalculated to approximately 0.1 µg/mL.

21.2 Laboratory-fortified matrix recovery

Recovery of analyte spiked onto the matrix is used as a surrogate measure of accuracy. It is determined by spiking known amounts of standards into tobacco and extracting the tobacco in the same way as for samples. Unspiked tobacco is also analysed. Recovery is calculated from the following formula:

$$\text{Recovery (\%)} = 100 \times (\text{analytical amount} - \text{unspiked amount} / \text{spiked amount})$$

Table 3. Sample recovery results for ammonia in tobacco filler

Spiked amount (mg)[a]	Analytical amount (mg)	Unspiked amount (mg)	Recovery (%)
0.0799 (low)	0.1751	0.0960	98.90
0.0999 (medium)	0.1944	0.0956	98.88
0.1199 (high)	0.2141	0.0953	99.10
			Mean: 98.96

[a] "Low", "medium" and "high" spiked amounts are about 80%, 100% and 120% of the unspiked amount.

21.3 Limit of detection (LOD) and limit of quantification (LOQ)

The LOD can be determined as three times the standard deviation of results obtained by analysing the lowest standard a minimum of 10 times over several days. The LOQ can be determined as 10 times the standard deviation of results obtained by analysing the lowest standard a minimum of 10 times over several days.

The LOD of ammonia in tobacco filler is 0.0033 mg/g, and the LOQ is 0.011 mg/g.

Alternatively, the lowest standard used can be taken as the LOQ.

22 REPEATABILITY AND REPRODUCIBILITY

An international collaborative study conducted in 2014-2015, involving testing of three reference cigarettes and two commercial brands by nine laboratories gave the following values for this method.

The difference between two single results found for matched cigarette tobacco filler samples by the same operator using the same apparatus within the shortest feasible time will exceed the repeatability, r, on average no more than once in 20 cases with normal, correct application of the method.

Single results for matched cigarette tobacco filler samples reported by two laboratories will differ by no more than the reproducibility, R, on average no more than once in 20 cases with normal, correct application of the method.

The test results were analysed statistically in accordance with ISO 5725-1 [2.3] and ISO 5725-2 [2.4] to give the precision data shown in Table 4. The ammonia concentrations in the commercial brands were within the range of those in the reference pieces.

To calculate r and R, one test result was defined as the average of seven replicates.

Table 4. Precision limits for determination of ammonia in tobacco from reference test pieces

Reference cigarette	n	m	Repeatability limit (mg/g), r	Reproducibility limit (mg/g), R
1R5F	9	1.214	0.002	0.153
3R4F	9	0.857	0.002	0.075
CM6	9	0.172	0.000	0.004
Commercial cigarette 1	9	0.383	0.001	0.010
Commercial cigarette 2	9	1.746	0.005	0.347

n, number of laboratories that participated; m, mean value of ammonia per cigarette

23 TEST REPORT

The following information shall be included in the test report:

(a) A reference to this method i.e. WHO TobLabNet SOP 07

(b) Date of receipt of the sample

(c) The results and its units

Annex 1. Typical calibration line and chromatograms obtained in the determination of ammonia in cigarette tobacco filler

Fig. 1a. Example of ammonia calibration line (without suppression)

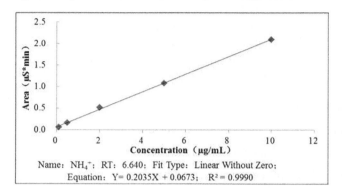

Name: NH_4^+; RT: 6.640; Fit Type: Linear Without Zero;
Equation: $Y = 0.2035X + 0.0673$; $R^2 = 0.9990$

Fig. 1b. Example of ammonia calibration line (with suppression)

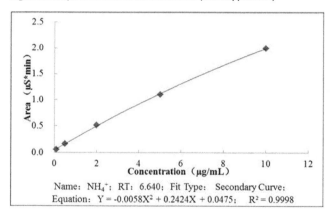

Name: NH_4^+; RT: 6.640; Fit Type: Secondary Curve;
Equation: $Y = -0.0058X^2 + 0.2424X + 0.0475$; $R^2 = 0.9998$

Fig. 2. Representative chromatogram of a standard solution of ammonia (about 5 μg/mL), with suppression

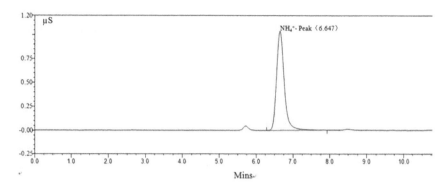

Fig. 3. Representative chromatogram of a sample solution of tobacco filler, with ammonia peak, with suppression

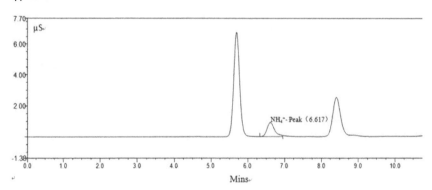

WHO TobLabNet
Official Method
SOP 08

Standard Operating Procedure for Determination of Aldehydes in Mainstream Cigarette Smoke under ISO and Intense Smoking Conditions

Method:	Determination of Aldehydes in Mainstream Cigarette Smoke Using ISO and Intense Smoking Conditions
Analytes:	Formaldehyde (CAS# 50-00-0) Acetaldehyde (CAS # 75-07-0) Acrolein (acrylaldehyde) (CAS # 107-02-8)
Matrix:	Mainstream Cigarette Smoke
Last update:	31 August 2018

No machine smoking regimen can represent all human smoking behaviour: machine smoking testing is useful for characterizing cigarette emissions for design and regulatory purposes, but communication of machine measurements to smokers can result in misunderstanding about differences between brands in exposure and risk. Data on smoke emissions from machine measurements may be used as inputs for product hazard assessment, but they are not intended to be nor are they valid as measures of human exposure or risks. Representing differences in machine measurements as differences in exposure or risk is a misuse of testing with WHO TobLabNet standards.

FOREWORD

This document was developed by members of the World Health Organization (WHO) Tobacco Laboratory Network (TobLabNet) as a standard operating procedure (SOP) for measuring aldehydes in mainstream cigarette smoke under ISO and intense smoking conditions.

INTRODUCTION

In order to establish comparable measurements for testing tobacco products globally, consensus methods are required for measuring specific contents and emissions of cigarettes. The Conference of the Parties to the WHO Framework Convention on Tobacco Control (WHO FCTC) at its third session in Durban, South Africa, in November 2008, recalling its decisions FCTC/COP1(15) and FCTC/COP2(14) on the elaboration of guidelines for implementation of Articles 9 (Regulation of the contents of tobacco products) and 10 (Regulation of tobacco product disclosures) of the WHO FCTC, noting the information contained in the report of the working group to the third session of the Conference of the Parties on the progress of its work ... requested the Convention Secretariat to invite WHO's Tobacco Free Initiative to, "validate, within five years, the analytical chemical methods for testing and measuring cigarette contents and emissions".[1]

Using the criteria for prioritization set at its third meeting in Ottawa, Canada, in October 2006, the working group on Articles 9 and 10 identified the following contents for which methods for testing and measurement (analytical chemistry) should be validated as a priority:

- nicotine
- ammonia
- humectants (propane-1,2-diol, glycerol (propane-1,2,3-triol) and triethylene glycol (2,2-ethylenedioxybis(ethanol)).

Measurement of these contents will require validation of three methods: one for nicotine, one for ammonia and one for humectants.

Using the criteria for prioritization set at the meeting in Ottawa mentioned above, the working group identified the following emissions in mainstream smoke for which methods for testing and measurement (analytical chemistry) should be validated as a priority:

- 4-(methylnitrosamino)-1-(3-pyridyl)-1-butanone (NNK)
- *N*-nitrosonornicotine (NNN)

---

[1] Decision, third session of the Conference of the Parties to the WHO FCTC; 2008 (http://apps.who.int/gb/fctc/PDF/cop3/FCTC_COP3_REC1-en.pdf).

- acetaldehyde
- acrolein (acrylaldehyde)
- benzene
- benzo[a]pyrene
- 1,3-butadiene
- carbon monoxide
- formaldehyde.

Measurement of these emissions with the two smoking regimens described below will require validation of four methods: one for tobacco-specific nitrosamines (NNK and NNN), one for benzo[a]pyrene, one for aldehydes (acetaldehyde, acryladehyde and formaldehyde) and one for volatile organic compounds (benzene, 1,3-butadiene and carbon monoxide).

The table below sets out the two smoking regimens for validation of the test methods referred above.

Smoking regimen	Puff volume (mL)	Puff frequency	Filter ventilation holes
ISO regimen: ISO 3308; Routine analytical cigarette smoking machine – definitions and standard conditions	35	Once every 60 s	No modifications
Intense regimen: Same as ISO 3308, but modified as indicated	55	Once every 30 s	All ventilation holes must be blocked 100% as described in WHO TobLabNet SOP 01.

This method SOP was developed to direct participating labs in the specifics of carrying out the analysis of aldehydes in mainstream cigarette smoke. For tobacco products other than cigarettes, specific parameters may have to be adjusted from those given in the document.

1. SCOPE

This standard operating procedure is suitable for the quantitative determination of the following three aldehydes in mainstream (MS) cigarette smoke: formaldehyde, acetaldehyde and acrolein (acrylaldehyde) to be analysed by liquid chromatography.

Note: Training in the use of the smoking machine and other analytical equipment is important for successful operation. For those not experienced in operating smoking machines or analyses using the analytical methods for measuring tobacco product emissions and contents, training should be obtained.

2. REFERENCES

2.1 ISO 3308: Routine analytical cigarette-smoking machine – Definitions and standard conditions. Geneva: International Organization for Standardization; 2012.

2.2 ISO 4387: Cigarettes – Determination of total and nicotine-free dry particulate matter using a routine analytical smoking machine. Geneva: International Organization for Standardization; 2000.

2.3 ISO 3402: Tobacco and tobacco products – Atmosphere for conditioning and testing. Geneva: International Organization for Standardization; 1999.

2.4 ISO 8243: Cigarettes – Sampling. Geneva: International Organization for Standardization; 2006.

2.5 ISO 5725-2: Accuracy (trueness and precision) or measurement methods and results – Part 2: Basic method for the determination of repeatability and reproducibility of a standard measurement method. Geneva: International Organization for Standardization; 1994.

2.6 Tobacco Laboratory Network. Standard Operating Procedure for Intense Smoking (WHO TobLabNet SOP 01). Geneva: World Health Organization; 2012 (http://apps.who.int/iris/bitstream/handle/10665/75261/9789241503891_eng.pdf;sequence=1, accessed 3 September 2018).

2.7 Tobacco Laboratory Network. Standard operating procedures for validation of analytical methods of tobacco product contents and emissions (WHO TobLabNet SOP 02). Geneva: World Health Organization., 2012 (http://apps.who.int/iris/bitstream/handle/10665/254998/9789241512060-eng.pdf?sequence=1, accessed 3 September, 2018).

2.8 Tobacco Laboratory Network, Standard Operating Procedure for the Determination of volatile organics in mainstream cigarette smoke under ISO and intense smoking conditions (WHO TobLabNet SOP 09). Geneva: World Health Organization; (in press).

2.9 Laboratory and Scientific Section. Guidelines on representative drug sampling. Vienna: United Nations Office on Drugs and Crime; 2009 (http://www.unodc.org/documents/scientific/Drug_Sampling.pdf).

2.10 ISO 5725-1. Accuracy (trueness and precision) of measurement methods and results—Part 1: General principles and definitions. Geneva: International Organization for Standardization; 1994.

2.11 Uchiyama S, Tomizawa T, Inaba, Y, Kunugita, N. Simultaneous determination of volatile organic compounds and carbonyls in mainstream cigarette smoke using a sorbent cartridge followed by two-step elution. J Chromatogr A. 2013; 1314:31-7. doi: 10.1016/j.chroma.2013.09.019.

2.12 Uchiyama, S, Hayashida, H, Izu, R, Inaba, Y, Nakagome, H, Kunugita N. Determination of nicotine, tar, volatile organic compounds and carbonyls in mainstream cigarette smoke using a glass filter and a sorbent cartridge followed by the two-phase/one-pot elution method with carbon disulfide and methanol. J Chromatogr A. 2015; 1426:48-55. doi.org/10.1016/j.chroma.2015.11.058

3. TERMS AND DEFINITIONS

Aldehydes: formaldehyde, acetaldehyde and acryladehyde

CX-572: carbon molecular sieves Carboxen 572.

DNPH-HCl: 2,4-Dinitrophenylhydrazine hydrochloride.

Intense Regime: parameters used to smoke tobacco products which include 55-mL puff volume, 30-sec puff interval, and 100% blocking of the filter ventilation holes.

ISO Regime: parameters used to smoke tobacco products which include 35-mL puff volume, 60-sec puff interval, and no blocking of the filter ventilation holes.

Laboratory sample: sample intended for testing in a laboratory, consisting of a single type of product delivered to the laboratory at one time or within a specified period.

Smoke trap: device for collecting such part of the smoke from a sample of cigarettes as is necessary for the determination of specified smoke components.

Test sample: product to be tested, taken at random from the laboratory sample. The number of products taken shall be representative of the laboratory sample.

Test portion: random sample from the test sample to be used for a single determination. The number of products taken shall be representative of the test sample.

Tobacco products: products entirely or partly made of the leaf tobacco as raw material which are manufactured to be used for smoking, sucking, chewing or snuffing (Article 1(f) of the WHO FCTC).

TPM: total particulate matter

4. METHOD SUMMARY

4.1 Mainstream cigarette smoke is performed according to the WHO Intense regime or the ISO regime. One cigarette is smoked per smoke trap, containing 300 mg CX-572 particles and a glass fibre filter pad.

4.2 Mainstream smoke total particulate matter from the cigarette test sample is trapped onto a smoke trap containing 300 mg CX-572 particles and a glass-fibre filter pad (commonly referred to as a glass fibre filter pad).

4.3 The aldehydes are extracted by adding a solution containing a mixture of carbon disulphide and methanol onto the Carboxen particles and glass fibre filter pad.

4.4 The extracted aldehydes are derivatized with 2,4-dinitrophenylhydrazine (DNPH).

4.5 The derivatized aldehydes are determined by high performance liquid chromatography (HPLC), equipped with ultraviolet (UV)/diode array detector (DAD).

4.6 This method described in this SOP simultaneously tests and measures the level of aldehydes (SOP 08) and volatile organics (SOP 09) produced in mainstream cigarette smoke.

4.7 The aldehyde chromatogram peak area is compared with a calibration curve created by analysis of standards with known concentrations of the aldehydes to determine the aldehyde content of each test portion.

5. SAFETY AND ENVIRONMENTAL PRECAUTIONS

5.1 Follow routine safety and environmental precautions, as in any chemical laboratory activity.

5.2 The testing and evaluation of certain products with this test method may require the use of materials or equipment that could be hazardous or harmful to the environment; this document does not purport to address all the safety aspects associated with its use. All persons using this method have the responsibility to consult with the appropriate authorities and to establish health and safety practices as well as environmental precautions in conjunction with any existing applicable regulatory requirements prior to its use.

5.3 Special care should be taken to avoid inhalation or dermal exposure to harmful chemicals. Use a chemical fume hood and wear an appropriate laboratory coat, gloves and safety goggles when preparing or handling undiluted materials, standard solutions, extraction solutions or collected samples.

6. APPARATUS AND EQUIPMENT

6.1 Equipment needed to condition cigarettes as specified in ISO 3402 [**2.3**].

6.2 Equipment needed to perform marking for butt length as specified in ISO 4387 [**2.2**].

6.3 Equipment needed to cover ventilation holes for the Intense Regimen as specified in WHO TobLabNet, Standard Operating Procedure for Intense Smoking (WHO TobLabNet SOP 01) [**2.6**].

6.4 Equipment needed to perform smoking of cigarettes as specified in ISO 3308 [**2.1**].

6.5 Analytical balance capable of measuring to at least four decimal places.

6.6 Pipettes and tips capable of accurately dispensing 40–5000 µl.

6.7 Volumetric flasks, 5 mL and 10 mL.

6.8 Magnetic stirring device.

6.9 Calibrated syringes or dispensers with the capacity of accurately transfer 2 to 20 mL of organic solvent or equivalent.

6.10 Vials or flasks with septum seal, capacity of at least 20 mL (linear) or 50 mL (rotary) or equivalent. Amber vials are recommended.

6.11 Liquid Chromatography (LC) autosampler vials, 2 mL or equivalent.

6.12 HPLC system, equipped with a UV/DAD, capable of measuring wavelength at 360 nm.

6.13 HPLC column capable of distinct separation of solvent peaks, DNPH peak and aldehyde peaks from other cigarette emissions components (e.g. Ascentis RP-Amide (150 mm × 4.6 mm, 3 μm).

6.14 Carboxen 572 particles, 20/45 mesh size.

6.15 CX-572 cartridge: either manually prepared 1 mL solid-phase extraction (SPE) cartridge filled with 300 mg of Carboxen 572 or preloaded CX-572 cartridge acquired by Sigma-Aldrich/Supelco (Cat. No. 54294-U).

6.16 Glass fibre filter pad as described in ISO 4387 [**2.2**]

6.17 Smoke trap set up consists of smoking machine holders attached with CX-572 (with or without cartridge) to be placed in front of glass fibre filter pad (See Figs. 1 to 5).

Note: Place the CX-572 cartridge in the holder facing the open side towards the cigarette. When positioned with the open end in the opposite direction, the content of the cartridge can be sucked into the smoking machine when puffs are taken. If necessary, a special stopper can be inserted into the cartridge as shown in Fig. 6.

Fig. 1a: Single port Borgwaldt smoking machine

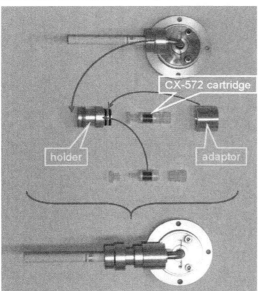

Fig. 1b: Linear Borgwaldt smoking machine

Fig. 2a and Fig. 2b illustrate the cigarette holder suitable for use in commercial rotary smoking machine (i.e. Cerulean and/or Borgwaldt smoking machine)

Figure 2a: with the CX-572 particles weigh directly onto the holder

Fig. 2b: with the CX-572 loaded cartridge inserted onto the holder

Figure 3: For Linear commercial smoking machine (i.e. Cerulean or Borgwaldt)

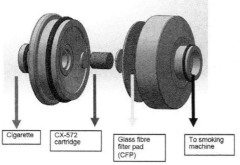

Note: CX-572 can be pre-packed into this holder or alternatively, a commercially available CX-572 cartridge can be used.

Fig. 4: For Linear Cerulean smoking machine (also applicable to Linear Borgwaldt or Rotary Cerulean or single channel smoking machine)

Fig. 5: For Linear Cerulean smoking machine (also applicable to Linear Borgwaldt smoking machine).

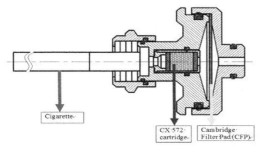

Fig. 6: Stoppers to be inserted into the CX-572 cartridges to prevent the content being sucked into the smoking machine.

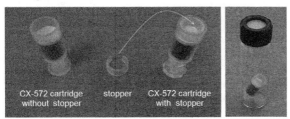

Note: Be sure to put the cartridge into a closed vessel (vial) directly after inserting the stopper to prevent contamination of the CX-572.

7. REAGENTS AND SUPPLIES

All reagents shall be of at least analytical reagent grade unless otherwise noted. When possible, reagents are identified by their Chemical Abstract Service (CAS) registry numbers.

7.1 2,4-dinitrophenylhydrazine hydrochloride (DNPH-HCl), > 98% (i.e. Tokyo Chemical Industry Co, Ltd part number D0846). CAS Number: 55907-61-4.

7.2 Formaldehyde-DNPH solution, 100 μg/ml (as aldehyde) (i.e. Sigma-Aldrich/Supelco art. nr. CRM47177 Supelco). CAS Number: 1081-15-8.

7.3 Acetaldehyde-DNPH solution 1000 μg/ml (as aldehyde) (i.e. Sigma-Aldrich/Supelco art. nr. CRM4M7340 Supelco). CAS Number: 1019-57-4.

7.4 Acrolein-DNPH solution, 1000 μg/ml (as aldehyde) (i.e. Sigma-Aldrich/Supelco art. nr. CRM47342 Supelco). CAS Number: 888-54-0.

7.5 Acetonitrile, HPLC-grade. CAS Number: 75-05-8.

7.6 Phosphoric acid 85%. CAS Number: 7664-38-2.

7.7 Carbon disulphide. CAS number: 75-15-0.

7.8 Methanol, anhydrous, analytical-grade. CAS number: 67-56-1.

7.9 Ethanol, >98%. CAS number: 64-17-5.

7.10 Carboxen-572 particles, 20/45 mesh size (i.e. Sigma-Aldrich/Supelco Cat. No. 11072-U).

8. PREPARATION OF GLASSWARE
Clean and dry glassware to ensure no contamination from residues.

9. PREPARATION OF SOLUTIONS

9.1 Derivatization reagent (DNPH-solution)

9.1.1 Weigh approximately 1 g of DNPH-HCL [7.1] in a volumetric flask of 50 mL.

9.1.2 Add 10 mL of phosphoric acid [7.6].

9.1.3 Dilute with acetonitrile to 50 mL [7.5].

9.1.4 Stir the solution continuously until a clear solution is obtained by using a magnetic stirring device. Store the solution at 4–10 °C, this solution is stable for one month.

9.2 Extraction solutions: There are two options to prepare the extraction solutions.

9.2.1 Option A: Carbon disulfide and methanol.

9.2.2 Option B: Carbon disulfide and methanol (1:4) mixed solution containing internal standard solution.

9.2.3 For analysis in combination with benzene and 1,3-butadiene as prescribed in WHO TobLabNet SOP09, add 20 mL of carbon disulfide and 500 μL or 250 μL of internal standard solution (benzene-d6 5 mg/mL) to the 100 mL volumetric flask and dilute with methanol.

9.2.3 Carbon disulfide and methanol (1:4) mixed solution

Add 20 mL of carbon disulfide to the 100 mL volumetric flask and dilute with methanol to 100 mL.

Note: The amount of extraction solution prepared can be adjusted based on the requirements of the analysis. The following are for illustrative purposes only.

10. PREPARATION OF STANDARDS

10.1 Aldehyde diluted mixed standard solution

10.1.1 Pipette 1 mL formaldehyde-DNPH [**7.2**], 1 mL acetaldehyde-DNPH [**7.3**] and 0.1 mL acrolein-DNPH [**7.4**] reference standards into a 10 mL volumetric flask.

10.1.1 Dilute to the mark with ethanol [**7.9**] and mix well.

10.1.1 Label and store at 4 ± 3 °C in amber vials.

10.2 Aldehyde final mixed standard solutions

10.2.1 Prepare final standard solutions according to Table 1.

10.2.2 Spike variable volumes of the diluted mixed standard solution [**10.1**] and 0.2 mL of DNPH solution [**9.1**] solution into 5 mL volumetric flasks (for rotary 0.4 mL DNPH solution into 10 mL).

10.2.3 Fill to the mark with ethanol [**7.9**] and mix well.

10.2.4 Label and store at 4 ± 3 °C.

10.2.5 Note: 10 mL is the final volume of [**10.1**] mixed standards prepared. Final volume of 5 mL (for linear) or 10 mL (for rotary) is the final volume of the working standards as indicated in Table 1, respectively. The final aldehyde concentrations in the standard solutions are shown in the tables below.

Table 1a: Linear smoking machine: concentrations of aldehydes in working standard solutions (mg/L)

Standard	Volume of mix standard solution (mL), V_{mix}	Final volume (mL), V_{tot}	Approximate aldehyde concentration in final mixed standard solution in mg/L for linear smoking machine, (mg/L)					
			Formaldehyde		Acetaldehyde		Acrolein	
			mg/L	µg/cig	mg/L	µg/cig	mg/L	µg/cig
1	0.05	5	0.1	10	1.0	100	0.1	10
2	0.25	5	0.5	50	5.0	500	0.5	50
3	0.50	5	1.0	100	10	1000	1.0	100
4	1.00	5	2.0	200	20	2000	2.0	200
5	2.00	5	4.0	400	40	4000	4.0	400

Table 1b: Rotary smoking machine: concentrations of aldehydes in working standard solutions (mg/L)

Standard	Volume of mix standard solution (mL), V_{mix}	Final volume (mL), V_{tot}	Approximate aldehyde concentration in final mixed standard solution in mg/L for linear smoking machine, (mg/L)					
			Formaldehyde		Acetaldehyde		Acrolein	
			mg/L	µg/cig	mg/L	µg/cig	mg/L	µg/cig
1	0.04	10	0.04	8	0.4	80	0.04	8
2	0.20	10	0.2	40	2	400	0.2	40
3	0.40	10	0.4	80	4	800	0.4	80
4	1.00	10	1.0	200	10	2000	1.0	200
5	2.00	10	2.0	400	20	4000	2.0	400

The range of standard solutions may be adjusted depending on the equipment used and samples to be tested, keeping in mind the possible influence on the sensitivity of the applied method.

All solvents and solutions must be adjusted to room temperature before use.

11. SAMPLING

11.1 Sample cigarettes according to ISO 8243 [**2.4**]. Alternative approaches may be used to obtain a representative laboratory sample in accordance with individual laboratory practice, or when required by specific regulation or the availability of samples.

11.2 Constitution of test sample

11.2.1 Divide the laboratory sample into separate units (e.g. packet, container), if applicable.

11.2.2 Take an equal amount of product for each test sample from at least $\sqrt{n}$ [**2.9**] of the individual units (e.g. packet, container). Note: Arbitrary sampling method is applied. $\sqrt{n}$ is one of the methods often used in practice and works well in many situations [**2.9**].

12. CIGARETTE PREPARATION

12.1 Condition all cigarettes to be smoked in accordance to ISO 3402 [**2.3**]

12.2 Mark cigarettes at a butt length in accordance with ISO 4387 [**2.2**] and WHO TobLabNet SOP 01 [**2.6**].

12.3 Prepare test samples to be smoked under intense smoking conditions as specified in WHO TobLabNet SOP 01 [**2.6**].

13. PREPARATION OF THE SMOKING MACHINE

13.1 Ambient conditions

The ambient conditions for smoking are specified in ISO 3308 [**2.1**].

13.2 Smoking machine specifications

Follow ISO 3308 [**2.1**] machine specifications, except for intense smoking as described in WHO TobLabNet SOP 01 [**2.6**].

13.3 Prepare the smoke trap as described above [**6.17**]. Prepare the smoke trap so that they are consistent as shown in Figs. 1–5 above.

14. SAMPLE GENERATION

14.1 Smoke the cigarette test samples, as specified in ISO 4387 [**2.2**] or in WHO TobLabNet SOP 01 [**2.6**], and collect the aldehydes of the mainstream smoke onto the smoke trap [**6.17**] containing the CX-572 cartridge [**6.15**] and glass fibre filter pad [**6.16**].

14.2 Include at least one reference test sample to be used for quality control, if applicable.

Table 2. Number of cigarettes to be smoked for one measurement in linear and rotary smoking machines. Refer to Appendix 2 for the appropriate smoking plan.

	ISO smoking regimen		Intense smoking regimen	
	Linear	Rotary	Linear	Rotary
No. of cigarettes per smoke trap	1	1	1	1
No. of smoke traps per result	3	3	3	3

14.3 Record the number of cigarettes and total puffs for each smoke trap.

14.4 After smoking the required number of test samples, perform one clearing puff and remove the smoke trap from the smoking machine.

14.5 One result is an average of three measurements.

14.6 Blank measurement is made on 10 puffs at least once per day before smoking cigarette samples.

Fig. 7: Overview of analytical procedures for simultaneous determination of aldehydes (SOP 08) and volatile organics (SOP 09) in mainstream cigarette smoke

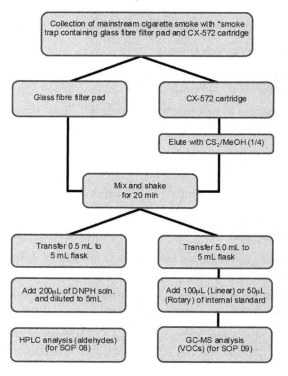

*It is an option to weigh the smoke trap before and after the smoking. The weight difference provides the TPM information as reference.

15. SAMPLE PREPARATION

Extraction (there are two options: individually add CS2 followed by MeOH (option A) or using a mixture of extraction solution (option B).

15.1 Option A (15.1.1 to 15.1.5)

15.1.1 Add 2 mL (for linear) or 4 mL (for rotary) of carbon disulfide [7.4] into the vial or flask [6.9]. Slowly add 8 mL (for linear) or 16 mL (for rotary) of methanol. DO NOT MIX the two solutions. CX-572 is added quickly into the two-phase solution.

15.1.2 After bubbling stop, swirl gently and place the glass fibre filter pad into the vial or flask.

15.1.3 Close the vial or flask.

15.1.4 Shake the mixture gently on a rotary shaker for 30 minutes at 120 cycles per minute.

15.1.5 Let the sample stand for at least 1 minute (or until the particles settle).

15.2 Option B (15.2.1 to 15.2.4)

15.2.1 Immediately transfer the Carboxen 572 particles and the glass fibre filter pad into the vial or flask [**6.9**].

15.2.2 Slowly add 10 mL (for linear) 20 mL (for rotary) of solvent mixture [**9.2.2**] into the vial or flask. Close the vial or flask immediately.

15.2.3 Shake the mixture gently on a rotary shaker for 30 minutes at 120 cycles or rounds per minute.

15.2.4 Let the sample stand for at least 1 minute (or until the particles settle).

15.3 Derivatization:

15.3.1 Let the sample stand for at least 1 minute (or until the particles settle).

15.3.2 Transfer 0.5 mL eluate using a syringe with needle slowly into a vial or flask [**6.10**].

15.3.3 Add 0.2 mL DNPH solution [**9.1**] and shake gently.

15.3.4 Let the solution stand for 10 minutes and dilute with ethanol [**7.9**] to 5 mL and mix.

15.3.5 Transfer a sufficient amount into an HPLC autosampler vial.

16. SAMPLE ANALYSIS

This method utilizes HPLC to quantify formaldehyde, acetaldehyde and acrolein in mainstream cigarette smoke. The analytes are resolved from other potential interferences on the column used. Comparison of the peak area of the unknowns with the peak area of known standard concentrations yields individual analyte concentrations.

16.1 HPLC operating conditions

This section illustrates an example of HPLC equipment and operating conditions.

HPLC Column:	Ascentis RP-Amide (150 mm × 4.6 mm, 3 µm) or equivalent
Injection Volume:	10 µL
Column oven temperature:	30 ± 5 °C (Alter as appropriate for the specific column used)
Degasser:	On

Binary pump flow: 1.0 mL/min
Mobile phase A: water
Mobile phase B: acetonitrile
Gradient: Shown in Table 3 below
Total analysis time: 45 mins
UV detection: 360 nm or maximum wavelength at 300-400 nm

Table 3. HPLC gradient for aldehydes separation

A: Water
B: acetonitrile

Time	A (%)	B (%)
0	55	45
10	55	45
25	0	100
30	0	100
31	55	45
40	55	45

Note: Adjustment of the operating parameters may be required, depending on the instrument and column conditions as well as the resolution of the chromatographic peaks.

16.2 General analytical information

16.2.1 For the conditions described here, the expected sequence of elution will be formaldehyde-DNPH, acetaldehyde-DNPH and acrolein-DNPH.

16.2.2 Differences in e.g.: flow rate, mobile phase concentration, age of the column, can be expected to alter retention times.

16.2.3 The sequence of determination of aldehydes will be in accordance with individual laboratory practice. This section gives an example.

16.2.4 Inject a blank solution (extraction solution) to check for any contamination in the system or reagents.

16.2.5 Inject a standard blank to verify the performance of the HPLC-DAD system.

16.2.6 Inject the calibration standards, the quality control and the samples.

16.2.7 Record the peak areas of formaldehyde-DNPH, acetaldehyde-DNPH and acrolein-DNPH.

Note: It is recommended that an aldehyde-DNPH mix solution containing other aldehydes that may occur in sample solutions be included in order to check if sufficient separation occurs. This is designed to prevent interference and falsely high results.

16.2.8 Plot the graph of the concentration of added formaldehyde, acetaldehyde and acrolein (X axis) in accordance with the peak area (Y axis).

16.2.9 The intercept should not be statistically significantly different from zero.

16.2.10 Linearity of the standard curve should extend over the entire standard range.

16.2.11 Calculate the linear regression (Y = a + bx) from these data, and use both the slope (b) and the intercept (a) of the linear regression. If the linear regression (R)2 is less than 0.99, the calibration should be repeated. If an individual calibration point differs by more than 10% from the expected value (estimated by linear regression), this point should be omitted.

16.2.12 The signal (peak area) obtained for all test portions must fall within the working range of the calibration curve; otherwise solutions should be adjusted as necessary.

Note: See Appendix 1 for representative chromatograms.

17. DATA ANALYSIS AND CALCULATIONS

17.1 Determine for each test portion the peak area response of the formaldehyde, acetaldehyde and acrolein.

17.2 Calculate the formaldehyde, acetaldehyde and acrolein concentration in mg/L for each test portion aliquot using the coefficients of the linear regression (mt = (Yt − a)/b).

Calculate the formaldehyde, acetaldehyde and acrolein content, mn, of the cigarette sample expressed in micrograms per cigarette, using the following equation:

$$M_n = (M_t - M_{bl}) * \frac{V_t * V_s}{V_e * N_c}$$

where:

- M_n is the calculated aldehyde content, in µg/cigarette
- M_t is the concentration of aldehyde in the test solution in mg/L
- M_{bl} is the concentration of the sample blank solution in mg/L
- V_t is the volume of the test solution in mL (5 mL for linear and 5 mL for rotary)
- V_s is the volume of the solvent (CS2 + MeOH) in mL (10 mL for linear or 20 mL for rotary)

V_e is the volume of the eluate used for derivatization (0.5 mL)

N_c is the number of cigarettes smoked onto one cartridge

Note: Alternative calculation procedures may be used if applicable.

18. SPECIAL PRECAUTIONS

18.1 After installing a new column, condition the column by injecting a sample, using the described instrument conditions. Injections should be repeated until the peak areas (or heights) of formaldehyde-DNPH, acetaldehyde-DNPH and acrolein-DNPH are reproducible.

19. DATA REPORTING

19.1 Report individual measurements for each of the samples evaluated.

19.2 Report results as specified by method specifications.

19.3 For more information, see WHO TobLabNet SOP 02 [**2.7**].

20. QUALITY CONTROL

20.1 **Control parameters**

Note: If the control measurements are outside the tolerance limits of the expected values, appropriate investigation and action must be taken.

Note: Additional individual laboratory quality assurance procedures should be carried out in compliance with the policies of the individual laboratory.

20.2 **Laboratory reagent blank**

To detect potential contamination during the sample preparation and analysis processes, include a laboratory reagent blank (LRB) as described in **16.2.4**. The blank consists of all reagents and materials used in performing the analysis on test samples, including the extraction of CX-572 and glass fibre filter pad, and is analysed as a test sample. The blank should be less than the limit of detection.

Note: In certain laboratories, a blank below the limit of detection is not feasible due to environmental conditions. In this case, a laboratory reagent blank shall be included into each test run and the sample results shall be corrected for the laboratory reagent blank.

20.3 **Quality control sample**

To verify consistency of the entire analytical process, analyse a reference cigarette in accordance with the practices of the individual laboratory.

20.4 Additional QC

Additional QC samples may be added as consistent with laboratory policy.

21. METHOD PERFORMANCE SPECIFICATIONS

21.1 Limit of Report (LOR)

The LOR is set to the lowest concentration of the calibration standards used, recalculated to µg/cigarette (i.e. 8–10 µg/cigarette for formaldehyde and acrolein and 80–100 µg/cigarette for acetaldehyde with 0.04–0.10 mg/l and 0.410–1.0 mg/L as lowest calibration solution respectively).

21.2 Lab fortified matrix recovery

Recovery of analyte spiked onto the matrix is used as a surrogate measure of accuracy. Recovery is determined by spiking known amounts of standards (after smoking) and extracting by the same method as used for samples. Unspiked samples are also analysed. The recovery is calculated by the following formula:

Recovery = 100*((analytical result-unspiked result)/spiked amount).

Table 4. Mean and recovery of analyte spiked onto the matrix

Formaldehyde			Acetaldehyde			Acrolein		
spiked amount (µg)	Mean (mg)	Recovery (%)	spiked amount (µg)	Mean (mg)	Recovery (%)	spiked amount (µg)	Mean (mg)	Recovery (%)
12.7	14.4	113.4	385.0	416.5	108.2	21.9	22.3	101.8
25.4	27.6	108.7	770.0	722.2	93.8	43.9	41.4	94.3
50.7	52.1	102.8	1540.0	1407.8	91.4	87.7	86.0	98.1

21.3 Analytical specificity

The retention time of the analyte of interest is used to verify the analytical specificity. An established range of the response of the component to that of the quality control cigarette smoke is used to verify the specificity of the results from an unknown sample.

21.4 Linearity

The aldehydes calibration curves established are linear over the standard concentration range from 0.04–4 mg/L for formaldehyde and acrolein and from 0.4–40 mg/L for acetaldehyde.

21.5 Possible interference

The presence of other aldehydes can cause interference, as the retention time might be similar to that of the aldehydes to be determined.

22. REPEATABILITY AND REPRODUCIBILITY

An international collaborative study conducted in 2015–2017 involving 10 laboratories and five samples (three reference cigarettes and two commercial brands), performed according to WHO TobLabNet SOP 02 [**2.7**] gave the following precision limits for this method.

The difference between two single results found on matched cigarette samples by the same operator using the same apparatus within the shortest feasible time interval will exceed the repeatability, r, on average not more than once in 20 cases in the normal and correct operation of the method.

Single results on matched cigarette samples reported by two laboratories will differ by more than the reproducibility, R, on average not more than once in 20 cases in the normal and correct operation of the method.

The test results were subjected to statistical analysis in accordance with ISO 5725-1 [**2.11**] and ISO 5725-2 [**2.5**] to give the precision data shown in tables 5 to 7.

For the purpose of calculating r and R, one test result under ISO smoking conditions was defined as the average of seven individual replicates of the mean yield of three smoke traps (one cigarette smoked per smoke trap) from linear smoking machines and three smoke traps for rotary smoking machines. Under Intense smoking conditions, one test result was defined as the average of seven individual replicates of the mean yield of three smoke traps from linear smoking machines and three smoke traps for rotary smoking machines. For more information, see reference [**2.6**].

Table 5a and 5b. Precision limits for determination of formaldehyde (µg/cigarette) in mainstream cigarette smoke of the reference test pieces.

Reference cigarette	Formaldehyde*: ISO smoking regimen, µg/cigarette			
	N	m_{cig}	r_{limit}	R_{limit}
1R5F	8	5.92	1.43	2.08
3R4F	8	22.87	3.05	13.34
CM6	8	48.94	4.54	37.53
Commercial Brand I	9	31.58	6.27	31.82
Commercial Brand II	9	31.66	5.98	24.74

Reference cigarette	Formaldehyde*: Intense smoking regimen, µg/cigarette			
	N	m_{cig}	r_{limit}	R_{limit}
1R5F	9	36.92	4.49	39.51
3R4F	6	72.99	5.38	7.62
CM6	6	84.83	6.76	12.17
Commercial Brand I	8	55.11	5.64	23.92
Commercial Brand II	9	58.05	7.67	30.90

Table 6a and 6b. Precision limits for determination of acetaldehyde (µg/cigarette) in mainstream cigarette smoke of the reference test pieces.

Reference cigarette	Acetaldehyde: ISO smoking regimen, µg/cigarette			
	N	m_{cig}	r_{limit}	R_{limit}
1R5F	9	156.21	26.00	36.80
3R4F	10	501.61	55.82	151.52
CM6	6	742.83	44.05	48.22
Commercial Brand I	9	641.25	80.07	295.24
Commercial Brand II	8	592.43	55.81	124.17

Reference cigarette	Acetaldehyde: Intense smoking regimen, µg/cigarette			
	N	m_{cig}	r_{limit}	R_{limit}
1R5F	8	866.60	90.51	168.60
3R4F	9	1054.47	114.99	290.90
CM6	6	961.75	62.96	78.09
Commercial Brand I	9	990.93	100.62	288.42
Commercial Brand II	9	930.38	124.48	334.17

Table 7a and 7b. Precision limits for determination of acrolein (acrylaldehyde) (µg/cigarette) in mainstream cigarette smoke of the reference test pieces.

Reference cigarette	Acrolein*: ISO smoking regimen, µg/cigarette			
	N	m_{cig}	r_{limit}	R_{limit}
1R5F	7	9.60	1.81	1.84
3R4F	7	48.12	4.77	19.29
CM6	9	67.77	9.70	37.30
Commercial Brand I	7	45.22	7.31	16.19
Commercial Brand II	9	57.31	7.45	54.40

Reference cigarette	Acrolein*: Intense smoking regimen, µg/cigarette			
	N	m_{cig}	r_{limit}	R_{limit}
1R5F	8	107.40	17.37	120.55
3R4F	9	137.25	24.49	138.17
CM6	9	121.01	26.93	108.33
Commercial Brand I	8	111.05	14.82	117.04
Commercial Brand II	9	120.16	27.97	163.28

Note:

• The method for this study is less optimal for collection of formaldehyde due to its highly volatile and hydrophilic nature [2.11].

• Due to i) the relative comparative repeatability (%r) is less than 30% ii) the low number of participating laboratories (≤10) and iii) the inter laboratory precision values are accompanied with relatively high variation, this method can be considered conditionally validated.

• On the basis of all the data obtained, this method shows that it fits for purpose in testing and measuring aldehydes in mainstream cigarette smoke and for a range of concentration of interest.

Appendix 1: Typical chromatograms obtained in the determination of aldehydes in mainstream cigarette smoke

Fig. A1: HPLC chromatogram of a typical combined carbonyl calibration standard using column Ascentis RP-Amide (150 mm × 4.6 mm, 3 μm).

Formaldehyde-DNPhydrazone (FA), acetaldehyde-DNPhydrazone (AA), acrolein-DNPhydrazone (ACR)

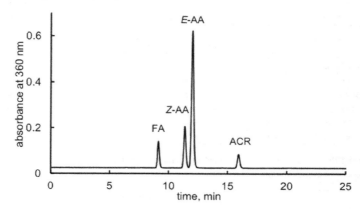

Fig. A2: HPLC chromatogram of carbonyls in DNPH extract of mainstream tobacco smoke using column Ascentis RP-Amide (150mm × 4.6mm, 3 μm)

Formaldehyde-DNPhydrazone (FA), acetaldehyde-DNPhydrazone (AA), acrolein-DNPhydrazone (ACR)

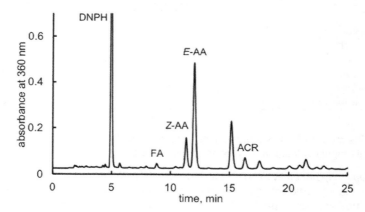

Appendix 2: Smoking Plans

The described smoking plans are based on five different samples (cigarettes brand/types) and seven replicates per sample. If a different number of samples or replicates is needed the plans should be adjusted accordingly.

Smoking plan for 20-port linear smoking machine (one cigarette per cartridge)

Day	1	2	3	4	5	6	7	8	9	10	11	12	13	14	15	16	17	18	19	20
1	A				B				C				D				E			
		A				B				C				D				E		
			A				B				C				D				E	
2		D				E				A				B				C		
			D				E				A				B				C	
				D				E				A				B				C
3	B				C					D				E					A	
		B				C					D				E					A
			B					D					E				A			
4	E				A				B				C				D			
		E				A				B				C				D		
			E				A				B				C				D
5	C			D			E				A					B				
		C			D			E					A					B		
			C			D								A					B	E
6				D				C			B			A			E			
	D			C						B			A				E			
		D			C						A				A			E		
7			C			B				A				E				D		
		C			B				A				E					D		
			C			B				A				E					D

· 361 ·

Smoking plan for rotary smoking machine (one cigarette per cartridge)

Day	Run	Sample	Day	Run	Sample	Day	Run	Sample	Day	Run	Sample
1	1	A	3	31	B	5	61	C	7	91	C
	2	A		32	B		62	C		92	C
	3	A		33	B		63	C		93	C
	4	B		34	C		64	D		94	B
	5	B		35	C		65	D		95	B
	6	B		36	C		66	D		96	B
	7	C		37	D		67	E		97	A
	8	C		38	D		68	E		98	A
	9	C		39	D		69	E		99	A
	10	D		40	E		70	A		100	E
	11	D		41	E		71	A		101	E
	12	D		42	E		72	A		102	E
	13	E		43	A		73	B		103	D
	14	E		44	A		74	B		104	D
	15	E		45	A		75	B		105	D
2	16	D	4	46	E	6	76	D			
	17	D		47	E		77	D			
	18	D		48	E		78	D			
	19	E		49	A		79	C			
	20	E		50	A		80	C			
	21	E		51	A		81	C			
	22	A		52	B		82	B			
	23	A		53	B		83	B			
	24	A		54	B		84	B			
	25	B		55	C		85	A			
	26	B		56	C		86	A			
	27	B		57	C		87	A			
	28	C		58	D		88	E			
	29	C		59	D		89	E			
	30	C		60	D		90	E			

WHO TobLabNet
Official Method
SOP 09

Standard Operating Procedure for Determination of Volatile Organics in Mainstream Cigarette Smoke under ISO and Intense Smoking Conditions

Method:	Determination of volatile organics in mainstream cigarette smoke under ISO and intense smoking conditions
Analytes:	Benzene (CAS# 71-43-2) 1,3-butadiene (CAS# 106-99-0)
Matrix:	Tobacco cigarette mainstream smoke particulate matter
Last update:	31 August 2018

No machine smoking regimen can represent all human smoking behaviour: machine smoking testing is useful for characterizing cigarette emissions for design and regulatory purposes, but communication of machine measurements to smokers can result in misunderstanding about differences between brands in exposure and risk. Data on smoke emissions from machine measurements may be used as inputs for product hazard assessment, but they are not intended to be nor are they valid as measures of human exposure or risks. Representing differences in machine measurements as differences in exposure or risk is a misuse of testing with WHO TobLabNet standards.

FOREWORD

This document was prepared by members of the World Health Organization (WHO) Tobacco Laboratory Network (TobLabNet) as an analytical method standard operating procedure (SOP) for measuring volatile organics in mainstream cigarette smoke under ISO and intense smoking conditions.

INTRODUCTION

In order to establish comparable measurements for testing tobacco products globally, consensus methods are required for measuring specific contents and emissions of cigarettes. The Conference of the Parties (COP) to the WHO Framework Convention on Tobacco Control (WHO FCTC) at its third session in Durban, South Africa, in November 2008, recalling its decisions FCTC/COP1(15) and FCTC/COP2(14) on the elaboration of guidelines for implementation of Articles 9 (Regulation of the contents of tobacco products) and 10 (Regulation of tobacco product disclosures) of the WHO FCTC, noting the information contained in the report of the working group to the third session of the Conference of the Parties on the progress of its work ... requested the Convention Secretariat to invite WHO's Tobacco Free Initiative to "validate, within five years, the analytical chemical methods for testing and measuring cigarette contents and emissions".[1]

Using the criteria for prioritization set at its third meeting in Ottawa, Canada, in October 2006, the working group on Articles 9 and 10 identified the following contents for which methods for testing and measurement (analytical chemistry) should be validated as a priority:

- nicotine
- ammonia
- humectants (propane-1,2-diol, glycerol (propane-1,2,3-triol) and triethylene glycol (2,2-ethylenedioxybis(ethanol)).

Measurement of these contents will require validation of three methods: one for nicotine, one for ammonia and one for humectants.

Using the criteria for prioritization set at the meeting in Ottawa mentioned above, the working group identified the following emissions in mainstream smoke for which methods for testing and measurement (analytical chemistry) should be validated as a priority:

- 4-(methylnitrosamino)-1-(3-pyridyl)-1-butanone (NNK)
- *N*-nitrosonornicotine (NNN)

---

[1] Decision, third session of the Conference of the Parties to the WHO FCTC; 2008 (http://apps.who.int/gb/fctc/PDF/cop3/FCTC_COP3_REC1-en.pdf).

- acetaldehyde
- acrylaldehyde (acrolein)
- benzene
- benzo[a]pyrene
- 1,3-butadiene
- carbon monoxide
- formaldehyde

Measurement of these emissions with the two smoking regimens described below will require validation of four methods: one for tobacco-specific nitrosamines (NNK and NNN), one for benzo[a]pyrene, one for aldehydes (acetaldehyde, acrolein and formaldehyde), one for volatile organic compounds (benzene, 1,3-butadiene) and one for carbon monoxide.

The table below sets out the two smoking regimens for validation of the test methods referred to above.

Smoking regimen	Puff volume (mL)	Puff frequency	Filter ventilation holes
ISO regimen: ISO 3308; Routine analytical cigarette smoking machine—definitions and standard conditions	35	Once every 60 s	No modifications
Intense regimen: Same as ISO 3308, but modified as indicated	55	Once every 30 s	All ventilation holes must be blocked 100% as described in WHO TobLabNet SOP 01.

This method SOP was developed to direct participating labs in the specifics of carrying out the analysis of volatile organic components in mainstream cigarette smoke. For tobacco products other than cigarettes, specific parameters may have to be adjusted from those given in the document

1. SCOPE

This method is suitable for quantitative determination of the following two volatile organics in mainstream cigarette smoke: benzene and 1,3-butadiene by combined gas chromatography mass spectrometry.

Benzene and 1,3-butadiene are potent carcinogens. Benzene and 1,3-butadiene are not originally present in tobacco leaves but are formed from the combustion of organic material during cigarette burning.

Note: Training in the use of the smoking machine and other analytical equipment is important for successful operation. For those not experienced in operating smoking machines or analyses using the analytical methods for measuring tobacco product emissions and contents, training should be obtained.

2. REFERENCES

2.1 ISO 3308: Routine analytical cigarette-smoking machine – Definitions and standard conditions. Geneva: International Organization for Standardization; 2012.

2.2 ISO 4387: Cigarettes—Determination of total and nicotine-free dry particulate matter using a routine analytical smoking machine. Geneva: International Organization for Standardization; 2000.

2.3 ISO 3402: Tobacco and tobacco products—Atmosphere for conditioning and testing. Geneva: International Organization for Standardization; 1999.

2.4 ISO 8243: Cigarettes—Sampling. Geneva: International Organization for Standardization; 2006.

2.5 ISO 5725-2: Accuracy (trueness and precision) or measurement methods and results—Part 2: Basic method for the determination of repeatability and reproducibility of a standard measurement method.

2.6 Tobacco Laboratory Network, Standard Operating Procedure for Intense Smoking (WHO TobLabNet SOP 01).). Geneva: World Health Organization; 2012 (http://apps.who.int/iris/bitstream/handle/10665/75261/9789241503891_eng.pdf;sequence=1, accessed 3 September 2018).

2.7 Tobacco Laboratory Network. Standard operating procedures for validation of analytical methods of tobacco product contents and emissions (WHO TobLabNet SOP 02). Geneva: World Health Organization., 2012 (http://apps.who.int/iris/bitstream/handle/10665/254998/9789241512060-eng.pdf?sequence=1, accessed 3 September, 2018).

2.8 Tobacco Laboratory Network. Standard Operating Procedure for the Determination of aldehydes in mainstream cigarette smoke under ISO and intense smoking conditions (WHO TobLabNet SOP 08). Geneva: World Health Organization; (jn press).

2.9 United Nations Office on Drugs and Crime. Guidelines on representative drug sampling. Vienna: Laboratory and Scientific Section; 2009 (http://www.unodc.org/documents/scientific/Drug_Sampling.pdf).

2.10 ISO 5725-1. Accuracy (trueness and precision) of measurement methods and results—Part 1: General principles and definitions. Geneva: International Organization for Standardization; 1994.

2.11 Uchiyama, S.; Tomizawa, T.; Inaba, Y.; Kunugita, N. Simultaneous determination of volatile organic compounds and carbonyls in mainstream cigarette smoke using a sorbent cartridge followed by two-step elution. J Chromatogr A, 2013. 1314:31-7. doi: 10.1016/j.chroma.2013.09.019.

2.12 Uchiyama, S.; Hayashida, H.; Izu, R.; Inaba, Y.; Nakagome, H.; Kunugita, N. Determination of nicotine, tar, volatile organic compounds and carbonyls in mainstream cigarette smoke using a glass filter and a sorbent cartridge followed by the two-phase / one-pot elution method with carbon disulfide and methanol. Journal of Chromatography A 2015. 1426:48-55. doi.org/10.1016/j.chroma.2015.11.058.

3. TERMS AND DEFINITIONS

CX-572: Carbon molecular sieves Carboxen 572

Intense Regime: Parameters for smoking tobacco products that include a 55-ml puff volume, a 30 s puff interval, 2 s puff duration and 100% blocking of filter ventilation hole

ISO regime: Parameters for smoking tobacco products that include a 35-ml puff volume, a 60 s puff interval, 2 s puff duration and no blocking of filter ventilation hole

Laboratory sample: Sample intended for testing in a laboratory, consisting of a single type of product delivered to the laboratory at one time or within a specified period.

Smoke trap: device for collecting such part of the smoke from a sample of cigarettes as is necessary for the determination of specified smoke components.

Test portion: Random portion from the test sample to be used for a single determination. The number of products taken shall be representative of the test sample.

Test sample: Product to be tested, taken at random from the laboratory sample. The number of products taken shall be representative of the laboratory sample.

Tobacco products: Products entirely or partly made of the leaf tobacco as raw material which are manufactured to be used for smoking, sucking, chewing or snuffing (Article 1(f) of the WHO FCTC)

TPM: total particulate matter

4. METHOD SUMMARY

4.1 Mainstream cigarette smoke is performed according to the WHO Intense regime or the ISO regime. One cigarette is smoked per smoke trap containing 300 mg CX-572 particles and a glass fibre filter pad.

4.2 Mainstream smoke total particulate matter from the cigarette test sample is trapped onto a smoke trap containing 300 mg CX-572 particles and a glass-fibre filter pad (commonly referred to as a glass fibre filter pad).

4.3 The volatile organic components are extracted by adding a solution containing a mixture of carbon disulphide and methanol onto the Carboxen particles and glass fibre filter pad.

4.4 The analytical determination of VOCs (1,3-butadiene and benzene) is performed by gas chromatography–mass spectrometry (GC-MS) with electron ionization mode.

4.5 The reconstructed ion chromatogram peak area ratio of native analyte to labelled internal standard is compared on a calibration curve created by analysis of standards with known concentrations of the benzene and 1,3-butadiene to determine

4.6 The benzene and 1,3-butadiene content of each test portion.

4.7 The method described in this SOP simultaneously tests and measures the level of aldehydes (SOP 08) and volatile organics (SOP 09) produced in mainstream cigarette smoke.

5. SAFETY AND ENVIRONMENTAL PRECAUTIONS

5.1 Follow routine safety and environmental precautions, as in any chemical laboratory activity.

5.2 The testing and evaluation of certain products with this test method may require the use of materials or equipment that could be hazardous or harmful to the environment; this document does not purport to address all the safety aspects associated with its use. All persons using this method have the responsibility to consult the appropriate authorities and to establish health and safety practices as well as environmental precautions in conjunction with any existing applicable regulatory requirements prior to its use.

5.3 Special care should be taken to avoid inhalation or dermal exposure to harmful chemicals. Use a chemical fume hood, and wear an appropriate laboratory coat, gloves and safety goggles when preparing or handling undiluted materials, standard solutions, extraction solutions or collected samples.

6. APPARATUS AND EQUIPMENT

6.1 Equipment needed to condition cigarettes as specified in ISO 3402 [**2.3**].

6.2 Equipment needed to mark butt length as specified in ISO 4387 [**2.2**].

6.3 Equipment needed to cover filter ventilation holes for the intense regimen as specified in WHO TobLabNet SOP 01 [**2.6**].

6.4 Equipment needed to perform smoking of cigarette products as specified in ISO 3308 [**2.1**].

6.5 Analytical balance capable of measurements to at least four decimal places.

6.6 Pipettes and tips capable of accurately dispensing volumes of 5–5000 µl.

6.7 Teflon-lined septa or equivalent.

6.8 Volumetric flasks 5 mL, 10 mL, 50 mL or equivalent.

6.9 Vials with septum seal, capacity of at least 20 mL (linear) or 50 mL (rotary) or equivalent.

6.10 Syringe, 1 mL and 5 mL or equivalent for transfer of organic solvent.

6.11 GC autosampler vials, 2 mL or equivalent.

6.12 GC coupled with mass spectrometer detector.

6.13 GC column capable of distinct separation of benzene, 1,3-butadiene and isotopically labelled benzene peaks from other cigarette emission components, e.g. InertCap AQUATIC-2 60 m x 0.25 mm i.d., d=1.4 µm by GL Sciences or equivalent.

6.14 Carboxen 572 particles, 20/45 mesh size.

6.15 CX-572 cartridge: either manually prepared SPE cartridge filled with 300 mg of Carboxen 572 particles or preloaded CX-572 cartridge acquired by Sigma Aldrich (Cat. No. 54294-U).

6.16 Glass fibre filter pad as described in ISO 4387 [**2.2**]

6.17 Smoke trap set up consists of smoking machine holders attached with CX-572 (with or without cartridge) to be placed in front of glass fibre filter pad. (See Figs. 1 to 5).

Note: Place the CX-572 cartridge in the holder facing the open side towards the cigarette. When positioned with the open end in the opposite direction, the content of the cartridge can be sucked into the smoking machine when puffs are taken. If necessary, special stopper can be inserted into the cartridge as shown in Fig. 6.

Fig. 1a: Single port Borgwaldt Smoking Machine

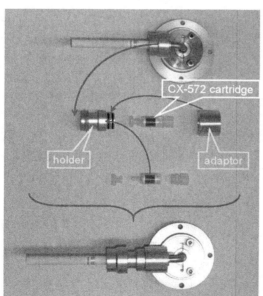

Fig. 1b: Linear Borgwaldt smoking machine

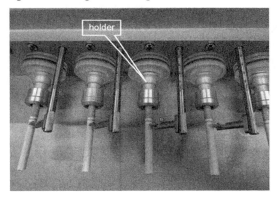

Fig. 2a and Fig. 2b illustrate the cigarette holder suitable for use in commercial rotary smoking machine (i.e. Cerulean and/or Borgwaldt smoking machine)

Fig. 2a: with the CX-572 particles weigh directly onto the holder

Fig. 2b: with the CX-572 loaded cartridge inserted onto the holder

Fig. 3: For Linear Cerulean smoking machine (also applicable to Linear Borgwaldt smoking machine)

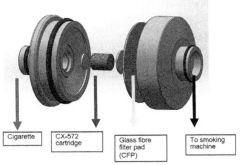

Note: CX-572 can be pre-packed into this holder or alternatively, a commercially available CX-572 cartridge can be used.

Fig. 4: For Linear Cerulean smoking machine (also applicable to Linear Borgwaldt or Rotary Cerulean or single channel smoking machine)

Fig. 5: For Linear Cerulean smoking machine (also applicable to Linear Borgwaldt smoking machine)

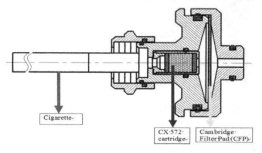

Fig. 6: Stoppers to be inserted into the CX-572 cartridges to prevent the contend to be sucked into the smoking machine

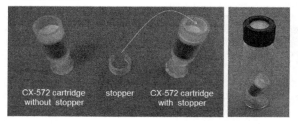

Note: Be sure to put the cartridge into a closed vessel (vial) directly after inserting the stopper to prevent contamination of the CX-572.

7. REAGENTS AND SUPPLIES

All reagents shall be of at least analytical reagent grade unless otherwise noted. When possible reagents are identified by their Chemical Abstract Service (CAS) registry numbers.

7.1 Benzene, analytical-grade. CAS number: 71-43-2.

7.2 Deuterium-labelled benzene (benzene-d6) purity>99%. CAS number: 1076-43-3.

7.3 1,3-butadiene standard solution (2 mg/ml) [CAS number:106-99-0]. (AccuStandard, S-406A-10X).

7.4 Carbon disulfide, analytical-grade. CAS number: 75-15-0.

7.5 Methanol, anhydrous, analytical-grade. CAS number: 67-56-1.

7.6 Carboxen 572 particles, 20/45 mesh size (Sigma-Aldrich/Supelco Cat. No. 11072-U).

8. PREPARATION OF GLASSWARE
Clean and dry glassware in a manner to avoid contamination from residues.

9. PREPARATION OF SOLUTIONS

9.1 Isotopically labelled internal standard solution (benzene-d6 5 mg/mL)

9.1.1 Weigh accurately 500 mg of benzene-d6 into a 100 mL volumetric flask added with *ca.* 80 mL methanol. Dilute to 100 mL with methanol.

9.1.2 Label and store at -20 oC ± 5 oC.

9.2 **Extraction solution: The extraction solutions can be prepared with the following two options:**

9.1.1 Option A: Carbon disulfide and methanol.

9.1.2 Option B: Carbon disulfide and methanol (1:4) mixed solution containing internal standard solution.

Add 20 mL of carbon disulfide and 500 µL or 250 µL of internal standard solution (benzene-d6 5 mg/mL) to the 100 mL volumetric flask and dilute with methanol.

The internal standard concentration of benzene-d6 is 25 µg/mL for linear or 12.5 µg/mL for rotary.

9.1.3 Carbon disulfide and methanol (1:4) mixed solution.

Add 20 mL of carbon disulfide to the 100 mL volumetric flask and dilute with methanol.

Note: The amount of extraction solution prepared can be adjusted based on the need of the analysis. The following are for illustrative purposes only.

10. PREPARATION OF STANDARDS
Note: can be adjusted accordingly with the condition that the standard range and internal standards concentration remains the same

10.1 **VOCs (benzene 80 µg/mL and 1,3-butadiene 200 µg/mL) stock solutions**

10.1.1 Benzene 2.0 mg/mL standard solution.

Weigh accurately 200 mg of benzene [**7.1**] into a 100 mL volumetric flask added with ca. 80 mL methanol. Dilute to 100 mL with methanol.

10.1.2 Pipette 0.8 mL of the benzene standard solution [**10.1**] and 2.0 mL of the 1,3-butadiene standard solution [**7.3**] into a 20 mL volumetric flask.

10.1.3 After addition of 4 mL of CS_2 [**7.4**], the solution is diluted to 20 mL with methanol [**7.5**].

10.1.4 Label and store at 4 °C ± 3 °C in amber vials.

10.2 VOCs working standard solutions

10.2.1 Depending on type of smoking machine used, pipette variable volumes of the stock solution [**10.1**] and solvent mixture [**9.2.3**] into gas chromatography (GC) autosampler vials according to Table 1a (linear) or 1b (rotary).

10.2.2 Add isotopically labelled internal standard [**9.1**] according to Table 1a (linear) or 1b (rotary) and mix well.

10.2.3 Label and store at 4 °C ± 3 °C.

Table 1a (option A). **Linear smoking machine: concentrations of VOCs in working standard solutions. Internal standard is added to each 5 mL standard solution.**

Standard	VOCs stock solution (mL) [10.1]	Total volume (mL)	Addition of internal standard (μl) [9.1]	concentration (mg/L) Benzene	1,3-Butadiene
1	0	5	100	0	0
2	0.50	5	100	8	20
3	1.00	5	100	16	40
4	1.50	5	100	24	60
5	2.00	5	100	32	80
6	4.00	5	100	64	160

Table 1b (option A). **Rotary smoking machine: concentrations of VOCs in working standard solutions. Internal standard is added to each 5 mL standard solution.**

Standard	VOCs stock solution (mL) [10.1]	Total volume (mL)	Addition of internal standard (μl) [9.1]	concentration (mg/L) Benzene	1,3-Butadiene
1	0	5	50	0	0
2	0.25	5	50	4	10
3	0.50	5	50	8	20
4	0.75	5	50	12	30
5	1.00	5	50	16	40
6	2.00	5	50	32	80

Table 1c (option B). Linear smoking machine: concentrations of VOCs in working standard solutions. Internal standard is involved in each 5 mL standard solution.

Standard	VOCs stock solution (mL) [10.1]	internal standard (μl) [9.1]	Total volume (mL)	Approximate VOCs concentration (mg/L)	
				Benzene	1,3-Butadiene
1	0.00	20	5	0	0
2	0.50	20	5	8	20
3	1.00	20	5	16	40
4	1.50	20	5	24	60
5	2.00	20	5	32	80
6	4.00	20	5	64	160

Table 1d (option B). Rotary smoking machine: concentrations of VOCs in working standard solutions. Internal standard is involved in each 5 mL standard solution.

Standard	VOCs stock solution (mL) [10.1]	internal standard (μl) [9.1]	Total volume (mL)	Approximate VOCs concentration (mg/L)	
				Benzene	1,3-Butadiene
1	0.00	10	5	0	0
2	0.25	10	5	4	10
3	0.50	10	5	8	20
4	0.75	10	5	12	30
5	1.00	10	5	16	40
6	2.50	10	5	40	100

The range of the working standard solutions may be adjusted depending on the equipment used and samples to be tested, keeping in mind the possible influence on the sensitivity of the applied method. All solvents and solutions must be adjusted to room temperature before use.

11. SAMPLING

11.1 Sample cigarettes according to ISO 8243 [2.4]. Alternative approaches may be used to obtain a representative laboratory sample in accordance with individual laboratory practice or when required by specific regulation or the availability of samples.

11.2 Constitution of test sample

11.2.1 Divide the laboratory sample into separate units (e.g. packet, container), if applicable.

11.2.2 Take an equal amount of product for each test sample from at least $\sqrt{n}$ [2.9] of the individual units (e.g. packet, container). Note: Arbitrary sampling method is applied. $\sqrt{n}$ is one of the method often used in practice and work well in many situations [2.9].

12. CIGARETTE PREPARATION

12.1 Condition all cigarettes to be smoked in accordance with ISO 3402 [**2.3**].

12.2 Mark cigarettes at a butt length in accordance with ISO 4387 [**2.2**] and WHO TobLabNet SOP 01 [**2.6**].

12.3 Prepare test samples to be smoked under Intense smoking conditions as specified in WHO TobLabNet SOP 01 [**2.6**].

13. PREPARATION OF THE SMOKING MACHINE

13.1 Ambient conditions

The ambient conditions for smoking are specified in ISO 3308 [**2.1**].

13.2 Smoking machine specifications

Follow ISO 3308 [**2.1**] machine specifications, except for intense smoking, as described in WHO TobLabNet SOP 01 [**2.6**].

Prepare the smoke trap as described above [6.17] so that they are consistent with Figs. 1 to 5.

14. SAMPLE GENERATION

14.1 Smoke the cigarette test samples, as specified in ISO 4387 [**2.2**] or in WHO TobLabNet SOP 01 [**2.6**] and collect the volatile organic compounds of the mainstream smoke onto the smoke trap [**6.17**] containing the CX-572 cartridge [**6.15**] and glass fibre filter pad [**6.16**].

14.2 Include at least one reference test sample to be used for quality control, if applicable.

14.3 The number of cigarettes to be smoked per measurement for linear and rotary smoking machines at ISO and intense smoking regimens are shown in Table 2.

Table 2. Number of cigarettes to be smoked for one measurement in linear and rotary smoking machines. Refer to Appendix 4 for the appropriate smoking plan.

	ISO smoking regimen		Intense smoking regimen	
	Linear	Rotary	Linear	Rotary
No. of cigarettes per smoke trap	1	1	1	1
No. of smoke traps per result	3	3	3	3

Note: Smoke 1 cigarette on the specified smoking machine such that breakthrough does not occur and the concentrations of benzene and 1,3- butadiene fall within the calibration range prepared for the analysis.

14.4 Record the number of cigarettes and total puffs for each smoke trap.

14.5 After smoking the required number of test samples, perform one clearing puff, and remove the smoke trap from the smoking machine.

14.6 One result is an average of three measurements.

14.7 Blank measurement is made on 10 puffs at least once per day before smoking cigarette samples.

Fig. 7: Overview of the analytical procedure for simultaneous determination of aldehydes (SOP 08) and volatile organics (SOP 09) in mainstream cigarette smoke

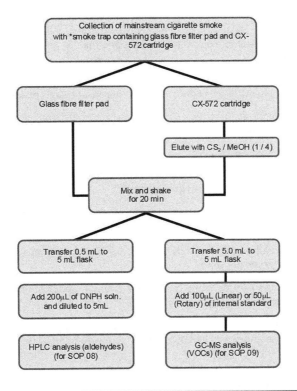

*It is an option to weigh the smoke trap before and after the smoking. The weight difference provides the TPM information as reference.

15. SAMPLE PREPARATION

Extraction (there are two options: individually add CS2 followed by MeOH (option A) or using a mixture of extraction solution (option B).

15.1 Option A (15.1.1 to 15.1.7)

15.1.1 Add 2 mL (for linear) or 4 mL (for rotary) of carbon disulfide [**7.4**] into the vial or flask [**6.9**]. Slowly add 8 mL (for linear) or 16 mL (for rotary) of methanol. DO NOT MIX the two solutions. CX-572 is added quickly into the two-phase solution.

15.1.2 After bubbling stop, swirl gently and place the glass fibre filter pad into the vial or flask.

15.1.3 Close the vial or flask.

15.1.4 Shake the mixture gently on a rotary shaker for 30 mins at 120 cycles per min.

15.1.5 Let the sample stand for at least one minute (or until the particles settle).

15.1.6 Transfer 5 mL of the supernatant solution into a vial and add 100 μl (for linear) or 50 μl (for rotary) of internal standard solution [**9.1**].

15.1.7 After mixed, transfer a 1 mL portion of the eluate solution to a 1.5 mL autosampler vial [**6.10**] and analyse the solution by GC/MS.

15.2 Option B [15.2.1 to 15.2.5]

15.2.1 Immediately transfer the Carboxen 572 particles and the glass fibre filter pad into the vial or flask [**6.9**].

15.2.2 Add slowly 10 mL (for linear) 20 mL (for rotary) of solvent mixture [**9.2.2**] into the vial or flask. Close the vial or flask immediately.

15.2.3 Shake the mixture gently on a rotary shaker for 30 mins at 120 cycles or rounds per min.

15.2.4 Let the sample stand for at least one minute (or until the particles settle).

15.2.5 After mixed, transfer a 1 mL portion of the eluate solution to a 1.5 mL autosampler vial [**6.10**] and analyse the solution by GC/MS.

16. SAMPLE ANALYSIS

This method utilizes gas chromatograph coupled with GC-MS to quantify benzene and 1,3-butadiene in mainstream cigarette smoke. The analytes are resolved from other potential interference on the column used. Comparison of the area ratio (native analyte area/isotope-labelled analyte area) of the unknowns with the concentration of known standard concentrations yields individual analyte concentrations.

16.1 Gas chromatography and mass spectrometer operating conditions

16.1.1 Gas chromatography conditions

Table 3. This section illustrates an example of GC equipment and its operating conditions.

GC Column	InertCap AQUATIC-2 60 m x 0.25 mm i.d., d=1.4 μm by GL Sciences
Injector temperature:	240 °C (iontrap: 200 °C)
Mode:	Constant flow
Flow rate:	1.0 mL/min (iontrap 1.0 mL/min)
Injection:	1 μL or 2 μL with split ratio 1:10 (Ion trap 1:5 after 1.5 mins 1:120)
Column temperature:	40 °C for 6 mins
	6 °C/min to 250 °C
Total run time	45 mins
Interface/Transfer line temperature:	180 °C (iontrap 200 °C)

- For the conditions described here, the expected sequence of elution will be 1,3-butadiene, carbon disulfide, benzene-d6 and benzene.

- It is expected that differences, such as in flow rate, mobile phase concentration, age of the column, will alter retention times.

- The elution order and retention times must be verified before analysis is begun.

- Using the above conditions, the expected total analysis time will be about 60 minutes.

Note: Adjustment of the operating parameters may be required, depending on the instrument and column conditions as well as the resolution of the chromatographic peaks.

16.1.2 Mass spectrometer conditions and peak identification

Table 4. The acquisition of the m/z signals is performed in full scan mode:

Ion source:	200 °C (ion trap 180°C)
Dwell time	50 msec
Ionization mode:	Electron Ionization (ionizing voltage at 70 eV)
Detection:	Full-scan m/z 30-300
Ion peak for identification	Benzene : m/z 78
	Benzene – d6: m/z 84
	1,3-Butadiene: m/z 54

16.2 VOCs Determination

16.2.1 Condition the system just prior to use by injecting two 1 μl or 2 μl aliquots of a sample solution as a primer.

16.2.2 Inject a check standard (blank with internal standard) under the same conditions as the samples to verify the performance of the GC-MS system.

16.2.3 Inject a blank (extraction solution) to check any contamination in the system or extraction solution or reagents.

16.2.4 Inject an aliquot of each working standard solution into the GC-MS.

16.2.5 Assess the retention times and relative response ratio (based on area counts) for the labelled standards. If the retention times are similar (±0.2 min) to the retention times and relative response ratio are within ± 20% of the typical responses from the previous runs, the system is ready to perform the analysis.

16.2.6 Inject the calibration standards, quality controls (QCs), and samples and determine peak areas and relative response ratio using the instrument software.

16.2.7 Record the peak areas of benzene, 1,3-butadiene and benzene-d6.

16.2.8 Calculate the ratio of the Benzene peak to the internal standard peak (RF = $A_{Benzene}/A_{Benzene-d6}$) for each of the Benzene and 1,3-butadiene standard solutions including the standard blanks.

16.2.9 Plot the graph of the concentration of added benzene (X axis) in accordance with the area ratios (Y axis).

16.2.10 The intercept should not be statistically significantly different from zero.

16.2.11 Linearity of the standard curve should extend over the entire standard range.

16.2.12 Calculate the linear regression (Y = a + bx) from these data, and use both the slope (b) and the intercept (a) of the linear regression.

If the linear regression R^2 is less than 0,99, the calibration should be repeated. If an individual calibration point differs by more than 10% from the expected value (estimated by linear regression), this point should be omitted.

16.2.13 Inject the QCs and samples and determine peak areas using the instrument software.

16.2.14 The relative response factors (peak area ratio) obtained for all test portions must fall within the working range of the calibration curve.

Note: See Appendix 1 for representative chromatograms.

17. DATA ANALYSIS AND CALCULATIONS

The slope and intercept are determined from the relative responses of the analyte and the labelled standard versus the relative concentrations of the analyte and the labelled standard.

17.1 Calculate the relative response ratio from the peak areas for each of the volatile organics for each of the calibration standards [**10.2.3**]:

$$RF = \frac{A_a}{A_{IS}}$$

where:

RF = relative response ratio

A_a = area of the target analyte

A_{IS} = area of the corresponding internal standard.

Plot a graph for each of the volatile organics, of the relative response factor, A_a/A_{IS}, (Y axis) versus concentration (X axis). Calculate a linear regression equation ($RF = kX + m$) from these data, and use both the slope (k) and the intercept (m) of the linear regression.

17.2 The linearity of the standard curve should cover the entire standard range.

17.3 The content M_{VOCs} of a given volatile organics of the target analyte (μg/cigarette) is determined from the calculated relative response ratio for the test portion, the slope and intercept obtained from the appropriate calibration curve, and the equation:

$$M_{VOCs} = \frac{RF - m}{k} \times \frac{V}{N}$$

where:

M_{VOCs} = calculated content, in μg/cigarette

V = volume of extraction solution (10 ml or 20 mL)

N = 1 (number of cigarette smoked onto each cartridge).

RF = Relative response ratio

m = intercept of the Y-axis of the calibration curve

k = slope of the calibration curve

Alternative calculation procedures may be used if applicable.

18. SPECIAL PRECAUTIONS

After installing a new column, condition it as specified by the manufacturer by injecting a tobacco sample extract under the specified instrument conditions. Injections should be repeated until the peak areas (or heights) of each of the volatile organics and the internal standards are reproducible.

19. DATA REPORTING

19.1 Report individual measurements for each of the samples evaluated.

19.2 Report results as specified by method specifications.

19.3 For more information, see WHO TobLabNet SOP 02 [**2.7**].

20. QUALITY CONTROL

20.1 Control parameters

Note: If the control measurements are outside the tolerance limits of the expected values, appropriate investigation and action must be taken.

Note: Additional laboratory quality assurance procedures should be carried out in compliance with the policies of the individual laboratory.

20.2 Laboratory reagent blank

To detect potential contamination during sample preparation and analysis, include a laboratory reagent blank, as described in [**16.2.3**]. The blank consists of all reagents and materials used in analysing test samples, including the extraction of CX-572 and glass fibre filter pad, and is analysed like a test sample.

The assessment of the blank should be less than the limit of detection.

20.3 Quality control sample

To verify the consistency of the entire analytical process, analyse a reference cigarette in accordance with the practices of the individual laboratory.

20.4 Additional QC

Additional QC samples may be added as consistent with laboratory policy.

21. METHOD PERFORMANCE SPECIFICATIONS

21.2 Limit of Report (LOR)

The LOR is set to the lowest concentration of the calibration standards used, recalculated to ng/cigarette (0.05 µmol/mL as the lowest calibration solution).

21.2 Laboratory fortified matrix recovery

Recovery of analyte spiked onto the matrix is used as a surrogate measure of accuracy. Recovery is determined by spiking known amounts of standards (after smoking) and extracting by the same method as used for samples. Unspiked samples are also analysed. The recovery is calculated from the following formula and as shown in Table 5 below:

Recovery = 100 × (analytical result − unspiked result/spiked amount)

Table 5. Mean and recovery spiked onto the matrix (to be added after the completion of the validation)

Benzene			1,3-butadiene		
Spiked amount (µg)	Mean (mg)	Recovery (%)	Spiked amount (µg)	Mean (mg)	Recovery (%)
16.9	18.4	108.9	10	9.4	94.0
33.8	37.0	109.5	20	18.6	93.0
67.6	75.2	111.2	40	41.9	104.8

21.4 Analytical specificity

GC-MS provides excellent analytical specificity. The retention time, the confirmation ion ratio, an established range of ratios of the response of the quantification ion to that of the confirmation ion and quality control samples are used to verify the specificity of the results for an unknown sample.

21.5 Linearity

The volatile organics calibration curves established are linear over the standard concentration range of 0.5–200 ng/mL.

21.6 Possible interference

There are no known components that can cause an interference by having both a similar retention time and confirmation ion pair ratio as volatile organics or the internal standards.

22. REPEATABILITY AND REPRODUCIBILITY

An international collaborative study conducted from 2015–2017, involving 10 laboratories and five samples (three reference cigarettes and two commercial brands), performed according to WHO TobLabNet SOP 02 (2.7) gave the following precision limits for this method.

The difference between two single results found on matched cigarette samples by the same operator using the same apparatus within the shortest feasible time interval will exceed the repeatability, r, on average not more than once in 20 cases in the normal and correct operation of the method.

Single results on matched cigarette samples reported by two laboratories will differ by more than the reproducibility, R, on average not more than once in 20 cases in the normal and correct operation of the method.

The test results were analysed statistically in accordance with ISO 5725-1 [**2.10**] and ISO 5725-2 [**2.5**] to give the precision data shown in Tables 6–7.

For the purpose of calculating r and R, one test result under ISO smoking conditions was defined as the average of seven individual replicates of the mean yield of three smoke traps (one cigarette smoked per smoke trap) from linear smoking machines and three smoke traps for rotary smoking machines. Under intense smoking conditions, one test result was defined as the average of seven individual replicates of the mean yield of three smoke traps from linear smoking machines and three smoke traps for rotary smoking machines. For more information, see reference [2.6].

Table 6a and 6b. Precision limits for determination of 1,3-butadiene (μg/cigarette) in mainstream cigarette smoke of the reference test pieces.

Reference cigarette	1,3-butadiene: ISO smoking regimen, μg/cigarette			
	N	m_{cig}	r_{limit}	R_{limit}
1R5F	8	11.06	1.96	4.27
3R4F	9	39.23	7.20	16.34
CM6	6	60.86	1.74	2.18
Commercial Brand I	8	72.71	8.03	42.79
Commercial Brand II	8	61.58	9.97	16.80

Reference cigarette	1,3-butadiene: Intense smoking regimen, μg/cigarette			
	N	m_{cig}	r_{limit}	R_{limit}
1R5F	7	82.26	8.11	8.63
3R4F	8	92.22	11.51	13.18
CM6	7	106.20	12.59	12.67
Commercial Brand I	8	153.96	17.20	70.04
Commercial Brand II	6	141.52	11.00	14.55

Table 7a and 7b. Precision limits for determination of Benzene (μg/cigarette) in mainstream cigarette smoke of the reference test pieces.

Reference cigarette	benzene: ISO smoking regimen, μg/cigarette			
	N	m_{cig}	r_{limit}	R_{limit}
1R5F	9	13.47	2.54	6.46
3R4F	9	44.50	6.94	10.15
CM6	8	72.86	5.94	11.40
Commercial Brand I	9	58.78	6.54	15.18
Commercial Brand II	9	49.94	8.83	15.37

Reference cigarette	benzene: Intense smoking regimen, μg/cigarette			
	N	m_{cig}	r_{limit}	R_{limit}
1R5F	8	85.59	8.17	16.68
3R4F	9	95.95	8.49	40.92
CM6	8	108.35	6.93	23.52
Commercial Brand I	8	122.00	7.21	48.76
Commercial Brand II	8	111.77	6.60	36.32

Appendix 1. Typical chromatograms obtained in the determination of volatile organics in mainstream cigarette smoke

Fig. A1. Representative chromatogram of standard 2[1]

Total ion chromatogram of a mixed standard solution. BD, 1,3-butadiene; BZ-d6, benzene-d6; BZ, benzene

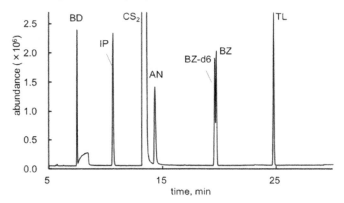

Fig. A2. Representative chromatogram of volatile organics in cigarette mainstream smoke[2]

Total ion chromatogram of a sample solution. BD, 1,3-butadiene; BZ-d6, benzene-d6; BZ, benzene

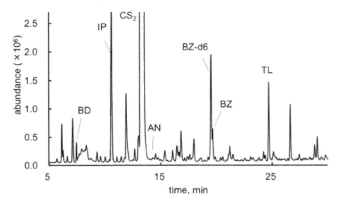

---

[1] Derived from an international collaborative study conducted in 2012 and provided by the National Institute of Public Health, Japan together with Centers for Disease Control and Prevention, China. For information only.

[2] Derived from an international collaborative study conducted in 2012 and provided by the National Institute of Public Health, Japan together with Centers for Disease Control and Prevention, China. For information only.

Appendix 2: Smoking Plans

The described smoking plans are based on five different samples (cigarettes brand/types) and seven replicates per sample. If a different number of samples or replicates is needed the plans should be adjusted accordingly.

Smoking plan for 20-port linear smoking machine (one cigarette per cartridge)

Day	1	2	3	4	5	6	7	8	9	10	11	12	13	14	15	16	17	18	19	20
1	A				B				C				D				E			
		A				B				C				D				E		
			A				B				C				D				E	
2		D				E				A				B				C		
			D				E				A				B				C	
				D				E				A				B				C
3	B				C				D				E				A			
		B				C				D				E				A		
			B				C				D				E				A	
4	E				A				B				C				D			
		E				A				B				C				D		
			E				A				B				C				D	
5	C				D				E				A				B			
		C				D				E				A				B		
			C				D				E				A				B	
6				D				C				B				A				E
	D				C				B				A				E			
		D				C				B				A				E		
7		C				B				A				E				D		
			C				B				A				E				D	
				C				B				A				E				D

Smoking plan for rotary smoking machine (one cigarette per cartridge)

Day	Run	Sample	Day	Run	Sample	Day	Run	Sample	Day	Run	Sample

1	1	A	3	31	B	5	61	C	7	91	C
	2	A		32	B		62	C		92	C
	3	A		33	B		63	C		93	C
	4	B		34	C		64	D		94	B
	5	B		35	C		65	D		95	B
	6	B		36	C		66	D		96	B
	7	C		37	D		67	E		97	A
	8	C		38	D		68	E		98	A
	9	C		39	D		69	E		99	A
	10	D		40	E		70	A		100	E
	11	D		41	E		71	A		101	E
	12	D		42	E		72	A		102	E
	13	E		43	A		73	B		103	D
	14	E		44	A		74	B		104	D
	15	E		45	A		75	B		105	D
2	16	D	4	46	E	6	76	D			
	17	D		47	E		77	D			
	18	D		48	E		78	D			
	19	E		49	A		79	C			
	20	E		50	A		80	C			
	21	E		51	A		81	C			
	22	A		52	B		82	B			
	23	A		53	B		83	B			
	24	A		54	B		84	B			
	25	B		55	C		85	A			
	26	B		56	C		86	A			
	27	B		57	C		87	A			
	28	C		58	D		88	E			
	29	C		59	D		89	E			
	30	C		60	D		90	E			

WHO TobLabNet
Official Method
SOP 10

Standard Operating Procedure for Determination of Nicotine and Carbon Monoxide in Mainstream Cigarette Smoke under Intense Smoking Conditions

Method:	Determination of Nicotine and Carbon Monoxide in Mainstream Cigarette Smoke under Intense Smoking Conditions
Analytes:	Nicotine (CAS# 54-11-5) Carbon Monoxide (CAS# 630-08-0)
Matrix:	Mainstream Cigarette Smoke
Last update:	21 September 2016

> No machine smoking regimen can represent all human smoking behaviour: machine smoking testing is useful for characterizing cigarette emissions for design and regulatory purposes, but communication of machine measurements to smokers can result in misunderstanding about differences between brands in exposure and risk. Data on smoke emissions from machine measurements may be used as inputs for product hazard assessment, but they are not intended to be nor are they valid as measures of human exposure or risks. Representing differences in machine measurements as differences in exposure or risk is a misuse of testing with WHO TobLabNet standards.

FOREWORD

This document was prepared by members of the World Health Organization (WHO) Tobacco Laboratory Network (TobLabNet) as a standard operating procedure (SOP) for measuring nicotine and carbon monoxide in mainstream cigarette smoke under intense smoking conditions.

INTRODUCTION

In order to establish comparable measurements for testing tobacco products globally, consensus methods are required for measuring specific contents and emissions of cigarettes. The Conference of the Parties (COP) to the WHO Framework Convention on Tobacco Control (WHO FCTC) at its third session in Durban, South Africa, in November 2008, recalling its decisions FCTC/COP1(15) and FCTC/COP2(14) on the elaboration of guidelines for implementation of Articles 9 (*Regulation of the contents of tobacco products*) and 10 (*Regulation of tobacco product disclosures*) of the WHO FCTC, noting the information contained in the report of the working group to the third session of the Conference of the Parties on the progress of its work .... requested the Convention Secretariat to invite WHO's Tobacco Free Initiative to.... validate, within five years, the analytical chemical methods for testing and measuring cigarette contents and emissions (FCTC/COP/3/REC/1).

Using the criteria for prioritization set at its third meeting in Ottawa, Canada, in October 2006, the working group on Articles 9 and 10 identified the following contents for which methods for testing and measurement (analytical chemistry) should be validated as a priority:

- nicotine
- ammonia
- humectants (propane-1,2-diol, glycerol (propane-1,2,3-triol) and triethylene glycol (2,2-ethylenedioxybis(ethanol)).

Measurement of these contents will require validation of three methods: one for nicotine, one for ammonia and one for humectants.

Using the criteria for prioritization set at the meeting in Ottawa mentioned above, the working group identified the following emissions in mainstream smoke for which methods for testing and measurement (analytical chemistry) should be validated as a priority:

- 4-(methylnitrosamino)-1-(3-pyridyl)-1-butanone (NNK)
- *N*-nitrosonornicotine (NNN)

- acetaldehyde
- acrylaldehyde (acrolein)
- benzene
- benzo[a]pyrene
- 1,3-butadiene
- carbon monoxide
- formaldehyde

Measurement of these emissions with the two smoking regimens described below will require validation of four methods: one for tobacco-specific nitrosamines (NNK and NNN), one for benzo[a]pyrene, one for aldehydes (acetaldehyde, acrolein and formaldehyde) and one for volatile organic compounds (benzene, 1,3-butadiene and carbon monoxide).

The table below sets out the two smoking regimens for validation of the test methods referred to above.

Smoking regimen	Puff volume (mL)	Puff frequency	Filter ventilation holes
ISO regimen: ISO 3308; Routine analytical cigarette-smoking machine — definitions and standard conditions	35	Once every 60 s	No modifications
Intense regimen: Same as ISO 3308, but modified as indicated	55	Once every 30 s	All ventilation holes must be blocked 100% as described in WHO TobLabNet SOP 01.

This method SOP was prepared to describe the procedure for the determination nicotine and carbon monoxide in mainstream cigarette smoke under intense smoking conditions.

1. SCOPE

This Standard Operating Procedure is suitable for the quantitative determination of nicotine and carbon monoxide in mainstream (MS) cigarette smoke. Nicotine will be analysed by gas chromatography and carbon monoxide by NDIR.

Note: Training in the use of the smoking machine and other analytical equipment is important for successful operation. For those not experienced in operating smoking machines or analysis using the analytical methods for measuring tobacco product emissions and contents, training should be obtained.

2. REFERENCES

2.1 *ISO 3308: Routine analytical cigarette-smoking machine — Definitions and standard conditions*

2.2 *ISO 4387: Cigarettes — Determination of total and nicotine-free dry particulate matter using a routine analytical smoking machine*

2.3 *ISO 3402: Tobacco and tobacco products — Atmosphere for conditioning and testing*

2.4 *ISO 10315: Cigarettes — Determination of nicotine in smoke condensates — Gas chromatographic method*

2.5 *ISO 8454: Cigarettes — Determination of carbon monoxide in the vapour phase of cigarette smoke — NDIR method*

2.6 *ISO 8243: Cigarettes — Sampling*

2.7 *ISO 5725-1: Accuracy (trueness and precision) of measurement methods and results — Part 1: General principles and definitions.*

2.8 *ISO 5725-2: Accuracy (trueness and precision) of measurement methods and results — Part 2: Basic method for the determination of repeatability and reproducibility of a standard measurement method.*

2.9 *World Health Organization Tobacco Laboratory Network, Standard Operating Procedure for Intense Smoking (WHO TobLabNet SOP_01).*

2.10 *World Health Organization Tobacco Laboratory Network, Standard Operating Procedure for validation of analytical methods of tobacco product contents and emissions. (WHO TobLabNet SOP_02).*

2.11 *United Nations Office on Drugs and Crime. Guidelines on representative drug sampling. Vienna: Laboratory and Scientific Section; 2009 (http://www.unodc.org/documents/scientific/Drug_Sampling.pdf).*

2.12 *World Health Organization. Method validation report of World Health Organization TobLabNet official method: Determination of tar, nicotine and carbon monoxide in mainstream cigarette smoke under intense smoking conditions. Geneva, Tobacco Laboratory Network, forthcoming.*

3. TERMS AND DEFINITIONS

3.1 *TPM*: total particulate matter.

3.2 *Smoke trap*: device for collecting such part of the smoke from a sample of cigarettes as is necessary for the determination of specified smoke components.

- **3.3** *Vapour phase*: portion of smoke which passes the particulate phase trap during smoking in accordance with ISO 4387 using a machine conforming to ISO 3308

- **3.4** *Clearing puff*: any puff taken after a cigarette has been extinguished or removed from the cigarette holder.

- **3.5** *Tobacco products*: Products entirely or partly made of leaf tobacco as raw material that are manufactured for smoking, sucking, chewing or snuffing (Article I(f) of the WHO FCTC).

- **3.6** *Intense Regime*: Parameters used to smoke tobacco products which include 55-mL puff volume, 30-second puff interval, and 100% blocking of the filter ventilation holes.

- **3.7** *ISO Regime*: Parameters used to smoke tobacco products which include 35-mL puff volume, 60-second puff interval and unblocked filter ventilation holes.

- **3.8** *Laboratory sample*: Sample intended for testing in a laboratory, consisting of a single type of product delivered to the laboratory at one time or within a specified period.

- **3.9** *Test sample*: Product to be tested, taken at random from the laboratory sample. The number of products taken shall be representative of the laboratory sample.

- **3.10** *Test portion*: Random sample from the test sample to be used for a single determination. The number of products taken shall be representative of the test sample.

4. METHOD SUMMARY

- **4.1** All samples are conditioned and marked according to ISO standard procedures.
- **4.2** Ventilation holes are blocked 100%.
- **4.3** Cigarettes are smoked according to ISO standard procedures with the exception of puff volume and puff frequency.
- **4.4** The vapour phase of the cigarette smoke is collected and carbon monoxide is measured using a non-dispersive infrared (NDIR) analyser.
- **4.5** The particulate phase of the cigarette smoke is collected on a trap for further analyses.
- **4.6** After extraction from the trap the nicotine content is analysed using gas chromatography.

5. SAFETY AND ENVIRONMENTAL PRECAUTIONS

5.1 Follow routine safety and environmental precautions, as in any chemical laboratory activity.

5.2 The testing and evaluation of certain products with this test method may require the use of materials or equipment that could be hazardous or harmful to the environment. This document does not purport to address all the safety aspects associated with its use. All persons using this method have a responsibility to consult the appropriate authorities and to establish health and safety practices, as well as environmental precautions, in conjunction with any existing applicable regulatory requirements prior to its use.

5.3 Special care should be taken to avoid inhalation or dermal exposure to harmful chemicals. Use a chemical fume hood, and wear an appropriate laboratory coat, gloves and safety goggles when preparing or handling undiluted materials, standard solutions, extraction solutions or collected samples.

6. APPARATUS AND EQUIPMENT
Usual laboratory apparatus, in particular:

6.1 Equipment needed to condition cigarettes as specified in ISO 3402 [**2.3**].

6.2 Equipment needed to mark butt length as specified in ISO 4387 [**2.2**].

6.3 Equipment needed to cover ventilation holes for the intense regimen as specified in WHO TobLabNet SOP 01 [**2.10**].

6.4 Equipment needed to perform smoking of cigarettes as specified in ISO 3308 [**2.1**].

6.5 Vapour phase collection system as specified in ISO 8454 [**2.5**].

6.6 Equipment needed to perform determination of carbon monoxide as specified in ISO 8454 [**2.5**].

6.7 Cambridge Filter Pad
Glass fibre filter pad as described in ISO 4387 (44 mm diameter for linear smoking machines, 92 mm diameter for rotary smoking machines) [**2.2**].

6.8 Extraction flasks: Erlenmeyer flasks (250 mL) or conical flasks (100 mL) with stoppers or other suitable flasks.

6.9 Wrist-action or circular shaker or equivalent device configured to hold extraction flasks in position.

6.10 GC equipped with flame ionization detector (FID) for nicotine determination.

6.11 GC columns capable of distinct separation of peaks for solvent, internal standard, nicotine and tobacco components (e.g. Varian WCOT Fused Silica, coating: CP-WAX 51, 25 m x 0.25 mm x 0.20 µm; DB-Wax, 30 m x 0.53 mm x 1 µm.).

6.12 Non-dispersive infrared (NDIR) analyser, selective and calibrated for the measurement of carbon monoxide in vapours and gases as specified in ISO 8454 [**2.5**].

6.13 Barometer, capable of measuring atmospheric pressures to the nearest 0.1 kPa ± 0.1 kPa.

6.14 Thermometer, capable of measuring temperature to the nearest 0.1 °C ± 0.1 °C.

7. REAGENTS AND SUPPLIES

All reagents shall be at least analytical reagent grade unless otherwise noted. When possible, reagents are identified by their Chemical Abstract Service (CAS) registry numbers.

7.1 Carrier gas: Helium [CAS Number: 7440-59-7] or nitrogen (CAS: 7727-37-9) of high purity (> 99.999%).

7.2 Auxiliary gases: Air and hydrogen [CAS Number: 1333-74-0] of high purity (> 99.999%) for the flame ionization detector.

7.3 Propan-2-ol [CAS Number: 67-63-0], with a maximum water content of 1.0 g/L.

7.4 -(-) Nicotine [CAS Number: 54-11-5] of known purity not less than 98%. Nicotine salicylate [CAS Number: 29790-52-1] of known purity not less than 98% may also be used. The laboratory may verify nicotine purity if necessary.

7.5 Internal Standard for GC analyses:
Nicotine determination:
n-heptadecane [CAS Number: 629-78-7] or quinaldine [CAS Number: 91-63-4] of purity not less than 99%.
Eicosane [CAS Number: 112-95-8], Isoquinoline [CAS Number: 119-65-3] or Quinolone [CAS Number: 91-22-5] of purity not less than 98% of mass fraction or other suitable alternatives may also be used.

7.6 Standard Carbon Monoxide (CO) gas mixtures:
At least three standard gas mixtures of accurately known concentrations within a relative error of 2%, covering the expected range in such a way as to avoid extrapolation of the calibration curve.

The volume percentages of CO in nitrogen are typically 1%, 3% and 5%. For low levels of CO, extension of the calibration is recommended, for instance by 0.25%.

8. PREPARATION OF GLASSWARE

Clean and dry glassware so as to prevent contamination from residues.

9. PREPARATION OF SOLUTIONS

9.1 Extraction solvent

Propan-2-ol [**7.3**] containing an appropriate concentration of internal standard.

Weigh 0.10 g of n-heptadecane [**7.5**] into a 1-litre volumetric flask.
Mix thoroughly and transfer the solution into a storage container equipped with features to prevent contamination.

Note: The concentration and/or type of internal standard may be adjusted, keeping in mind the possible effect of internal standards on the sensitivity and selectivity, as well as the linear range of the method.

10. PREPARATION OF STANDARDS

10.1 Nicotine determination

10.1.1 Nicotine stock solution (10 mg/mL)
Weigh approximately 1000 mg of nicotine to 0.1 mg accuracy into a 100 mL volumetric flask and dilute to volume with the extraction solvent.
Mix thoroughly and store between 0 °C and 4 °C and exclude light.

Solvent and solutions stored at low temperatures shall be allowed to equilibrate to (22 ± 2) °C before use.

10.1.2 Nicotine working standard solutions
Pipette various volumes of the nicotine stock solution [**10.1.1**] into separate 100 mL volumetric flasks as indicated in Table 1 and dilute to volume with extraction solvent.

Store between 0 °C and 4 °C and exclude light. Solvent and solutions stored at low temperatures shall be allowed to equilibrate to (22 ± 2) °C before use.

Note: When, in addition to the nicotine determination, a water determination is performed simultaneously on a single GC, the working standard solutions of nicotine and water can be combined.

10.1.3 The final nicotine concentrations of standards (mg/L) are determined from:

$$\text{Final concentration mg/L} = \frac{x \times y}{10} \times \text{purity of standard}$$

where;

x   is the original weight (in mg) of nicotine as weighed in **10.1.1**.
y   is the volume of stock standard solution as pipetted in **10.1.2**.

The final nicotine concentrations in the standard solutions are shown in Table 1.

Table 1. Concentrations of nicotine in standard solutions

Standard	Volume of nicotine stock solution (10 mg/mL)	Volume of internal standard solution (mL)	Total volume (mL)	Approximate nicotine concentration in final standard solution (mg/L)
1	0.2		100	20
2	2	Not applicable, included in extraction solution	100	200
3	4		100	400
4	6		100	600
5	8		100	800

The range of the standard solutions may be adjusted, depending on the equipment used and the samples to be tested, keeping in mind the possible effect on the sensitivity of the method.

11. SAMPLING

Sample cigarettes according to ISO 8243 [**2.6**]. Alternative approaches may be used to obtain a representative laboratory sample in accordance with individual laboratory practice or when required by specific regulation or the availability of samples.

11.2 Constitution of test sample

11.2.1 Divide the laboratory sample into separate units (e.g. packet, container), if applicable.

11.2.2 Take an equal amount of product for each test sample from at least $\sqrt{n}$ [**2.11**] of the individual units (e.g. packet, container).

12. CIGARETTE PREPARATION

12.1 Condition all cigarettes to be smoked in accordance to ISO 3402 [**2.3**]

12.2 Mark cigarettes at a butt length in accordance with ISO 4387 [**2.2**] and World Health Organization Tobacco Laboratory Network, Standard Operating Procedure for Intense Smoking (WHO TobLabNet SOP_01) [**2.9**].

12.3 Prepare test samples to be smoked under intense smoking conditions as specified in World Health Organization Tobacco Laboratory Network, Standard Operating Procedure for Intense Smoking (WHO TobLabNet SOP_01) [**2.9**].

13. PREPARATION OF THE SMOKING MACHINE

13.1 Ambient conditions

The ambient conditions for smoking are specified in ISO 3308 [**2.1**].

13.2 Smoking machine specifications

Follow ISO 3308 [**2.1**] machine specifications, except for intense smoking as described in World Health Organization Tobacco Laboratory Network, Standard Operating Procedure for Intense Smoking (WHO TobLabNet SOP_01) [**2.9**].

14. SAMPLE GENERATION

Smoke a sufficient number of cigarettes on the specified smoking machine such that breakthrough does not occur and the concentrations of nicotine and carbon monoxide fall within the calibration range prepared for the analysis

14.1 Smoke the cigarette test samples and collect the TPM, as specified in ISO 4387 [**2.2**] or in WHO TobLabNet SOP_01 [**2.9**].

14.2 Collect the vapour phase in a suitable collection system [**6.5**] as specified in ISO 8554 [**2.5**].

14.3 Include at least one reference test sample to be used for quality control, if applicable.

14.4 When testing sample types for the first time, breakthrough should be evaluated. The number of cigarettes might have to be adjusted to prevent breakthrough of the filter pad [**6.7**]. If the TPM exceeds 600 mg for a 92 mm filter pad or 150 mg for a 44 mm filter pad, the number of cigarettes smoked onto each pad must be decreased.

14.5 The number of cigarettes to be smoked per measurement for linear and rotary smoking machines at ISO and intense smoking regimens are shown in Table 2.

Table 2. Number of cigarettes to be smoked for one measurement in linear and rotary smoking machines.

	ISO smoking regimen		Intense smoking regimen	
	Linear	Rotary	Linear	Rotary
No. of cigarettes per smoke trap	5	20	3	10
No. of smoke traps per result	4	1	7	2

14.6 After smoking an individual cigarette, remove the cigarette butt and take one clearing puff for each trap. After smoking the required number of test samples per trap, perform five clearing puffs for each trap and remove the smoke trap from the smoking machine.

14.7 Record the number of cigarettes and total puffs for each smoke trap, including clearing puffs.

14.8 Perform a blank measurement, at least once per day, by placing a Cambridge Filter Pad [**6.7**] in the smoking machine area during a smoke run.

15. SAMPLE PREPARATION

15.1 After collection of TPM, transfer each Cambridge filter (CF) pad [**6.7**] from the holder (for linear — three cigarettes smoked per 44 mm CF pad; for rotary — 10 cigarettes smoked per 92 mm CF pad) into a clean, dry, 100 mL conical flask for 44 mm disks and a 250 mL conical flask for 92 mm discs with glass stopper [**6.8**].

15.2 Wipe the front half of the holder twice using a ¼ clean 44 mm CF pad for linear smoking machines and twice using a clean 44 mm CF pad for rotary smoking machines and add these to the respective flask.

15.3 Add 20 mL extraction solvent [**9.1**] for 44 mm discs and 50 mL extraction solvent [**9.1**] for 92 mm discs into each flask. Ensure that the discs are fully covered.

15.4 Shake the flasks for about 30 minutes on the shaker [**6.9**] at about 200 rpm. Adjust the time or speed to prevent disintegration of the pads.

16. SAMPLE ANALYSIS

16.1 TPM weight
The method for determining TPM weight is described in ISO 4387 [**2.2**].

16.2 Nicotine Content
The method for determining nicotine in mainstream cigarette smoke is described in ISO 10315 [**2.4**].

Suitable GC–FID operating conditions: example
Injector temperature: 250 °C
Detector temperature: 250 °C
Carrier gas: Helium, or nitrogen at a flow rate of 25 mL/min
Injection: 1 µL or 2 µL
Column temperature: 170 °C (Isothermal)

Note: The operating parameters may have to be adjusted to the instrument and column conditions and the resolution of the chromatographic peaks.

16.3 Carbon Monoxide (CO) Content
The method for determining carbon monoxide in the vapour phase of smoke is described in ISO 8454 [**2.5**].

17. DATA ANALYSIS AND CALCULATIONS

17.1 Nicotine Content

17.1.1 For each sample extract, calculate the ratio of the nicotine response to the internal standard response from the peak area.

17.1.2 Calculate the nicotine concentration in mg/L in each sample extract using the coefficients of the linear regression

$$m_t = \frac{(y_t - a)}{b}$$

where;

m_t is the concentration of nicotine in the sample extract in mg/L

Y_t is the ratio of nicotine response to the internal standard response from the peak area

a is the intercept of the linear regression obtained from the standard calibration curve

b is the slope of the linear regression obtained from the standard calibration curve

17.1.3 The nicotine content for each sample extract, in mg per cigarette, is calculated as follows:

$$\text{Nicotine content (mg/cigarette)} = \frac{m_t \times V}{1000 \times N}$$

where;

m_t :   Concentration of nicotine in the sample extract, in mg/L.

N :   Number of cigarettes smoked through each smoke trap (three for linear; 10 for rotary).

V :   Volume of extraction solvent in which the contents of the extract was dissolved (20 mL for linear; 50 mL for rotary).

17.2 Carbon Monoxide (CO) content

17.2.1 The average volume of carbon monoxide per cigarette is given by the following equation:

$$V_{as} = \frac{C \times V \times N \times p \times T_0}{S \times 100 \times p_0 \times (t + T_0)}$$

where;

V_{as} is the average volume of carbon monoxide per cigarette, in mL

C is the percentage by volume of carbon monoxide observed;

V is the puff volume, in mL;

N is the number of puffs in the measured sample (including clearing puffs);

p	is the ambient pressure, in kPa;	
p_o	is the standard atmospheric pressure, in kPa	
S	is the number of cigarettes smoked;	
T	is the ambient temperature, in °C;	
T_o	is the temperature for the triple point of water, in Kelvin.	

17.2.2 The average mass of carbon monoxide per cigarette is given by the following equation

$$m_{cig} = V_{as} \times \frac{M_{co}}{V_m}$$

where;

V_{as}	is the average volume of carbon monoxide per cigarette, in mL
m_{cig}	is the average mass of carbon monoxide per cigarette, in mg
M_{co}	is the molar mass of carbon monoxide, in g/mol;
V_m	is the molar volume of an ideal gas, in L/mol

18. SPECIAL PRECAUTIONS

18.1 After installing a new column, condition it as specified by the manufacturer by injecting a tobacco sample extract under the specified instrument conditions. Injections should be repeated until the peak areas (or heights) for the analyte of interest and internal standard are reproducible.

18.2 When the peak areas (or heights) for the internal standard are significantly higher than expected, it is recommended that the tobacco sample be extracted without internal standard in the extraction solution.
This makes it possible to determine whether any component co-elutes with the internal standard, which would cause artificially lower values for nicotine.

19. DATA REPORTING

19.1 Report individual measurements for each of the samples evaluated.

19.2 Report results as specified by method specifications.

19.3 For more information, see World Health Organization Tobacco Laboratory Network, Standard Operating Procedures for validation of analytical methods of tobacco product contents and emissions. (WHO TobLabNet SOP_02). [**2.10**].

20. QUALITY CONTROL

20.1 Control Parameters

Note: If the control measurements are outside the tolerance limits of the expected values, appropriate investigation and action must be taken.

Note: Additional individual laboratory quality assurance procedures should be carried out if necessary in order to comply with the policies of individual laboratories.

20.2 Laboratory Reagent Blank (LRB)

To detect potential contamination during sample preparation and analysis, include a laboratory reagent blank. The blank consists of all reagents and materials used in analysing test samples and is analysed like a test sample. The blank should be assessed in accordance with the practices of individual laboratories.

20.3 Quality Control Sample

To verify consistency of the entire analytical process, analyse a reference cigarette (or an appropriate quality control sample) in accordance with the practices of the individual laboratories.

21. METHOD PERFORMANCE SPECIFICATIONS

21.1 Limit of Reporting (LOR)

The limit of reporting is set to the lowest concentration of the calibration standards used, recalculated to mg per cigarette. (e.g. for the linear smoking machine the lowest nicotine calibration standard of 20 mg/L corresponds to 0.1 mg per cigarette).

21.2 Analytical Specificity

For gas chromatographic determinations (nicotine), the retention time of the analyte of interest is used to verify the analytical specificity. An established range of ratios of the response of the component to that of the internal standard component of a quality control cigarette is used to verify the specificity of the results for an unknown sample.

21.2 Linearity

Nicotine Content: The nicotine calibration curves established are linear over the standard calibration range of 20–800 mg/L.

21.3 Possible interferences

The presence of eugenol can cause interference, as its retention time is similar to that of nicotine. This interference is most likely to occur with samples containing clove. The laboratory may need to resolve this by adjusting the analytical instrument parameters.

22. REPEATABILITY AND REPRODUCIBILITY

A global collaborative study conducted in 2007, involving 14 laboratories and three reference cigarettes (1R5F, 3R4F and CM6) and two commercial brands, was conducted to validate a method for assessing TNCO in mainstream cigarette smoke following ISO standards using an intense smoking regimen [**2.12**]. The precision limits for this method are indicated in Table 3 and Table 4.

The difference between two single results found for matched cigarette samples by the same operator using the same apparatus within the shortest feasible time will exceed the repeatability, r, on average not more than once in 20 cases in the normal, correct application of the method.

Single results for matched cigarette samples reported by two laboratories will differ by more than the reproducibility, R, on average no more than once in 20 cases with normal, correct application of the method.

The test results were analysed statistically in accordance with ISO 5725-1 [**2.7**] and ISO 5725-2 [**2.8**] to give the precision data.

Table 3. Precision limits for determination of Carbon monoxide (CO) (mg/cig) in cigarette

Reference Cigarettes	N	m_{cig}	r_{limit}	R_{limit}
1R5F	13	27.39	2.53	4.62
3R4F	13	32.19	2.24	2.97
CM6	13	26.87	1.40	2.91

Table 4. Precision limits for determination of Nicotine (mg/cig) in cigarette

Reference Cigarettes	N	m_{cig}	r_{limit}	R_{limit}
1R5F	13	1.02	0.08	0.17
3R4F	13	1.91	0.14	0.31
CM6	13	2.73	0.20	0.42

Appendix 1. Typical chromatograms obtained in the analysis of nicotine by GC method

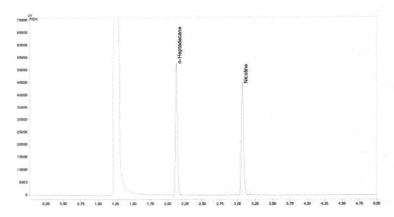

Figure 1. Example of a chromatogram of a nicotine standard solution

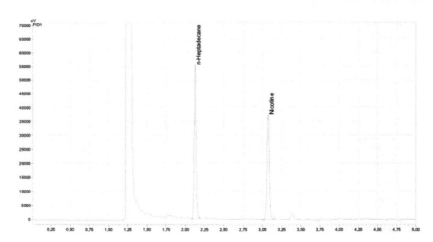

Figure 2. Example of a chromatogram of a nicotine sample solution